James Furman Kemp

The Ore Deposits of the United States

James Furman Kemp

The Ore Deposits of the United States

ISBN/EAN: 9783337186722

Printed in Europe, USA, Canada, Australia, Japan

Cover: Foto ©ninafisch / pixelio.de

More available books at **www.hansebooks.com**

THE

ORE DEPOSITS

OF THE

UNITED STATES

BY

JAMES F. KEMP, A.B., E.M.

PROFESSOR OF GEOLOGY IN THE SCHOOL OF MINES, COLUMBIA COLLEGE

NEW YORK
THE SCIENTIFIC PUBLISHING COMPANY
1893

COPYRIGHTED, 1893,
BY
THE SCIENTIFIC PUBLISHING COMPANY.

In Memory of

JOHN STRONG NEWBERRY[*]

FIRST PROFESSOR OF GEOLOGY IN THE SCHOOL OF MINES, COLUMBIA COLLEGE

THIS BOOK IS RESPECTFULLY INSCRIBED BY

HIS OLD STUDENT AND FRIEND

THE AUTHOR

[*] Dr. NEWBERRY died while the book was in press.

PREFACE.

The following pages presuppose for their comprehension some acquaintance with geology and mineralogy. The materials for them have been collected and arranged in connection with lectures on economic geology, first at Cornell University and later at the School of Mines, Columbia College. To the descriptions of others the author has endeavored to add, as far as possible, observations made by himself in travel during the last ten years. The purpose of the book is twofold, and this fact has been conscientiously kept in view. It is, on the one hand, intended to supply a condensed account of the metalliferous resources of the country, which will be readable and serviceable as a text-book and work of reference. For this reason every effort has been put forth to make the bibliography complete, so that, in cases where fuller accounts of a region are desired, the original sources may be made available in any good library. But, on the other hand, it has also been the hope and ambition of the author to treat the subject in such a way as to stimulate investigation and study of these interesting phenomena. If, by giving an extended view over the field, and by making clear what our best workers have done in late years toward explaining the puzzling yet vastly important questions of origin and formation, some encouragement may be afforded those in a position to observe and ponder, the second aim will be fulfilled. In carrying out this purpose, the best work of recent investigators on the origin and changes of rocks, especially as brought out by microscopic study, has been kept constantly in mind, and likewise in the artificial production of the ore and gangue minerals. So much unsound and foolish theorizing has been uttered and believed about ores, that too much care cannot be exercised in basing explanations on reasonable and right foundations.

Acknowledgments are due to many friends for encouragement, suggestion, and criticism. To Prof. Henry S. Williams, now of

Yale, but late of Cornell, whose interest made the book possible, these are especially to be made. On particular regions much advice has been obtained from Dr. W. P. Jenney, for which the author is grateful. In the same way Prof. H. A. Wheeler of St. Louis, Prof. R. A. F. Penrose of Chicago, and several other friends have contributed. Dr. R. W. Raymond suggested the method of enumerating the paragraphs. It has the advantages of being elastic and of showing at once in what part of the book any paragraph is situated.

The geologists of the United States Geological Survey, who have been engaged in the study of our great mining regions, especially in the West, have laid the whole scientific world under a debt of gratitude, and in this country have probably been the most potent influences toward right geological conceptions regarding ores. Of writers abroad, Von Groddeck has been a means of inspiration to all readers of German who have interested themselves in this branch of geology. The writer cannot well forbear acknowledging their influence.

Should errors be noted by any reader, the author will be very appreciative of the kindness if his attention is called to them.

J. F. KEMP.

COLUMBIA COLLEGE,
IN THE CITY OF NEW YORK.

TABLE OF CONTENTS.

	PAGE
PREFACE	iii
LIST OF ILLUSTRATIONS	xi
LIST OF ABBREVIATIONS	xv

PART I.—INTRODUCTORY.

CHAPTER I.—GENERAL GEOLOGICAL FACTS AND PRINCIPLES.
    The two standpoints of geology, 3–4; the scheme of classification, 4–5; classification of rocks, 6; brief topographical survey of the United States, 6–7; brief geological outline, 7–10; the forms assumed by rock masses, 10, 11..........................1–11

CHAPTER II.—ON THE FORMATION OF CAVITIES IN ROCKS.
    By local contraction, 12, 13; by more extensive movements, 13–17; faults, 17–20; secondary modifications of cavities, 20–22..12–22

CHAPTER III.—THE MINERALS IMPORTANT AS ORES; THE GANGUE MINERALS, AND THE SOURCES WHENCE BOTH ARE DERIVED.
    The minerals, 23; source of the metals, 23–27..............23–27

CHAPTER IV.—ON THE FILLING OF MINERAL VEINS.
    Résumé, 28; methods of filling, 28, 29; ascension by infiltration, 29, 32; lateral secretion, 30, 31; replacement, 33, 34.......28–34

CHAPTER V.—ON CERTAIN STRUCTURAL FEATURES OF MINERAL VEINS.
    Banded structure, 35–37; clay selvage, 37; pinches, swells, and lateral enlargements, 37; changes in the character of the vein filling, 38; secondary alteration of the minerals in veins, 38–40; electrical activity, 40, 41..35–41

CHAPTER VI.—THE CLASSIFICATION OF ORE DEPOSITS, A REVIEW AND A SCHEME BASED ON ORIGIN.
    Statement of principles, 42–44; schemes involving only the classification of veins by Von Weissenbach, 44; Von Cotta, 44, 45; Le Conte, 45; general schemes based on form: Von Cotta and

Prime, 46, 47; Lottner-Serlo, 47; Koehler, 47; Callon, 47; schemes partly based on form, partly on origin: J. D. Whitney, 48; J. S. Newberry, 48; J. A. Phillips, 49; schemes largely based on origin: J. Grimm, 50; A. von Groddeck, 51; R. Pumpelly, 51, 52; schemes entirely based on origin: H. S. Munroe, 52, 53; J. F. Kemp, 53-55; remarks on the above, and discussion of methods of formation, 56-62; bibliography, 62-65...........................42-65

PART II.—THE ORE DEPOSITS.

CHAPTER I.—THE IRON SERIES (IN PART).—INTRODUCTORY REMARKS ON IRON ORES.—LIMONITE.—SIDERITE.

General literature, 69-70; table of analyses, 70; general remarks, 70-73; Limonite, **Example 1,** Bog Ore, 73-75; Example 2, Brown Hematite, not Siluro-Cambrian, 75-79; Example 2*a*, Siluro-Cambrian Limonites, 79-84; origin of same, 84-85. Analyses of limonites, 86. Siderite or Spathic Ore, introductory, 86; Example 3, Clay **Ironstone,** 86-87; Example 3*a*, Black-band, 87-89; Example 4, Burden mines, 89-91; Example 5, Roxbury, Conn., 91.........69-91

CHAPTER II.—THE IRON SERIES, CONTINUED.—HEMATITE, RED AND SPECULAR.

Introductory remarks, 92; **Example 6,** Clinton Ore, 92-98; Example 7, Crawford Co., Mo., 99; Example 8, Jefferson Co., N. Y., 99; Example 9, Lake Superior Hematites, 100-113; introductory, 100-102; Marquette district, 102-106; Menominee, 107-108; Penokee-Gogebic, 108-109; Vermilion Lake, 110-111; Mesabi, 111-113; Example 10, James River, Va., 113; Example 11, Pilot Knob., Mo., 114-116; **Example 11***a***,** Iron Mountain, Mo., 116-118; **Analyses of Hematites,** 118...92-119

CHAPTER III.—MAGNETITE AND PYRITE.

Example 12, Magnetite beds, 120-127; Adirondack region, 120-122; New York, New Jersey, and Pennsylvania, 123-125; North Carolina and Virginia, 125; Colorado, 126; California, 126-127; Example 13, Cornwall, Penn., 127; Example 14, Iron Co., Utah, 128; **Example 15,** Magnetite Sands, 128; origin of Magnetite deposits, 129-130; distribution of Phosphorus, 130; Analyses of Magnetites, 131; Pyrite, 131-132; Example 16, Pyrite Beds, 131-132; Statistics of iron ores, 133........................120-133

CHAPTER IV.—COPPER.

Table of analyses of copper ores, 134; Example 16, continued, Pyrite Beds, 134-135; Ore Knob, N. C., 135; Spenceville, Cal., 136; Example 17, Butte, Mont., 136-137; Gilpin Co., Colo., 138; Llano Co., Texas, 139; Example 18, Keweenaw Point, Mich., 139-143; origin of the copper, 141-142; Example 19, St. Genevieve, Mo., 143-144; Example 20, Arizona copper, 144-151; Morenci, 145; Bisbee, 148; Globe district, 150; Santa Rita, N. M., 150-151; Black

Range, 151; Copper Basin, 151; Example 20g, Crismon-Mammoth, Utah, 152; Sunrise, Wyo., 152; Example 21, **copper ores in Triassic or** Permian sandstone, 152-154; Eastern **States,** 152-153; **Western States,** 154; statistics of **copper,** 155 134-155

CHAPTER V.—LEAD ALONE.
    Introductory **and analyses,** 156; Example 22, **Atlantic Border, St. Lawrence Co., N. Y.,** 156; **Massachusetts, Connecticut, and Eastern New York,** 157, 158; **Southeastern Pennsylvania,** 157; Davison Co., **N. C.,** 157; Example 23, **Southeast Missouri,** 158-159; **statistics of lead,** 160................................. 156-160

CHAPTER VI.—LEAD AND ZINC.
    Example 24, the **Upper Mississippi** Valley, 161-164; **Example** 24a, **Washington Co.,** 165; Example 24b, **Livingston Co., Ky.,** 165; Example 25, **Southwest Missouri,** 166-172; **Example 26, Wythe Co., Va.,** 172.. 161-173

CHAPTER VII.—ZINC ALONE OR WITH METALS OTHER THAN LEAD.
    Introduction, tables of analyses, 174; Example 27, **Saucon Valley, Penn.,** 174-175; **Example 28, Franklin Furnace and Sterling, N. J.,** 175-179; statistics of zinc, 180..................... 174-180

CHAPTER VIII.—LEAD AND SILVER.
    Introduction, 181; Rocky Mountain Region and **Black Hills,** 181-193; **New** Mexico, 181-182; **Example** 29, **Kelley Lode,** 181; Colorado, 182-191; **Example 30, Leadville,** 182-185; **Example** 30a, **Ten Mile, Summit Co.,** 185-186; Example 30b, **Monarch** district, 186; Example 30c, **Eagle River,** 186; Example 30d, **Aspen,** 186 and 188-190; Example 30e, **Rico,** 190; Example 31, **Red Mountain,** 190; **South Dakota, Example** 30f, **191; Montana, Idaho,** 191-192; **Example 32, Glendale, 191;** Example 32a, **Wood River, 191-192; Example 33, Wickes,** 192; Example 34, Cœur d'Alene, 192; **region of the Great Basin,** 192-198; Utah, Example 35, Bingham **and Big and Little Cottonwood, 192-194; Example** 35a, **Tooele Co., 194; Example** 35b, **Tintic; Example** 30g, **Horn Silver, 195; Example** 33a, **Carbonate Mine, Beaver Co., 195;** Example 32b, **Cave Mine, Beaver Co., 195;** Nevada, **Example 36, Eureka,** 196-197; **Arizona, California,** 198... 181-198

CHAPTER IX.—SILVER AND GOLD.—INTRODUCTORY: EASTERN SILVER MINES AND THE ROCKY MOUNTAIN REGION OF NEW MEXICO AND COLORADO.
    Introduction, 199; Examples 37-42 defined, 199-200; silver minerals, 200; Example 22a, Atlantic **Border,** 201; **Example 42, Silver Islet,** 201; region of **the** Rocky Mountains **and the Black Hills,** 202-215; New Mexico, **geology,** 202-203; **mines,** 203-204; **Colorado, geology,** 204-205; **San Juan region,** 205-210; **Gunnison**

viii TABLE OF CONTENTS.

district, 211; Eagle Co., 211; Summit Co., 211; Park, Chaffee, Rio Grande, and Codejos counties, 212; Custer Co., 212-213; Gilpin, Clear Creek, and Boulder counties, 214 199-215

CHAPTER X.—SILVER AND GOLD, CONTINUED.—ROCKY MOUNTAIN REGION, WYOMING, THE BLACK HILLS, MONTANA, AND IDAHO.
Wyoming, 216; the Black Hills, 216-218; Montana, geology, 218; Madison, Beaverhead, and Jefferson counties, 219; Silver Bow Co., 220-221; Deer Lodge and Lewis and Clarke counties, 221; Missoula Co., 222; Idaho, geology, 222; Custer, Boisé, Alturas, and other counties, 223....................................216-223

CHAPTER XI.—SILVER AND GOLD, CONTINUED.—THE REGION OF THE GREAT BASIN, IN UTAH, ARIZONA, AND NEVADA.
Utah, geology, 224; Ontario and other mines, 225-226; Silver Reef, 226-227; Arizona, geology, 227; northern counties and the Silver King mine, 228; Tombstone, 229; Pima and Yuma counties, 229; Nevada, geology, 230; Lincoln Co., 230; Ney and White Pine counties, 231; Lander and other counties, 232; the Comstock Lode, 233-237...224-237

CHAPTER XII.—THE PACIFIC SLOPE.—WASHINGTON, OREGON, AND CALIFORNIA.
Washington, geology, 238; mines, 239; Oregon, geology, 239; gold quartz and placers, 240; Example 44a, Port Orford, 240; California, geology, 241; Calico district, 242-243; Example 44, auriferous gravels, 243-248; river gravels, 243-245; high or deep gravels, 245-248; Example 45, gold quartz veins, 248-251.............238-251

CHAPTER XIII.—GOLD ELSEWHERE IN THE UNITED STATES AND CANADA.
Example 45a, Southern States, 252; Example 45b, Ishpeming, Mich., 253; Alaska, geology, 253-254; Example 46, Douglass Island, 254-255; Example 45c, Nova Scotia, 255; Example 45d, gold elsewhere in Canada, 256; statistics, 256-257 252-257

CHAPTER XIV.—THE LESSER METALS.—ALUMINIUM, ANTIMONY, ARSENIC, BISMUTH, CHROMIUM, MANGANESE.
Aluminium, 258; antimony, 259-260; Example 47, including California, Nevada, Arkansas, and New Brunswick, 259; Example 48, Utah, Iron Co., 259-260; arsenic, 260; bismuth, 260-261; chromium, 261; Example 49, chromite in serpentine, 261; manganese, 262; Example 50, manganese ores in residual clay, 262-266; statistics, 266..251-266

CHAPTER XV.—THE LESSER METALS, CONTINUED.—MERCURY, NICKEL AND COBALT, PLATINUM, TIN.
Mercury, 267-269; Example 50, New Almaden, 267, 268; Exam-

ple 50*a*, Sulphur Bank, 268; Example 50*b*, Steamboat Springs, 268–269; nickel and cobalt, 269–272; platinum, 272; tin, 273–274; Example 51, Black Hills, 273 267–273

CHAPTER XVI.—CONCLUDING REMARKS.
Summation of such general geological relations among North American ore deposits as can be detected 275–277

ADDENDA.—Wadsworth's scheme of classification of ore deposits, 279; J. P. Kimball on the genesis of spathic ores, 279; bog ores of the Three Rivers district, Province of Quebec, 280; iron ores of Arkansas, 280; titaniferous magnetites, 281; the origin of magnetite as segregated veins, 282; Cambrian at St. Genevieve, Mo., 283; C. R. Boyd on zinc mines at Austinville, Va., 283; W. Lindgren on the American and Yuba rivers, California, 283; Posepny's great paper on the origin of ores, 283; Fuchs et De Launay, Traité des Gîtes Minéraux et Métallifères, Paris, 1893, 285............278–285

LIST OF ILLUSTRATIONS.

FIGS.		PAGE
1.	Illustration of rifting in granite at Cape Ann, Mass. After R. S. Tarr	12
2.	Open fissure in the Aubrey limestone (Upper Carboniferous) 25 miles north of Cañon Diablo Station, on the A. & P. R. R., Arizona. Photographed by G. K. Gilbert, 1892	15
3.	Normal fault at Leadville, Colo. After A. A. Blow	17
4.	Reversed fault at Holly Creek, near Dalton, Ga. After C. W. Hayes	18
5.	Illustration of one vein faulting another at Newman Hill, near Rico, Colo. After J. B. Farish	20
6.	Banded vein at Newman Hill, near Rico, Colo. After J. B. Farish.	36
7.	Illustration of the oxidized zone, or gossan, the zone of enrichment, and the unchanged sulphides, at Ducktown, Tenn. After A. F. Wendt	39
8.	Section of the Hurst limonite bank, Wythe Co., Va., illustrating the replacement of shattered limestone with limonite and the formation of geodes of ore. After E. R. Benton	76
9.	Geological section of the Amenia mine, Dutchess Co., N. Y., illustrating a Siluro-Cambrian limonite deposit. After B. T. Putnam	80
10.	View of the Siluro-Cambrian brown hematite bank at Baker Hill, Ala. From the *Engineering and Mining Journal*	83
11.	Map and sections of the Burden Spathic ore mines. After J. P. Kimball	90
12.	Clinton ore, Ontario, Wayne Co., N. Y. After C. H. Smyth, Jr.	93
13.	Clinton ore, Clinton, N. Y. After C. H. Smyth, Jr	94
14.	Clinton ore, Eureka mine, Oxmoor, Ala. After C. H. Smyth, Jr	95
15.	Cross section of the Sloss mine, Red Mountain, Ala. From the *Engineering and Mining Journal*	95
16.	Map of the vicinity of Birmingham, Ala. After W. P. Barker	96
17.	Open cut in the Republic mine, Marquette range, showing a horse of jasper. From a photograph by H. A. Wheeler	103
18.	Cross sections to illustrate the occurrence and associations of iron ore in the Marquette district, Michigan. After C. R. Van Hise.	105
19.	Plan of Ludington ore body, Menominee district, Michigan. After P. Larsson	108

xli

LIST OF ILLUSTRATIONS.

| FIGS. | | PAGE |

20. Cross section of the Colby mine, Penokee-Gogebic district, Michigan, to illustrate occurrence and origin of the ore. After C. R. Van Hise.. 109
21. General cross section of ore body at Biwabik, Mesabi Range, Minn. After H. V. Winchell... 112
22. Cross section of Pilot Knob, Mo. From a drawing by W. B. Potter. 114
23. View of open cut at Pilot Knob, Mo., showing the bedded character of the iron ore. From a photograph by J. F. Kemp. 115
24. View of Iron Mountain, Mo. From a photograph by H. A. Wheeler 116
25. Cross section of Iron Mountain, Mo., showing the knob of porphyry, with the veins of ore, the conglomerate, etc. After W. B. Potter... 117
26. View of open cut and underground work in Mine 21, Mineville, near Port Henry, N. Y. From a photograph by J. F. Kemp... 121
27a and 27b. Model of the Tilly Foster ore body. After F. S. Ruttmann and the model itself... 124
28. Cross section of the magnetite ore body at Cornwall, Penn. After Bailey Willis.. 127
29. Illustration of overlapping lenses of pyrite. After A. F. Wendt.. 132
30. Cross section of the Bob-tail mine, Central City, Colo. After F. M. Endlich... 138
31. Geological sections of Keweenaw Point, Mich., near Portage Lake and through Calumet. After R. D. Irving...................... 140
32. Cross section of the St. Genevieve copper mine, illustrating the relations of the ore. After F. Nicholson..................................... 144
33. Section at the St. Genevieve mine, illustrating the intimate relations of ore and chert. After F. Nicholson 144
34. Geological map of the Morenci or Clifton copper district of Arizona. After A. F. Wendt... 145
35. Vertical section of Longfellow Hill, Clifton district, Arizona. After Wendt... 146
36. Horizontal sections of Longfellow ore body. After Wendt. 146
37. Geological section of the Metcalf mine, Clifton district, Arizona. After Wendt... 147
38. Section of Copper Queen ore body, Bisbee district, Arizona. After A. F. Wendt.. 148
39. View of the Copper Queen mine, Bisbee district, Arizona. From a photograph by James Douglass.. 149
40. Cross section of the Schuyler copper mine, N. J. After N. H. Darton.. 153
41. Gash veins, fresh and disintegrated. After T. C. Chamberlain... 162
42. Idealized section of "flats and pitches," forms of ore bodies in Wisconsin. After T. C. Chamberlain..................................... 163
43. Vertical section of a typical zincblende ore body, near Webb City, Mo. After C. Henrich 168
44. View of the Motley mine, Webb City, Mo. After a photograph by W. P. Jenney.. 169

LIST OF ILLUSTRATIONS.

FIGS.		PAGE
44a.	View of the Bertha zinc mines, Wythe Co., Va. From a photograph by A. E. W. Miller	173
45.	Section at Franklin Furnace, N. J., showing the geological relations of the franklinite ore body. After F. L. Nason	176
46.	Section of the White Cap chute, Leadville, showing the geological relations of the ore, and its passage into unchanged sulphides in depth. After A. A. Blow	184
47.	Geological section of the Eagle River mines, Colo. After E. E. Olcott	187
48.	Geological section at Aspen, Colo. After A. Lakes	188
49.	View of the Bunker Hill and Sullivan mines, Wardner, Idaho. From a photograph loaned by E. E. Olcott	193
50.	Section at Eureka, Nev. After a plate by J. S. Curtis	197
51.	Geological cross sections of strata and veins at Newman Hill, near Rico, Colo. After J. B. Farish	208
52.	Geological cross sections of strata and veins at Newman Hill. After J. B. Farish	209
53.	View of Lower Creede, Colo. From the *Engineering and Mining Journal*	210
54.	Geological section of the Black Hills. After Henry Newton	217
55.	Cross section of vein at the Alice mine, Butte, Mont. After W. P. Blake	220
56.	Two sections of the argentiferous sandstone at Silver Reef, Utah. After C. M. Rolker	226
57.	Section of Comstock Lode. After G. F. Becker	233
58.	Geological section of the Calico district, California. After W. Lindgren	242
59.	View of the Union diggings, Columbia Hill, Nevada Co., California. From a photograph	243
60.	View of the Timbuctoo diggings, Yuba Co., California. From a photograph	244
61.	Generalized section of a deep gravel bed, with technical terms. After R. E. Browne	246
62.	Section of Forest Hill Divide, Placer Co., California, to illustrate the relations of old and modern lines of drainage. After R. E. Browne	247
63.	Sections of the Crimora manganese mine, Virginia. After C. E. Hall	263
64.	Geological sections illustrating the formation of the manganese ores in Arkansas. After R. A. F. Penrose	264
65.	The Turner mine, Batesville region, Arkansas. After R. A. F. Penrose	265
66.	Section of the Great Western cinnabar mine. After G. F. Becker	268
67.	Horizontal section of the Etta granite knob, Black Hills. After W. P. Blake	273

ABBREVIATIONS.

A. A. A. S.—Proceedings of the American Association for the Advancement of Science.
A. G. or *Amer. Geol.*—*American Geologist*, Minneapolis, Minn.
A. J. S. or *Amer. Jour. Sci.*—*American Journal of Science*, also known as *Silliman's Journal*. Fifty half-yearly volumes make a series. The *Journal* is now (1893) in the third series. In the references the series is given first, then the volume, then the page.
Ann. des Mines.—*Annales des Mines*. Paris, France.
Bost. Soc. Nat. Hist.—See Proceedings of same.
Bull. Geol. Soc. Amer. or *G. S. A.*—Bulletin of the Geological Society of America.
Bull. Mus. Comp. Zoöl.—Bulletin of the Museum of Comparative Zoölogy, Harvard University. Cambridge, Mass.
B. und H. Zeitung.—*Berg- und Huettenmännische Zeitung*. Leipzig, Germany.
M. E.—Transactions of the American Institute of Mining Engineers.
Neues Jahrb.—Neues Jahrbuch für Mineralogie, Geologie und Palaeontologie, often called Leonhard's Jahrbuch. Stuttgart, Germany.
Oest. Zeit. f. Berg- u. Huett.—*Oesterreichische Zeitschrift für Berg- und Huettenwesen*. Vienna, Austria.
Phil. Magazine.—*Philosophical Magazine*. Edinburgh, Scotland.
Proc. Amer. Acad.—Proceedings of the American Academy of Arts and Sciences. Boston, Mass.
Proc. and Trans. N. S. Inst. Nat. Sci.—Proceedings and Transactions of the Nova Scotia Institute of Natural Science. Halifax, Nova Scotia.
Proc. Bost. Soc. Nat. Hist.—Proceedings of the Boston Society of Natural History. Boston, Mass.
Proc. Colo. Sci. Soc.—Proceedings of the Colorado Scientific Society. Denver, Colo.
Raymond's Reports.—Mineral Resources West of the Rocky Mountains. Washington, 1867–1876. The first two volumes were edited by J. Ross Browne, the others by R. W. Raymond.

Trans. Min. Asso. and Inst., Cornwall.—Transactions of the Mining Association and Institute of Cornwall. Tuckingmill, Camborn, England.

Trans. N. Y. Acad. of Sci.—Transactions of the New York Academy of Sciences, formerly the Lyceum of Natural History.

Zeit. d. d. g. Ges.—*Zeitschrift der deutschen geologischen Gesellschaft.* Berlin, Germany.

Zeitsch. f. B., H. und S. im P. St.—*Zeitschrift für Berg-, Huetten-, und Salinenwesen im Preussischen Staat.* Berlin, Germany.

Zeitschr. f. Krys.—*Zeitschrift für Krystallographie.* Munich, Germany.

The remaining abbreviations are deemed self-explanatory. The numbering of the paragraphs is on the following principle: The first digit refers invariably to the part of the book, the second two digits to the chapter, and the last two to the paragraph of the chapter.

PART I.
INTRODUCTORY.

CHAPTER I.

GENERAL GEOLOGICAL FACTS AND PRINCIPLES.

1.01.01.* In the advance of geological science the standpoints from which the strata forming the earth's crust are regarded necessarily change, and new points of view are established. In the last two years two have become especially prominent, and there are now two sharply contrasted positions from which to obtain a conception of the structure and development of the globe. The first is the physical, the second the biological. We may, for example, consider the surface of the earth as formed by rocks, differing in one part and another, and these different rocks or groups of rocks are known by different names. The names have no special reference to the animal remains found in them, but merely indicate that series of related strata form the surface in particular regions. On the other hand, the rocks are also regarded as having been formed in historical sequence, and as containing the remains of organisms characteristic of the period of their formation. They illustrate the development of animal and vegetable life, and in this way afford materials for historical-biological study. In the original classification the biological and historical considerations are all-important. But when once the rocks are placed in their true position in the scale, and are named, these considerations, for many purposes, no longer concern us. The formations are regarded simply as members in the physical constitution of the outer crust. The International Geological Congress held in Berlin in 1885 expressed these different points of view in two parallel and equivalent series of geological terms, which are tabulated on p. 4. They are now very

---

* The numbers at the beginning of the paragraphs are so arranged that the first figure denotes the part of the book, the next two figures the chapter, and the last two the paragraph. Thus 1.06.21 means Part I., Chapter VI., Paragraph 21 under Chapter VI.

generally adopted. For clearness in illustration, the equivalent terms employed by Dana are appended.

Biological Terms.	Physical Terms.	Dana's Terms.	Illustrations.
Era.	Group.	Time.	Paleozoic.
Period.	System.	Age.	Devonian.
Epoch.	Series.	Period.	Hamilton.
Age.	Stage.	Epoch.	Marcellus.

The United States Geological Survey divides as follows: Era and System, Period and Group, Epoch and Formation. In considering the ore deposits of the country, we employ only the physical terms. We understand, of course, the chronological position of the systems in the historical sequence, but it is of small moment in this connection what may be the forms of life inclosed in them. The purely physical character of the rocks—whether crystalline or fragmental; whether limestone, sandstone, granite, or schists; whether folded, faulted, or undisturbed—are the features on which we lay especial stress. In all the periods the same sedimentary rocks are repeated, and in the hand specimen it is often impossible to distinguish those of different ages from one another. The classification, briefly summarized, is as follows:

1.01.02. ARCHÆAN GROUP.—I. Laurentian System. II. Huronian System. Additional subdivisions have been introduced by Canadian and Minnesota geologists (Animikie, Montalban, etc.), and there is a tendency to group all the later schistose members, especially in the region of the Great Lakes, under the name Algonkian. (See discussion under Example 9.)

PALEOZOIC GROUP.—III. Keweenawan System. (This may belong with the Archæan.) IV. Cambrian System: (*a*) Georgian Stage; (*b*) Acadian Stage; (*c*) Potsdam Stage. V. Lower Silurian System. (*A*) Canadian Series: (*a*) Calciferous Stage; (*b*) Chazy Stage. (This will probably experience revision.) (*B*) Trenton Series: (*a*) Trenton Stage; (*b*) Utica Stage; (*c*) Cincinnati or Hudson River Stage. VI. Upper Silurian System. (*A*) Niagara Series: (*a*) Medina Stage; (*b*) Clinton Stage; (*c*) Niagara Stage. (*B*) Salina Series. (*C*) Lower Helderberg Series. (*D*) Oriskany Series. (The Oriskany may be made the base of the Devonian.) VII. Devonian System. (*A*) Corniferous Series: (*a*) Cauda-Galli Stage; (*b*) Schoharie Stage; (*c*) Corniferous Stage. (*B*) Hamilton Series: (*a*) Marcellus Stage; (*b*) Hamilton Stage; (*c*) Genesee

Stage. (*C*) Chemung Series : (*a*) Portage Stage ; (*b*) Chemung Stage. (*D*) Catskill Series. VIII. Carboniferous System. (*A*) Subcarboniferous or Mississippian Series. (*B*) Carboniferous Series. (*C*) Permian Series.

MESOZOIC GROUP.—IX. Triassic System. X. Jurassic System. IX. and X. are not sharply divided in the United States, and we often speak of Jura-Trias. A stratum of gravel and sand, along the Atlantic coast, that contains Jurassic fossils has been called the Potomac formation by McGee. XI. Cretaceous System. Subdivisions differ in different parts of the country. Atlantic Border : (*a*) Raritan Stage ; (*b*) New Jersey Greensand Stage. Gulf States : (*a*) Tuscaloosa Stage ; (*b*) Eutaw Stage ; (*c*) Rotten Limestone Stage ; (*d*) Ripley Stage. Rocky Mountains : (*a*) Comanche Stage ; (*b*) Dakota Stage ; (*c*) Benton Stage ; (*d*) Niobrara Stage ; (*e*) Pierre Stage ; (*f*) Fox Hills Stage ; (*g*) Laramie Stage. Stages (*c*), (*d*), (*e*), and (*f*) are sometimes collectively called the Colorado Stage. Pacific coast : (*a*) Shasta Stage ; (*b*) Chico Stage ; (*c*) Tejon Stage.

CENOZOIC GROUP.—XII. Tertiary System. (*A*) Eocene or Alabama Series : Gulf States, (*a*) Claiborne Stage ; (*b*) Jackson Stage ; (*c*) Vicksburg Stage. Western States, (*a*) Puerco Stage ; (*b*) Wasatch Stage ; (*c*) Wind River Stage ; (*d*) Bridger Stage ; (*e*) Uintah Stage. (*B*) Miocene or Yorktown Series, including perhaps the Sumter of the Atlantic Border. On the Pacific Slope it has the following : (*a*) White River Stage ; (*b*) John Day Stage ; (*c*) Loup Fork Stage. (*C*) Pliocene Series. (Of doubtful American determination.) XIII. Quaternary System. (*A*) Glacial Series. (*B*) Champlain Series. (*C*) Terrace Series. (*D*) Recent Series. Pleistocene is sometimes employed as a name for the early Quaternary, especially south of the Glacial Drift. The Quaternary granitic sands and clay of the coast below the terminal moraine have been called by McGee the Appomattox and Columbia formations.

Other terms are also often used, especially when we do not wish to speak too definitely. "Formation" is a word loosely employed for any of the above divisions. "Terrane" is used much in the same way, but is rather more restricted to the lesser divisions. A stratum is one of the larger sheet-like masses of sedimentary rock of the same kind ; a bed is a thinner subdivision of a stratum. "Horizon" serves to indicate a particular position in the geological column ; thus, speaking of the Marcellus Stage, we say that shales of this horizon occur in central New York.

1.01.03. The rock species themselves are classified into three great groups—the Igneous, the Sedimentary, and the Metamorphic.

The Igneous (synonymous terms, in whole or in part: massive, eruptive, volcanic, plutonic) include all those which have solidified from a state of fusion. They are marked by three types of structure—the holo-crystalline, the porphyritic, and the glassy, depending on the circumstances under which they have cooled. Under the first type of structure come the granites, syenites, diorites, gabbros, diabases, and peridotites; under the second, quartz-porphyries, rhyolites, porphyries, trachytes, porphyrites, andesites, and basalt; under the third, pitchstone, obsidian, and other glasses.

The Sedimentary rocks are those which have been deposited in water. They consist chiefly of the fragments of pre-existing rocks and the remains of organisms. They include gravel, conglomerate, breccia, sandstone,—both argillaceous and calcareous,—shales, clay, limestone, and coal. In volcanic districts, and especially where the eruptions have been submarine, extensive deposits of volcanic lapilli and fine ejectments have been formed, called tuffs. With the sedimentary rocks we place a few that have originated by the evaporation of solutions, such as rock salt, gypsum, etc.

The Metamorphic rocks are usually altered and crystallized members of the sedimentary series, but igneous rocks are known to be subject to like change, especially when in the form of tuffs. They are all more or less crystalline, more or less distinctly bedded or laminated, of ancient geological age or in disturbed districts. They include gneiss, crystalline schists, quartzite, slate, marble, and serpentine.

After a brief topographical survey, we shall employ the above terms to summarize the geological structure of the United States. The several purely artificial territorial divisions are made simply for convenience. Nothing but intelligent travel will perfectly acquaint one with the topographical and geological structure of the country, and in this connection Macfarlane's "Geological Railway Guide" and a geological map are indispensable.

1.01.04. On the east we note the great chain of the Appalachians, with a more or less strongly marked plain between it and the sea. This is especially developed in the south, and is now generally called the Coastal Plain. It is of late geological age and contains the pine barrens and seacoast swamps. The Appalachians themselves consist of many ridges, running on the north into the White

Mountains, the Green Mountains, and the Adirondacks. Farther south the Highlands of New York and New Jersey, the South Mountain of Pennsylvania, the Alleghanies, the Blue Ridge, and the other southern ranges make up the great eastern continental mountain system. In western New York and Ohio we find a rolling, hilly country; in Kentucky and Tennessee, elevated tableland, with deeply worn river valleys. Indiana, Illinois, Iowa, and Missouri contain prairie and rolling country, more broken in southern Missouri by the Ozark uplift. In Michigan, Wisconsin, and Minnesota the surface is rolling and hilly with numerous lakes. In Arkansas, Louisiana, and Mississippi there are bottom lands along the Mississippi and Gulf, with low hills back in the interior. Across Arkansas and Indian Territory runs the east and west Ouachita uplift. West of these States comes the great billowy prairie region, and then the chain of the Rocky Mountains, consisting of high, dome-shaped peaks and ridges, with extended elevated valleys (the parks) between the ranges. Some distance east of the main chain are the Black Hills, made up of later concentric formations around a central, older nucleus, and also the extinct volcanic district of the Yellowstone National Park. In western Colorado, Utah, and New Mexico, between the Rocky Mountains and the Wasatch, is the Colorado plateau, an elevated tableland. This is terminated by the north and south Wasatch range and is traversed east and west by the Uintah range. West of this lies the region called the Great Basin, characterized by alkaline deserts, and subordinate north and south ranges of mountains. Next comes the chain of the Sierra Nevada, and lying between it and the Coast range is the great north and south valley of California. This rises in the comparatively low Coast range, which slopes down to the Pacific Ocean. To the north, these mountains extend into eastern Oregon and Washington, with forests and fertile river valleys. These topographical features are important in connection with what follows, for the reason that the ore deposits especially favor mountainous regions. Mountains themselves are due to geological disturbances—upheaval, folding, faulting, etc.—and are often accompanied by great igneous outbreaks. They therefore form the topographical surroundings most favorable to the development of cavities, waterways, and those subterranean, mineral-bearing circulations which would fill the cavities or replace the rock with useful minerals.

1.01.05. GEOLOGICAL OUTLINE. I. *New England, New York, New Jersey, and Eastern Pennsylvania District.*—In New England and northern New York the Archæan is especially developed, forming the White Mountains, the Adirondacks, and the Highlands of New York and New Jersey. These all consist of granite and other igneous rock, of gneiss, and of crystalline schists. There are also great areas of metamorphic rocks whose true age may be later. The Green Mountains are formed of such, and were elevated at the close of the Lower Silurian. In New England there are small, scattered exposures of the undoubted Paleozoic (Devonian, Carboniferous). In eastern New York, and to some extent in New Jersey and eastern Pennsylvania, the entire Paleozoic, except the Carboniferous, is strongly developed. Up and down the coast there are narrow north and south estuary deposits of red Jura-Trias sandstone, which are pierced by diabase eruptions. The Cretaceous clays are strong, and Tertiary strata occur at Martha's Vineyard, in Massachusetts, while over all as far south as Trenton is found the glacial drift. Between the Archæan ridges of the Highlands and the first foldings of the Paleozoic on the west is found the so-called Great Valley, which also runs to the south and is a very important topographic and geologic feature. It follows the outcrop of the Siluro-Cambrian limestones, to whose erosion it is due.

II. *Eastern-Middle and Southeastern Coast District.*—The low plains of the coast are formed by Quaternary, Tertiary, and Cretaceous, consisting of gravel, sand, shell beds, and clay. Inland there are exposures of Jura-Trias, as in the north. The Archæan crystalline rocks are also seen at numerous points not far from the ocean. Florida is largely made up of limestones, with a mantle of calcareous sand.

III. *Alleghany Region and the Central Plateau.*—The Appalachian mountain system, from New York to Alabama, consists principally of folded Paleozoic (largely Carboniferous), with Archæan ridges on its eastern flank. There is an enormous development of folds, with northeast and southwest axes. On the west they are succeeded by the plateau region of Kentucky and Tennessee, chiefly Paleozoic. Along central latitudes the Archæan does not again appear east of the Mississippi.

IV. *Region of the Great Lakes.*—In Michigan, Wisconsin, and Minnesota the Archæan rocks are extensively developed, both Laurentian and Huronian. Around Lake Superior are found the igneous and sedimentary rocks of the Keweenawan, followed by the lower Paleozoic. Lake Michigan and Lake Huron are surrounded by Silurian, Devonian, and Carboniferous; Lake Erie, by

Devonian; Lake **Ontario**, by Silurian. **Running south through Ohio**, we find an **important fold** known **as the Cincinnati uplift, with a** north and **south axis.** It was elevated **at the close of the Lower Silurian.** In the lower peninsula of **Michigan and in eastern Ohio and western Pennsylvania the Carboniferous is extensively** developed.

V. *Mississippi Valley.*—The **head waters** of the Mississippi **are in the** Archæan. It then passes **over Cambrian and Silurian strata in** Minnesota, Wisconsin, **northern Iowa**, and Illinois, which in these States lie **on the flanks of the** Archæan " **Wisconsin Island"** of central Wisconsin. **These are** succeeded by subordinate **Devonian, and in southern Iowa,** Illinois, and Missouri **by Carboniferous.** In southern Missouri the Lower Silurian **forms the west bank. Thence to the Gulf the** river flows on estuary deposits of Quaternary **age,** with Tertiary and Cretaceous farther inland.

VI. *Gulf Region.*—The **Gulf States along the water front are formed** by the Quaternary. **This is soon succeeded inland by** very **extensive** Tertiary beds, which are **the principal formation represented.**

VII. *Prairie Region.*—West **of the Paleozoic rocks of the** States bordering on the Mississippi is found **a great strip of Cretaceous** running **from the Gulf of Mexico to and across British America, and bounded on** the west **by the foothills of the Rocky Mountains. A few Tertiary** lake deposits **are found in it.** Quite **extensive Triassic rocks** are developed on the **south.** The surface **is a gradually rising** plateau to the Rocky Mountains.

VIII. *Region of the Rocky Mountains, the Black Hills, and the Yellowstone National Park.*—**The Rocky** Mountains rise from **the prairies in long north and south ranges,** consisting of Archæan **axes with the** Paleozoic in relatively **small** amount, but with abundant **Mesozoic on the** east and west **flanks. In the parks** are found **lake deposits of Tertiary age.** There are also great bodies **of igneous rocks, which attended the** various **upheavals.** The principal **upheavals began at the close of the** Cretaceous. **The outlying Black Hills consist of an** elliptical Archæan **core, with concentric Paleozoic and Mesozoic strata** laid up around it. **The National Park** consists **chiefly** of igneous (volcanic) **rocks in enormous** development.

IX. *Colorado Plateau.*—**The Rocky** Mountains **shade out on the west into a great elevated** plateau, extending to central Utah, **where it is cut off by the north and** south chain of the Wasatch.

The Uintah Mountains are an east and west chain in its northern portion. The rocks on the north are chiefly Tertiary, with Mesozoic and Paleozoic in the mountains. To the south are found Cretaceous and Triassic strata, with igneous rocks of great extent. The principal upheaval of the Wasatch began at the close of the Carboniferous and seems still to be in progress.

X. *Region of the Great Basin.*—Between the Wasatch and the Sierra Nevada ranges is found the Great Basin region, once lake bottoms, now very largely alkaline plains of Quaternary age. The surface is diversified by subordinate north and south ranges, formed by great outflows of eruptive rocks, and by tilted Paleozoic. The ranges are extensively broken and the stratified rocks often lie in confused and irregular positions. There is no drainage to the ocean.

XI. *Region of the Pacific Slope.*—The depression of the Great Basin is succeeded by the heights of the Sierra Nevada. On the west the Sierras slope down into the Central Valley of California. The flanks are largely metamorphosed Jurassic and Cretaceous rocks with great developments of igneous outflows. The surface rises again in the Coast ranges, which slope away farther west to the ocean. In addition to the Jurassic and Cretaceous, the Tertiary and Quaternary are also developed, and in the Coast ranges are many outflows of igneous rock. The principal upheaval of the Sierra Nevada began at the close of the Jurassic, that of the Coast range at the close of the Miocene Tertiary.

XII. *Region of the Northwest.*—Washington and Oregon, along the coast, are formed by Cretaceous and Tertiary strata similar to California. But inland, immense outpourings of igneous rocks cover the greater portion of both States and extend into Idaho. On the north the Carboniferous is extensive, running eastward into Montana. Quaternary and Tertiary lake deposits are also not lacking.

1.01.06. *On the Forms Assumed by Rock Masses.*—All sedimentary rocks have been originally deposited in beds, approximately horizontal. They are not of necessity absolutely horizontal, because they may have been formed on a sloping bottom or in a delta, in both of which cases an apparent dip ensues. We find them now, however, in almost all cases changed from a horizontal position by movements caused primarily by the compressive strain in the earth's crust. Beds thus assume folds known as monoclines, anticlines, and synclines.

GENERAL GEOLOGICAL FACTS AND PRINCIPLES. 11

A monocline is a terrace-like dropping of a bed without changing the direction of the dip. There is usually a zone, more or less shattered, along the folded portion, and such a zone may become a storage receptacle. Monoclines of a gentle character in Ohio, which have been detected by Orton in studies of natural gas, have been called "arrested anticlines." An anticline is a convex fold with opposing dips on its sides, while a syncline is a concave fold with the dips on its sides coming together. We speak of the axis of a fold, and this marks the general direction of the crest or trough. The axis is seldom straight for any great distance. Folds are often broken and faulted across the strike of their axes, and this causes what is called a "pitch" of the axes and makes the original dips run diagonally down on the final one. Folds are the primary cause of the phenomena of dip and strike. Horizontal beds have neither. A dome-like elevation of beds, with dips radiating in every direction from its summit, is called a quaquaversal, but it is a rare thing. An anticline or syncline with equal dips on opposite sides of its axes is called a normal fold. If the dip is steeper on one side than on the other, it is an overthrown fold; if the sides are crushed together, it is a collapsed or sigmoid fold.

Igneous rocks are in the form of sheets (the term "bed" should be restricted to sedimentary rocks), knobs or bosses, necks, laccolites, and dikes. A sheet is the form naturally assumed by surface flows, and by an igneous mass which has been intruded between beds. It has relatively great length and breadth as compared with its thickness, and coincides with its walls in dip and strike. A knob, or boss, is an irregular mass, of approximately equal length and breadth, which may be related in any way to the position of its walls. Such masses are often left projecting by erosion. A neck is the filled conduit of a volcano, which sometimes remains after the overlying material has been denuded. A laccolite is a lenticular sheet which has spread between beds radially from its conduit, and thus has never reached the surface, unless revealed by subsequent erosion. A dike is a relatively long and narrow body of igneous rock which has been intruded in a fissure. It is analogous to a vein, but the term "vein" ought not to be applied to an undoubtedly igneous rock. Some granitic mixtures, however, of quartz, feldspar, and mica, leave us yet in uncertainty as to whether they are dikes or veins. (See Example 56.) From the above it will be seen at once that bosses, knobs, and necks may be practically indistinguishable.

CHAPTER II.

ON THE FORMATION OF CAVITIES IN ROCKS.

1.02.01. *By Local Contraction.*—In the contraction caused by cooling, drying, or hardening, both igneous and sedimentary rocks break into more or less regular masses along division planes, called joints, or diaclases. Numerous cracks and small cavities are thus formed. Basaltic columns, or the prismatic masses, formed by the separation, in cooling and consolidating, of the heavier basic rocks, along planes normal to the cooling surface, are good illustrations of the first. Larger manifestations of them often become filled with zeolites, calcite, and other secondary minerals. Granitic

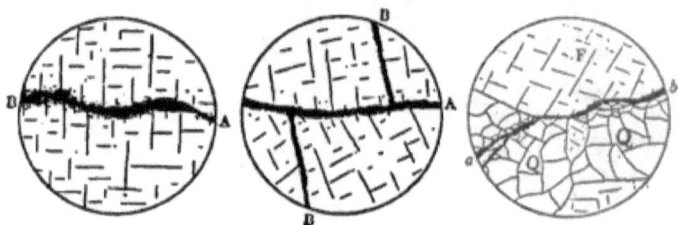

FIG. 1.—*Illustration of rifting in granite at Cape Ann, Mass. After R. S. Tarr.*

rocks and porphyries break up less regularly from the same cause, but still exhibit prismoids and polygonal blocks and benches. (J. P. Iddings' paper on "The Columnar Structure in the Igneous Rocks on Orange Mountain, N. J.," *Amer. Jour. Sci.*, III., xxxi. 320, is an excellent discussion.) Large cracks have been referred to this cause, which have afterward formed important receptacles for ores. (See Example 11*a*.) Very small microscopic cracks may occasion lines of weakness and brecciation which are not readily apparent. They afford a cleavage, called rifting, but are not well understood. (See R. S. Tarr, "On Rifting in Cape Ann Granites," *Amer. Jour. Sci.*, April, 1891.) Joints are generally prominent in sedimentary rocks, and probably afford many of the regular planes of separation which are often seen crossing one or several beds. They are chiefly due to drying and consolidation. Both the joints formed by cooling and those formed by drying may be afterward modified or increased by rock movements, so that it may be a matter of difficulty to decide between the two forms of origin. The undulatory tremors of an earthquake have been cited, with great reason, by W. O. Crosby as a prolific cause of joints.*

1.02.02. *Cavities Formed by More Extensive Movements in the Earth's Crust.*—The strains induced by the compression in the outer portion of the earth are by far the most important causes of fractures. The compression develops a tangential stress which is resisted by the archlike disposition of the crust. (By the term "crust" is simply meant the outer portion of the globe without reference to the character of the interior.) Where there is insufficient support, gravity causes a sagging of the material into synclinals, which leave salient anticlinals between them. Where the tangential strain is also greater than the ability of the rocks to resist, they are upset and crumpled into folds from the thrust. Both kind of folds are fruitful causes of fissures, cracks, and general shattering, and every slip from yielding sends its oscillations abroad, which cause breaks along all lines of weakness. The simplest result, either from sagging or from thrust, is a fissure, on one of whose sides the wall has dropped, or on the other of which it has risen, or both, as will be more fully described under "Faults." If the rocks are firm and quite thickly bedded, as is the case with limestones and quartzites, the separation is cleanly cut; but if they are softer and more yielding, they are sheared downward on the stationary or lifting side, and upward on the one which relatively sinks. Such fissures may pass into folds along their strike, as at Leadville, Colo.

---

* In addition to the usual text-books the following references may be consulted: W. O. Crosby, "On the Origin of Jointed Structures," *Boston Soc. Nat. Hist.*, XXII., October, 1882, p. 72; *Amer. Jour. Sci.*, III., xxv. 476; G. K. Gilbert, "On the Origin of Jointed Structures," *Amer. Jour. Sci.*, III., xxiv. 50, and xxvii. 47; J. H. Kinahan, *Valleys, and their Relations to Fissures, Fractures, and Faults*, London, Trübner & Co. See also a short letter in the *Amer. Jour. Sci.*, III., xxiv., p. 68, on the "Origin of Jointed Structures;" J. Leconte, "Origin of Jointed Structure in Undisturbed Clay and Marl Deposits," *Amer. Jour. Sci.*, III., xxiii. 233; W. J. McGee, "On Jointed Structure," *Amer. Jour. Sci.*, III., xxv. 152, 476.

1.02.03. A phenomenon which is especially well recognized in metamorphic regions, and which is analogous to those last cited, is furnished by the so-called "shear zones." A faulting movement, or a crush, may be made apparent by changes in mineralogical composition and structure. Massive diabases, for instance, pass into hornblende-schists or amphibolites for limited stretches. Garnets and other characteristically metamorphic minerals appear, and pyroxenes alter to amphiboles. Strains are manifested in the optical behavior of the minerals in thin sections of specimens taken from such localities. These crushed strips, or shear zones, may be formed with very slight displacement, but they afford favorable surroundings for the formation of ore bodies. This conception of the original condition of a line of ore deposition is a growing favorite with recent writers, and combined with the idea of replacement is often applicable. (See Example 17, Butte, Mont.) Fahlbands, which are very puzzling problems, may have originated as shear zones.

1.02.04. A more extended effect is produced by the monocline, which has a double line of shattered rock marking both the crest and the foot of its terrace. Anticlines and synclines occasion the greatest disturbances. Comparatively brittle materials like rocks cannot endure bending without suffering extended fractures. When strained beyond their limit of resistance, along the crest of an anticline and in the trough of a syncline, cracks and fractures are formed which radiate from the axis of each fold. As these open upward and outward in anticlines, they become the easiest points of attack for erosion, so that it is a very common thing to find a stream flowing in a gorge, which marks the crest of an anticline, while synclinal basins are frequently left to form the summits of ridges, as is so markedly the case in the semi-bituminous coal basins of Pennsylvania. It is quite probable, however, that the anticline may have been leveled off at this fissured crest because it was upheaved under water and became exposed at its vulnerable summit to wave action.

Ore deposits may collect in these fissured strips, of which the lead and zinc mines of the upper Mississippi Valley (Example 24) are illustrations. Such fissures are peculiar in that they exhibit no displacement. The accompanying figure is from a photograph of a gaping crack in the Aubrey (Upper Carboniferous) limestone, twenty-five miles north of Cañon Diablo station, Ariz., on the Atlantic and Pacific Railroad. It was caused by a low anticlinal roll, and contained water about one hundred feet below the top. Its reproduction of the conditions of a vein, with horse, pinches and swells, devious course, and all, is striking. The photograph was made by Mr. G. K. Gilbert, of the United States Geological Survey, and to his courtesy its use is due.

FIG. 2.—*Open fissure in the Aubrey limestone (Upper Carboniferous), 25 miles north of Cañon Diablo station, Ariz., on the A. & P. R. R. Photographed by G. K. Gilbert, 1892.*

While it is true that in many regions the folds and fractures have resulted in this simple way, and exhibit the unmistakable course through which they have passed, yet geological structure is by no means always so clear. Extended disturbances, great faults and displacements, combined with folds and the intrusion of igneous rocks, have often so broken up a district that it is a matter of much difficulty to trace out the course through which it has passed. Subsequent erosion, or the superposition of heavy beds of gravel or forest growths, etc., may so obstruct observation even of the facts as to add to the obscurity. The expense of making and the consequent scarcity of accurate contour maps to assist in such work are other obstacles. The profound dynamic effects wrought by mountain-making processes, although in individual cases producing only the simpler phenomena already cited, yet in general are much more extensive, and must be considered in the study of many large districts. When folds are the result of compression or thrust, the dynamic effects are more marked than in those formed by sagging. Faults are larger and more abundant. When sedimentary beds have been laid down along an older axis of granite or some equally resistant rock and the thrust crowds the beds against this axis, the conditions are eminently favorable to great fracturing and disturbance. The flanks of the Rocky Mountains furnish such examples.

1.02.05. There are also great lines of weakness in the outer portion of the earth, which seems to have been the scene of faulting movements from a very early period. Thus on the western front of the Wasatch Mountains, in Utah, is a great line of weakness, that was first faulted, as nearly as we can discover, in Archæan times, and has suffered disturbances even down to the present. A few instances of actual movements within recent years have been recorded. In 1889 a sudden small fold and fissure developed under a paper mill near Appleton, Wis., and heaved the building four and a half inches. (See F. Cramer, "Recent Rock Flexure," *Amer. Jour. Sci.*, III., xxxix. 220.) This occurred in what was regarded a settled region and one not liable to disturbance.

1.02.06. Wherever igneous rocks form relatively large portions of the globe they necessarily share extensively in terrestrial disturbances. Not being often in sufficiently thin sheets, they rarely furnish the phenomena of dip and strike. Folds are largely wanting. They are replaced by faults and shattering. The fissures thus formed are at times of great size and indicate important movements. The Comstock Lode fissure is four miles long and in the central part exhibits a vertical displacement of three thousand feet. (See 2.11.19.) Such fissures seldom occur alone, but minor ones are found on each side and parallel with the main one.

1.02.07. The intrusion of igneous dikes may start earthquake vibrations which fracture the firm rock masses. Fissures caused in this way radiate from the center of disturbance or else appear in concentric rings. The violent shakings which so often attend great volcanic eruptions, and the sinking of the surface from the removal of underlying molten material, all tend to form cracks

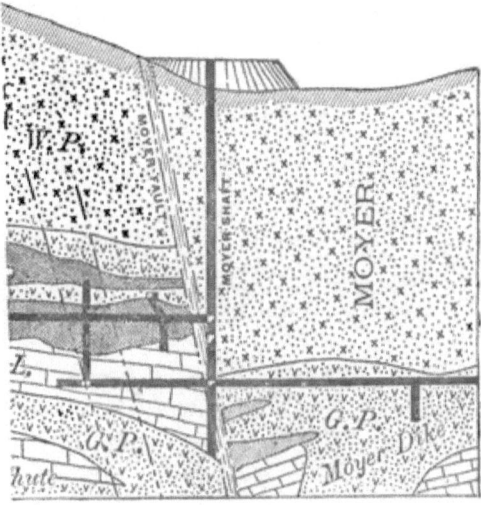

FIG. 3.—*Normal fault at Leadville, Colo. After A. A. Blow.*

and cavities. They are possible causes which may well be borne in mind in the study of an igneous district.

1.02.08. *Faults.*—When fractures have been formed by any of the means referred to above, and the opposite walls slip past each other, so as not to correspond exactly at all horizons, they are called "faults," a term which indicates this lack of correspondence.

The separation is chiefly due to the relative slipping down or sinking of one side. The distance through which this has taken place is called the amount of displacement, or throw. Faults are most commonly inclined to the horizon, so that there is both a vertical and a horizontal displacement. What would be the dip of an inclined stratum is in a fault now generally known as the "hade," although the word formerly had a different meaning. Experience has shown that where beds or veins encounter faults and operations are brought to a standstill, the continuation is usually found as follows, according to Schmidt's law. If the fault dips or hades **away** from the workings, the continuation is down the hade; if it dips toward the workings, it should be followed upward. Such a fault is called a normal fault, and is illustrated in the figure on p. 17, after A. A. Blow. This is a natural result of the drawing apart of the two sides. The least supported mass would slip down **on the one** which has the larger base. Less commonly the **opposite movement** results. Thus when the fault is due to compression, the beds pass each other in the reverse direction, and what **is** called a reverse fault results. The accompanying cut illustrates a very extended one in the southern Appalachians. While we would naturally think of a reversed fault as resulting

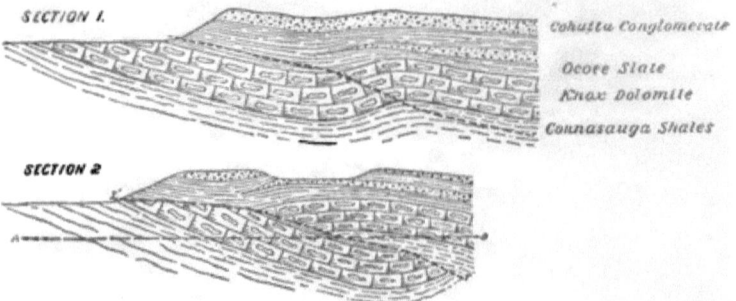

FIG. 4.—*Reversed fault at Holly Creek, near Dalton, Ga. After C. W. Hayes,* **Bull. G. S. A.,** *Vol. II., Pl. 3, p. 152.*

from a compressive strain, in that in this case the lower wedged-shaped portion **would be forced under** the upper one, yet normal faults **can likewise,** in instances, be explained by compression. If we **consider the** fault to be caused **by** the vertical thrust or component, that would always be present **in the compression of a** completely supported arch, **this** would tend to heave upward the **portion next the fissure, that** had the larger base. Along an inclined fracture such portion is manifestly the under one. It is also **important to note whether the fault** plane, both in normal and in **reversed** faults, cuts inclined **beds in** the direction of the dip or **across it.** (See Margerie and Heim, *Dislokation der* **Erdrinde,** Zurich, 1888.)

1.02.09. **The movement of the walls on each** other produces **grooves** and polished **surfaces called slickensides,** or slips. They are usually covered with a layer of serpentine or talc or some such secondary product. The strain caused by the movement may in rare instances leave the slips in such a state of tension that when, from any cause such as excavation, the pressure is relieved, they will scale off with a small explosion. (See A. Strahan, "On Explosive Slickensides," *Geological Magazine*, iv. 401, 522.) Observations on the directions of slips may, in cases of doubt, throw some additional light on the direction of the movement which caused the fault. But the best guide in stratified rocks is a knowledge of the succession of the beds as revealed by drill cores or excavations. Attempts have been made to deduce mathematical formulas for the calculation of the amount of downthrow or upthrow, and when sufficient data are available, as is often the case in coal seams, this may be done. The methods depend on the projection of the planes in a drawing, on the principles of analytical geometry, and on the calculation of the displacements by means of spherical trigonometry. (See G. Koehler, *Die Störungen der Gänge, Flötze und Lager*, Leipzig, 1886; William Engelmann. A translation by W. B. Phillips, entitled, "Irregularities of Lodes, Veins, and Beds," appeared in the *Engineering and Mining Journal*, June 25, 1887, p. 454, and July 2, 1887, p. 4. A very excellent paper, having a quite complete bibliography, is F. T. Freeland's "Fault Rules," *M. E.*, June, 1892.) Prof. Hans Hoefer has called attention to the fact that in faulting there is frequently a greater displacement in one portion of the fissure than in a neighboring part, and even a difference of hade. This causes a twisting or circular movement of one wall on the other, and needs to be allowed for in some calculations. (*Oestereich. Zeitschrift für Berg- und Huettenwesen*, Vol. XXIX. An abstract in English is given by R. W. Raymond, *M. E.*, February, 1882.) In the *Engineering and Mining Journal* for April and May, 1892, a quite extended discussion of faults by several prominent American mining engineers and geologists is given, apropos of the question raised by Mr. J. A. Church as to whether fissure veins are more regular on the dip or on the strike. In a relatively uniform massive rock the regularity should be greater on the dip, but in inclined and diversified stratified rocks too many variables enter to warrant any sweeping assertions. In soft rocks like shales the fissure may become so split into small stringers as to be valueless. Again, in very firm rock, where there is little drawing apart, the fissure may be very tight. In the veins of Newman Hill, near Rico, Colo. (see 2.09.10), the fissure is so narrow above a certain stratum as practically to fail. Quartzite is a favorable rock for such effect. Despite all rules, faults are often causes of great uncertainty, annoyance, and expensive exploration.

1.02.10. If a number of faults succeed one another in a short distance they are called "step faults." An older and completed vein may also be faulted by one formed and filled later. In such a case the continuous one is the younger. The figure below will illustrate each case. At the intersection of the two, the later vein is often richer than in other parts.

1.02.11. If a faulted series of rocks is afterward tilted and eroded, so as to expose a horizontal section across the strike of the faulting plane, an apparent horizontal fault may result ; or if the erosion succeeds normal faulting and lays bare two unconformable beds each side of the fissure, a lack of correspondence in plan as

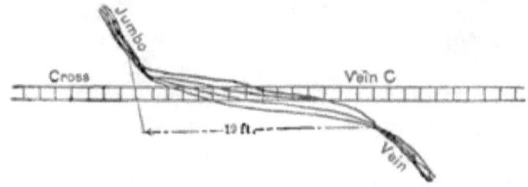

Fig. 5.—*Illustration of one vein faulting another at Newman Hill, near Rico, Colo. After J. B. Farish, Proc. Colo. Sci. Soc., April 4, 1892.*

well as in section may be seen. Faulting fractures are seldom straight ; on the contrary, they bend and corrugate. When the walls slip past each other, they often stop with projection opposite projection, and depression opposite depression. These irregularities cause pinches and swells in the resulting cavity, and constitute one of the commonest phenomena of veins. Fissures also gradually pinch out at their extremities, or break up into various ramifications that finally entirely cease. They also pass into folds, as stated above.

1.02.12. *Secondary Modifications of Cavities.*—Fractures and cavities of all sorts speedily become lines of subterranean drainage. The dissolving power of water, and to a much smaller degree its eroding power, serve to modify the walls very greatly. An enlargement may result, and what was perhaps a small joint or fissure may become a waterway of considerable size. This is especially true in limestones, in which great caverns (like the

Mammoth Cave and Luray's Cave) are excavated. Caves are, however, almost always due to surface waters, and do not extend below the permanent water level unless they have been depressed after their formation. (See J. S. Curtis, *Monograph VII., U. S. Geol. Survey*, Chap. VIII.)

The solvent action of water is vastly augmented by the carbonic acid which it gathers from the atmosphere, and this is the chief cause of the excavations wrought by it in limestones. Pure cold water has comparatively small dissolving and almost no erosive power. It has also been advocated that various acids which result from the decay of vegetable matter aid in such results. (A. A. Julien, *Amer. Asso. Adv. Sci.*, 1879, p. 311.) This may be true, but in general carbonic acid is the chief agent. Iron in minerals falls an easy prey, as well as calcium, and is dissolved out in large amount. (See Example 1.) When charged with alkaline carbonates, water has the power to attack other less soluble minerals, such as quartz and the silicates, and by such action the walls of a cavity in the crystalline rocks may be much affected.

1.02.13. Waters percolating to great depths in the earth, or circulating in regions of igneous disturbances, become highly heated, and this too at great pressure. Under such circumstances the solvent action is very strongly increased, and all the elements present in the rock-making minerals are taken into solution. Alkaline carbonates are formed in quantity; silica is easily dissolved; alkaline sulphides result in less amount; and even the heaviest and least tractable metals enter into solution, either in the heated waters themselves, or in the alkaline liquors formed by them. The action on the walls of cavities and courses of drainage is thus profound, and accounts for the frequent decomposed character of the walls and the general lack of sharpness in their definition. The vast amount of siliceous material, etc., deposited by hot springs and geysers is additional evidence of its importance.

Magnesia is one of the alkaline earths readily taken into solution by carbonated waters, and when such waters again meet limestone the effect is often very great, and constitutes one of the most important methods of the formation of cavities. Solutions of magnesium carbonate, on meeting calcium carbonate, effect a partial exchange of the former for the latter. This leaves the rock a double carbonate of calcium and magnesium, which is the composition of the mineral and rock dolomite. The process is therefore called dolomitization. (See Example 25.) It may bring about a general shrinkage of eleven or **twelve per cent**. In any extended **thickness of strata this would cause vast shattering** and porosity. As an illustration of its results, the following analyses of normal, unchanged Trenton limestone of Ohio, and of well drillings **from the** porous, gas-bearing, dolomitized **portions of the same, are given.** They are taken from **a paper by Edward Orton.** (*Amer. Manuf. and Iron World*, Pittsburg, **Dec. 2, 1887.)**

			$CaCO_3$.	$MgCO_3$.	$Fe_2O_3.Al_2O_3$.	SiO_2.
Unchanged	Trenton	limestone...	79.30	0.92	7.00	12.00
"	"	"	...82.36	1.67	0.58	12.34
Dolomitized	"	"	...53.50	43.50	1.25	1.70
"	"	"	...51.78	36.80		

1.02.14. **Late studies in ore** deposits by Posepny, **Curtis, and** Emmons indicate also that solutions of metallic ores **may** effect an interchange of their contents with **the carbonate of calcium or magnesium,** in limestones and dolomites, leaving an ore **body in place of the** rocks. **This change is** effected molecule by **molecule, and is spoken of as a** metasomatic interchange or replacement. **(See Example 30.)** By "metasomatic" is meant an **interchange of substance without, as** in pseudomorphs, an **imitation of form. Alteration of the metallic ores may follow and occasion cavities from shrinkage. (See Example 36, and** Curtis, on Eureka, Nev., *Monograph VIII., U. S. Geol. Survey,* Chap. VIII.)

CHAPTER III.

THE MINERALS IMPORTANT AS ORES; THE GANGUE MINERALS, AND THE SOURCES WHENCE BOTH ARE DERIVED.

1.03.01. The minerals which form the sources of the metals are almost without exception included in the following compounds: the sulphides, the related compounds of arsenic and antimony, oxides and oxidized compounds such as hydrous oxides, carbonates, sulphates, phosphates, and silicates, and one or two compounds of chlorine. A few metals occur in the native state. All the other mineral compounds such as a chromate or two, a bromide or iodide, etc., are rarities. It may be said that nine tenths of the productive ores are sulphides, oxides, hydroxides, carbonates, and native metals. The ores of each metal are subsequently outlined before its particular deposits are described.

1.03.02. The most common gangue mineral is quartz, while in less amount are found calcite, siderite, barite, fluorite, and in places feldspar, pyroxene, hornblende, rhodonite, etc. The silicates are chiefly present where the gangue is a rock and the ore is disseminated through it. All the common rocks serve in this capacity in one place or another.

1.03.03. *Source of the Metals.*—The metallic contents of the minerals which constitute ores must logically be referred to a source, either in the igneous rocks or in the ocean. If the nebular hypothesis expresses the truth,—and it is the best formulation that we have,—all rocks, igneous, sedimentary, and metamorphic, must be traced back to the original nebula. This, in cooling, afforded a fused magma, which chilled and assumed a structure analogous to the igneous rocks with which we are familiar. Igneous rocks must thus necessarily be considered to have furnished by their erosion and degradation the materials of the sedimentary rocks; while igneous and sedimentary have alike afforded the substances whose alterations have produced the metamorphic rocks. It may also be true that eruptive rocks, especially when basic, have been formed, by the oxidation and combination with silica, of inner metallic portions of the earth, for such is one of our most reasonable explanations of volcanic phenomena, suggested alike by the composition of basalts, by the high average specific gravity of the globe, and by analogy with meteorites.

1.03.04. As opposed to this conception, there are those who would derive the metallic elements of ores from the ocean, in which they have been dissolved from its earliest condensation. Thus it is said that substantially all the metals are in solution in sea water. From the sea they are separated by organic creatures, it may be, through sulphurous precipitation, attendant on the decay of their dead bodies. The accumulations of the remains of organisms bring the metals into the sedimentary strata. Once thus entombed, circulation may concentrate them in cavities. When present in igneous rocks, the latter are regarded as derived from fused sediments. If the metallic contents of sedimentary rocks do not come from the ocean in this way, the igneous rocks as outlined above are the only possible source. No special mention is here made of the metamorphic rocks, because in their original state they are referable to one or the other of the two remaining classes. But it is not justifiable, in the absence of special proof, to consider them altered sediments, any more than altered igneous rocks, and it is doubtless true that the too generally and easily admitted sedimentary origin for our gneisses and schists has materially hindered the advance of our knowledge of them in the last forty years.

1.03.05. Microscopic study of the igneous rocks has shown that, with few exceptions, the rock-making minerals separate from a fused magma on cooling and crystallizing, in a quite definite order.[1] Thus the first to form are certain oxides, magnetite, specular hematite, ilmenite, rarely chromite and picotite, a few silicates, unimportant in this connection (zircon, titanite), and the sulphides pyrite and pyrrhotite. Next after these metallic oxides, etc., the heavy, dark-colored, basic silicates, olivine, biotite, augite, and hornblende, are formed. All these minerals are characterized by high percentages of iron, magnesium, calcium, and aluminum. They are very generally provided with inclusions of the first set. Following the bisilicates in the order of crystallization,

---

[1] H. Rosenbusch, "Ueber das Wesen der Körnigen und Porphyrischen Structur bei Massengesteine," *Neues Jahrbuch*, 1882, ii., I.

come the feldspars, and after these, when some residual silica remains uncombined, it separates as quartz.

1.03.06. If we regard the igneous rocks as the source, the metallic elements are thus to be ascribed to the first and second series of crystallizations, while the elements of the gangue minerals are derived from the last three. It is a doubtful point whether the less common metals, such as copper, silver, and nickel, enter into the composition of the dark silicates as bases, replacing the iron, alumina, lime, etc., or whether they are present in them purely as inclusions of the first series. F. Sandberger[1] argues in support of the first view, but his critics, notably A. W. Stelzner, cast doubt upon his conclusions on the ground that his chemical methods were indecisive. The case is briefly this: Sandberger, as an advocate of views which will be subsequently outlined, separated the dark silicates of a great many rocks. By operating on quantities of thirty grams he proved the presence in them of lead, copper, tin, antimony, arsenic, nickel, cobalt, bismuth, and silver, and considered these metals to act as bases. The weak point of the demonstration consists in dissolving out from the powdered silicate any possible inclusions. There seems to be no available solvent which will take the inclusions and be without effect on the silicates. This is the point attacked by the critics, and apparently with reason. It is, however, important to have shown the presence of these metals, even though their exact relations be thus doubtful. Quite recently in a series of "Notes on Chilean Ore Deposits" (Tschermaks *Min. und Petrog.*, Mitth. XII., p. 195) Dr. Möricke mentions native gold in pearlstone (obsidian) from Guanaco, in skeleton crystals in the glass, as inclusions in perfectly fresh plagioclase and sanidine crystals, and in spherulites. The existence of silver in quartz-porphyry has been demonstrated in this country by J. S. Curtis, at Eureka, Nev.;[2] both the precious metals have been shown by G. F. Becker to be in the diabase near the Com-

---

[1] The principal paper of Professor Sandberger is his "Untersuchungen über Erzgänge," 1882, abstracted in the *Engineering and Mining Journal*, March 15, 22, and 29, 1884; but a long series of others might be cited in which the investigations, notably at Pribram, Bohemia, are interpreted as indicated above A. W. Stelzner, *B. and H. Zeit.*, xxxix., No. 3. *Zeitsch. d. d. g. Gesell.*, xxxi. 644. "Die Lateral-secr-tions-Theorie, etc." Reprint Freiberg, 1889.

[2] *Monograph VII., U. S. Geol. Survey.*

stock Lode;[1] and, by the same investigator, antimony, arsenic, lead, copper, gold, and silver were proved to be contained in the granite near Steamboat Springs, Nev.[2] S. F. Emmons has also shown that the porphyries at Leadville contain appreciable, though small, amounts of silver.[3] Of forty-two specimens tested, thirty-two afforded it; of seventeen tested for lead, fourteen yielded results. Undoubtedly the multiplication of tests will show similar metallic contents in other regions. Thus the augite of the eastern Triassic diabase will probably yield copper, for this metal is abundant in connection with the outflows.

1.03.07. That the metals are so generally combined with sulphur in ore deposits seems to be due to the extended distribution of this element, and to its being a vigorous precipitating agent of nearly all the metals at the temperatures and pressures near the surface. Sulphur is widespread as pyrite, an original mineral in many igneous rocks, and one much subject to alteration; while sulphuretted hydrogen is common in waters from sedimentary rocks, and is a very general result of organic decomposition. Natural gas and petroleum from limestone receptacles almost always contain it. (See, in this connection, J. F. Kemp, "The Precipitation of Metallic Sulphides by Natural Gas," *Engineering and Mining Journal*, Dec. 13, 1890.) Many sulphides, too, are soluble under the pressures and temperatures prevailing at great depths, but are deposited spontaneously at the pressures and temperatures prevailing at or near the surface.

1.03.08. Where veins occur in igneous rocks the bases for gangue minerals have been obtained from the rock-making silicates. Calcium is afforded by nearly all the important ones; silicon is everywhere present; barium has been proved in many feldspars, in small amount; and magnesia is present in many pyroxenes and amphiboles. Of the sedimentary rocks, limestone of course affords unlimited calcium, and recently Sandberger reports that he has identified microscopic crystals of barite in the insoluble residues of one. (*Sitzungsberichte d. Math. phys. Classe d. k. bayer, Akad. d. Wiss.*, 1891, xxi. 291.) This is of interest, as barite is such a common gangue in limestone.

1.03.09. It may be remarked that the natural formation of both

---

[1] *Monograph III., U. S. Geol. Survey.*
[2] *Monograph XIII., U. S. Geol. Survey.*
[3] *Monograph XII., U. S. Geol. Survey.*

ore and gangue minerals has doubtless proceeded in nature with great slowness, and from very dilute solutions. Both classes exhibit a tendency to concentrate in cavities, even from a widely dispersed condition through great masses of comparatively barren rock. The formation may have proceeded when the walls were far below their present position with regard to the surface, so that to those inclined a wide latitude for speculation on origin is afforded. It is also possible that in the earlier history of the globe circulations were more active than they are now—a line of argument on which a conservative writer would hesitate to enlarge.

CHAPTER IV.

ON THE FILLING OF MINERAL VEINS.

1.04.01. Bearing in mind what precedes, the preliminaries for the discussion of mineral veins are set in order. We have traced the formation of cavities by the shrinkage of rock masses in cooling or drying, by the movements and disturbances of the earth's crust (which are far the commonest and most important causes), and by dolomitization. The enlargement of such cavities by subterranean circulations followed, and the general effect of waters, cold and heated. The sources of the elements of the useful minerals were pointed out so far as known. All these general and indisputable truths assist in the drawing of right conclusions. It should be emphasized, as will appear later, that mineral veins or cavity fillings do not embrace all metalliferous deposits. On the contrary, the deposits which either form beds by themselves, or which are disseminated through beds of barren rock and are of the same age with them, do not enter into the discussion. They are characterized by being younger than their foot walls and older than the hanging. Their geological structure is far simpler, and, as will appear in the discussion of particular examples, the working out of their origin does not so often carry the investigator into the realms of speculation and hypothesis. And yet it is not to be inferred from the prominence here given to the discussion of veins that bedded deposits yield to them, in any degree, in importance. Iron ores, for instance, are often in beds.

1.04.02. *Methods of Filling.*—Methods of filling were summed up a very long time ago by Von Herder and Von Cotta,[1] as follows: 1. Contemporaneous formation. 2. Lateral secretion. 3. Descension. 4. Ascension by (*a*) infiltration, or (*b*) sublimation with steam, or (*c*) by sublimation as gas, or (*d*) by igneous injection. To these should be added the more recent theory of (5) re-

---

[1] *Erzlagerstätten*, 2d ed., 1859, Vol. I., p. 172.

placement, which, however, is rather a method of precipitation than of derivation. No one longer believes in contemporaneous formation, and descension has an extremely limited, if, indeed, any application. Ascension by sublimation as gas or with steam, or by igneous injection, has few, if any, supporters. The discussion is practically reduced to lateral secretion and to ascension by infiltration.

1.04.03. By lateral secretion is understood the derivation of the contents of a vein from the wall rock. The wall rock may vary in character along the strike and in depth. Three interpretations may be made, two of which approach a common middle ground with ascension by infiltration. It may first be supposed that the vein has been filled by the waters near the surface which are known to be soaking through all bodies of rock, even where no marked waterway exists, and which seep from the walls of any opening that may be afforded. Being at or within comparatively short distances of the surface, the waters are not especially heated. As they emerge to the oxidizing and evaporating influence of the air in the cavity, their burden of minerals is deposited as layers on the walls. The second interpretation supposes the walls to be placed during the time of the filling at considerable depth below the surface, so that the percolating waters are brought within the regions of elevated temperature and pressure. Essentially the same action takes place as in the first case. The third interpretation increases the extent of the rock leached. Thus if a mass of granite incloses a vein and extends to vast depths, we may suppose the waters to come from considerable distances, and to derive their dissolved minerals from a great amount of rock of the same kind as the walls. Portions of this may even be in the regions of high temperature, while the place of precipitation is nearer the surface. These last two interpretations have much in common with the theory of ascension by infiltration, and on this common middle ground lateral-secretionists and infiltration-ascensionists may be in harmony.

1.04.04. *Ascension by Infiltration.*—The theory of infiltration by ascension in solution from below considers that ore-bearing solutions have come from the heated zones of the earth, and that they rise through cavities, and at diminished temperatures and pressures deposit their burdens. No restriction is placed on the source from which the mineral matter has been derived. Indeed, beyond that it is "below," and yet within the limits reached by waters, all of which have descended from the surface, and that the metals have been gathered up from a disseminated condition in rocks, —igneous, sedimentary, and metamorphic,—no more definite statement is possible. This theory is of necessity largely speculative, because the materials for its verification are beyond actual investigation.

1.04.05. In favor of lateral secretion the following arguments may be advanced. I. According to Sandberger, actual experience with the conduits, either natural or artificial, of mineral springs, shows that a deposit seldom, if ever, gathers in a moving current. It is only when solutions come to rest on the surface and are exposed to oxidation and evaporation that precipitation ensues. Deposits in veins have therefore formed in standing waters, whose slight overflow or evaporation would be best compensated by the equally slight and gradual inflow from the walls. If in hot springs there were a strong and continuous flow from below and discharge from above, the mineral matter would reach the surface. (Sandberger, *Untersuchungen über Erzgänge*, Heft I.) Hence the deposit would be more likely to gather by the slow infiltrations from the wall rock, which would stand in cavities like a well. We have, however, some striking instances of deposits in artificial conduits.

Prof. H. S. Munroe has called the writer's attention to a case recently met by him. The fourteen-inch column pipe of a pump at the Indian Ridge Colliery, Shenandoah, Penn., which was raising ferruginous waters, became reduced in diameter to five inches within two years by the deposit of limonite. The same amount of water was forced through the five-inch as through the fourteen-inch. By figuring out the stroke and cylinder contents, it was found that in the clear pipe the water moved 162 feet per minute, and in the contracted pipe 1268 feet. And yet the deposit gathered. The conditions necessitated the continuous action of the pump, and it was not idle over two hours in each two months of that period. The boiler feed-pipes of steamers plying on the Great Lakes also become coated with salts of lime. Years ago a disastrous boiler explosion occurred from the virtual stoppage of the feed by this precipitation.

1.04.06. II. If a vein were opened up, in mining, which ran through two different kinds of rock, and if in the one rock one kind of ore and gangue minerals were found, and in the other a different set, the wall rock would clearly have some influence.

Thus in a mine at Schapbach, in the Black Forest, investigated by Sandberger, a vein ran through granite and gneiss. The mica of the granite contained arsenic, copper, cobalt, bismuth, and silver, but no lead. The principal ore in this portion was gray copper. The mica of the gneiss contained lead, copper, cobalt, and bismuth, and the vein held galena, chalcopyrite, and a rare mineral, schapbachite, containing bismuth and silver, but probably a mixture of several sulphides. No two ores were common to both parts of the vein. Another well-established foreign illustration is at Klausen, in the Austrian Tyrol. Lead, silver, and zinc occurred in the veins where they cut diorite and slates, but copper where mica schist and felsite formed the walls. In America there are a number of similar cases. At the famous Silver Islet Mine[1] on Lake Superior the vein runs through unaltered flags and shales, and then crosses and faults a large diorite dike. When the diorite forms the walls, the vein carries native silver and sulphides of lead, nickel, zinc, etc., but where the flags form the walls, the vein carries only barren calcite. Along the edges of the estuary Triassic sandstones of the Atlantic border, where they adjoin Archæan gneiss, a number of veins are found carrying lead minerals, while in the sandstones near the well-known diabase sheets and dikes are others carrying copper ores. It was early remarked by J. D. Whitney that the lead was usually associated with the gneiss, the copper with the diabase.

1.04.07. From instances like these it is inferred that the ores were derived each from its own walls, and by just such a leaching action by cold surface waters as is outlined above. As opposed to this, it has usually been claimed that each particular wall exerted a peculiar selective and precipitating action on the metals found adjacent to it and none on the others ; so that if a solution arose carrying both sets, each came down in its particular surroundings, while the others escaped. Dr. W. P. Jenney has called the writer's attention to such a case. The Head Centre mine, in

---

[1] W. M. Courtis, " On Silver Islet," *Engineering and Mining Journal*, Dec. 21, 1878. *M. E.*, V. 474.

E. D. Ingall, *Geol. Survey of Canada*, 1887-88, p. 27, H.

F. A. Lowe, " The Silver Islet Mine," etc., *Engineering and Mining Journal*, Dec. 16, 1882, p. 321.

T. MacFarlane, "Silver Islet," *M. E.*, VIII. 226. *Canadian Naturalist*, IV. 37.

McDermott, *Engineering and Mining Journal*, January, 1877.

the Tombstone district, Arizona, is on a vein which pierces slates, and in one place forty feet of limestone. In the slates it carried high-grade silver ores, with no lead, but in the limestone, lead-silver ores. A rock like limestone might well exercise a precipitating action, which, however, we cannot attribute to rocks composed of the more inert silicates. Again, it has been said that the solutions coming from below have varied in different portions of the vein or at different periods. An earlier opening would thus be filled with one ore, a later opening with another. This is hypothetical, but has been advanced for Klausen by Posepny. (*Archiv f. Praktische Geologie*, p. 482.) A further general objection to the first interpretation of lateral secretion is the weak dissolving power of cold surface waters, and this is a very serious one.

1.04.08. As opposed to the second interpretation, it may be advanced that precipitation in a cavity at a great depth would be retarded by the heat and the pressure, to just that extent to which solution in the neighboring walls would be aided. The temperature and pressure being practically the same, the tendency to remain in solution would be great until the minerals had reached the upper regions and filled the cavity by ascension. Under such circumstances ores would only be deposited below, by some such action as replacement. To the third interpretation no theoretical objections can be made.

1.04.09. *Infiltration by Ascension.*—On the side of infiltration by ascension, if two veins or sets of veins were found in the same wall rock, but with different kinds of ores and minerals, the conclusion would be irrefutable that the respective solutions which formed them had come from two different sources below. Thus at Butte, Mont., there is a great development of a dark, basic granite. It contains two series of veins, of which the southern produces copper sulphides in a siliceous gangue, the northern sulphides of silver, lead, zinc, and iron, also in a siliceous gangue, but abundantly associated with manganese minerals, especially rhodonite. No manganese occurs in the copper belt, nor is any copper found in the silver belt. Such results could originate only in different, deep-seated sources. Again, at Steamboat Springs, Nev., and Sulphur Bank, Cal., the hot springs are still in action and are bringing their burdens of gangue and ore to the surface. The former has afforded a long series of metals, the latter chiefly cinnabar. G. F. Becker[1] has shown that the cinnabar probably comes up in solution with alkaline sulphides.

---

[1] G. F. Becker, "Natural Solutions of Cinnabar, Gold, and Associated

1.04.10. *Replacement.*—The conception of replacement is one that has been applied of late years by some of the most reliable observers. About 1873 it appears to have been first extensively developed by Franz Posepny, an Austrian geologist, in relation to certain lead-silver deposits at Raibl, in the Province of Kaernthen. About the same time it was suggested by Pumpelly, then State Geologist of Missouri, to Adolf Schmidt, who was engaged in studying the iron deposits of Pilot Knob and Iron Mountain (see Examples 11 and 11*a*), and by Schmidt it was considered applicable to them. ("Iron Ores and Coal Fields," *Missouri Geol. Survey*, 1873.) Some ten years later J. S. Curtis based his explanation of the formation of the Eureka (Nev.) lead-silver deposits on the same idea, and according to Emmons (1886) it holds good for Leadville. R. D. Irving, who credited Pumpelly with bringing it to his attention, published in 1886 an explanation of the hematite beds of the Penokee-Gogebic range (Example 9*e*), in which the idea is applied, and Van Hise has since elaborated it. In the process of replacement no great cavity is supposed to exist previously. There is little, in fact, but a circulation or percolation of ore-bearing solutions which exchange their metallic contents, molecule by molecule, for the substance of the rock mass. We would not expect the ore body to be as sharply defined against the walls as when it filled a fissure, but rather to fade into barren material. Thus rock may be impregnated but not entirely replaced, and, while apparently unchanged, yet carry valuable amounts of ore. Some of the ores of Aspen, Colo. (Example 30*d*), are at times only to be distinguished by assay from the barren limestones. Yet decomposition may bring out the limits of each.

1.04.11. The chemistry of the replacement process is none too well understood, but it presents fewer difficulties when applied to a soluble rock, like limestone or dolomite, than when rocks composed of silicates and quartz have given away to ores. Acid solutions would readily yield to calcium carbonate; but if the metals are present as sulphates, some reducing agent, such as organic matter, is necessary in order to change the metallic sulphate to sulphide.[1] Or else, if the metallic sulphides come up in solution with alkaline sulphides, some third agent is needed to remove the calcium car-

---

Sulphides," *Amer. Jour. Sci.*, III., xxxiii. 199; *Eighth Ann. Rep. Director U. S. Geol. Survey*; *Monograph XIII., U. S. Geol. Survey*, p. 965.

[1] Compare S. F. Emmons, "On the Replacement of Leadville Limestones and Dolomites by Sulphides," *Monograph XII., U. S. Geol. Survey*, p. 563.

bonate, *pari passu*, just before the metallic sulphide is precipitated. It must be confessed that for enormous bodies of ore, like those of Leadville, **the small amount of** organic matter present seems hardly equal to the task assigned it, and the delicate balance of the latter case—causing deposition to tread so closely on the heels of rock removal, in order to avoid assuming an extended cavity—makes it appear that the entire chemistry of the **process is** perhaps hardly understood.

1.04.12. When silicate rocks are replaced, leaving a siliceous gangue, the process may have been somewhat **as suggested** by R. C. Hills for the mines of the Summit district, Rio Grande County, **Colorado.** (See *Proc. Colo. Sci. Soc.*, Vol. I., p. 20.) Alkaline solutions **remove silica and** have slight action on silicates, but solutions **acid with** sulphuric acid attack silicates, such **as feldspar and biotite,** remove the alumina or change it to kaolin, and cause **the separation of free silica.** In the alteration products abundant opportunity would be afforded for the precipitation of sulphides, which **would in** part at least replace the rock. Along a crack or line of **drainage** definite walls would **thus** easily fade out. Such phenomena are afforded by **innumerable ore deposits (see R. W.** Raymond, discussion of S. F. Emmons' "**Notes on the Geology** of Butte, Mont.," *M. E.*, July, 1877), and often **come under the notice of** every one **familiar with mining.** Yet **we cannot but hope** that in the **future our knowledge** of the chemical reactions involved will be **increased.**

It may again be stated **that the formation of ore deposits has proceeded with** great slowness, and the solutions bringing the met**als have been,** beyond question, very dilute. The extremely small **amounts of the** metals that **have been** detected in relatively large **amounts** of igneous rocks, even by the most refined analytical methods, **have necessarily made** the progress of solution **a** protracted one. **Curtis records some** careful **observations** on the growth of aragonite at Eureka, Nev., **where he found that in** three weeks, **so long as wet by a** drop **of water, the** crystals increased in one case **as a maximum, five eighths of an inch,** and in another three eighths. **But this** was where the whole inclosing mass of rock consisted of the **compound deposited.**

CHAPTER V.

ON CERTAIN STRUCTURAL FEATURES OF MINERAL VEINS.

1.05.01. *Banded Structure.*—Mineral veins sometimes exhibit a banded structure, by which is understood the arrangement of the ore and gangue in parallel layers that correspond on opposite walls. They are most conspicuous where the walls are well defined. The solutions which have brought the minerals have varied from time to time, and the precipitated coatings correspond to these variations. They alternate from gangue to ore, it may be, several times repeated. The ore may be in small scattered masses preserving a distinct lineal arrangement in the midst of the barren quartz, calcite, barite, fluorite, siderite, etc., or itself be so abundant as to afford a continuous parallel streak. The commonest ores so observed are pyrite, chalcopyrite, galena, blende, and the various sulphides of silver. The veins of the Reese River district, in Nevada, furnish good illustrations of alternating ruby silver ores and quartz. Those of Gilpin County, Colorado (Example 170), afford alternations of pyrite, chalcopyrite, and gangue. (See figures in Endlich's report, *Hayden's Survey*, 1873, p. 280.) The Bassick Mine, in Colorado, has pebbles remarkably coated. The figure on p. 36 shows a vein at Newman Hill, near Rico, Colo.

Banded veins, however, except of a rude character, are not common in this country. They have received much more attention in Germany, where, especially near Freiberg, they show remarkable perfection. The famous Drei Prinzen Spat Vein, figured by Von Weissenbach and copied in many books, has ten corresponding alternations of six different minerals on each wall.

1.05.02. A line of cavities, or vuggs, is often seen at the central portion of a vein, into which crystals of the last formed minerals emerge, forming a comb (see p. 36). These may project into each other and interlock,—especially if quartz,—forming a comb in comb. The same may occur between side layers. These cavities are a most prolific source of finely crystallized minerals. If, after the fissure—perhaps at the time small—has become once filled, subsequent movements take place, it may strip the vein from one wall and cause a new series of minerals to be deposited, with the previously formed vein on one side and the wall rock on the other. This occasions unsymmetrical fillings. But it may also happen that, with otherwise symmetrical fillings, one layer may be lacking on one side or the other. Where portions of the wall rock have been torn off by the vein matter in these secondary movements, they may be buried in the later deposited vein filling, and form great masses of barren rock called horses. The vein then forks around them. If the ore and the gangue have partly replaced the wall in deposition,

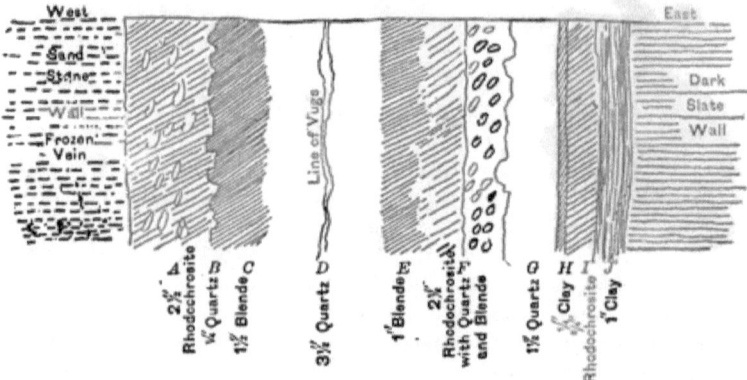

Fig. 6.—*Banded vein at Newman Hill, near Rico, Colo. After J. B. Farish, Proc. Colo. Sci. Soc., April 4, 1892; Engineering and Mining Journal, Aug. 20, 1892.*

unchanged masses of wall may also become inclosed and afford horses of a different origin. An originally forked fissure gives an analogous result.

It is a curious fact that veins are often most productive just at the split. If the masses are small, or if the vein fills a shattered strip and not a clean fissure, or if it occupies an old volcanic conduit, deposition and replacement may surround unchanged cores of wall rock with concentric layers of ores and minerals. Thus the Bassick, at Rosita, Colo., referred to above, consists of rounded cores of andesite, inclosed in five concentric layers of metallic sulphides. The Bull Domingo, in the same region, exhibits shells of galena and quartz mantling nodules of gneiss. Such cores strongly resemble rounded, water-worn boulders, a similarity which has suggested some rather improbable hypotheses of deposition.

1.05.03. *Clay Selvage.*—An extremely common feature is a band of clay, most often between the vein matter and the wall. This is called a selvage, gouge, flucan, clay seam, or parting. It may come in also between layers of different minerals, and even rests as a mantle on the crystals which line cavities. It is at times the less soluble portion left by the decay and removal in solution of wall rock (residual clay), at times the comminuted material resulting from the friction of moving walls (attrition clay), and again it may be taken up by currents and redeposited from bodies of the first two sorts. Such layers of clay, being wellnigh impervious to water, may have exercised an important influence in directing the subterranean circulations. (See Becker on the Comstock Lode, 2.11.19.)

1.05.04. *Pinches, Swells, and Lateral Enrichments.*—The swells and pinches of veins have been referred to above and explained. Aside from these thicker portions of the ore, it is often seen that the richer or even the workable bodies follow certain more or less regular directions, forming so-called "chutes." They probably correspond to the courses taken and followed by the richer solutions. J. E. Clayton observed that they follow the directions of the slips, or striæ, of the walls rather more often than not, and in the west this disposition or tendency is called Clayton's law. Chute is sometimes spelled "shoot" or "shute." Chimney and ore-course are synonyms of chute. Bonanza is used, especially on the Comstock Lode, to indicate a localized, rich body of ore.

Lateral enrichments are caused by the spreading of the ore-bearing currents sidewise from the vein, and often along particular beds of rock, which they may replace more or less with ore. Beds of limestone—it may be quite thin, when in a series composed of shales or sandstones—are favorite precipitants, and from such lateral enlargement the best returns may be obtained. The valuable ore bodies of Newman Hill, near Rico, Colo., whose interesting description by J. B. Farish has already been several times cited, are found as lateral enrichments along a bed of limestone less than three feet thick and embedded in shales. Above the limestone the veins practically cease. Lateral enrichments may closely resemble bedded deposits if the supply fissures are relatively small. This interpretation is placed by W. P. Jenney on the disseminated lead ores of southeastern Missouri (2.15.09), and he has suggested the expressive term "melon vein," thus comparing them with a vine and its melons.

1.05.05. *Changes in Character of Vein Filling.*—In discussing the influence of wall rock the changes that occur in veins were briefly mentioned. But even where the walls remain uniform there is always variation in contents, and of course in value, from point to point. Ore, gangue, horses, and walls alternate both longitudinally and in depth, and such changes must be allowed for and averaged by keeping exploration well in advance of excavation. Even a series of parallel veins may all prove fickle. In illustration of the above the Marshall tunnel of Georgetown, Colo., may be cited. It cut twelve veins below their actual workings, and every one was barren at the tunnel though productive above. (J. J. Stevenson, Wheeler's Survey, *Geology*, Vol. III., p. 351.)

1.05.06. *Secondary Alteration of the Minerals in Veins.*—It has already been stated that the chief ore minerals in vein fillings are sulphides. Where these lie above the line of permanent subterranean water they are exposed to the oxidizing and hydrating action of atmospheric waters, which, falling on the surface, percolate downward. The ores are thus subjected to alternating soakings and dryings which encourage alteration. The sulphides change to sulphates, carbonates, oxides, or hydrous forms of the same, and the metallic contents are in part removed in the acid waters which are also formed. Pyrite, which is the most widespread of the sulphides, becomes limonite, staining everything with its characteristic color. Galena becomes cerusite or anglesite. Blende affords calamine and smithsonite. Copper ores, of which the usual one is chalcopyrite, change to malachite, azurite, chrysocolla, cuprite, and melaconite, and to the sulphide chalcocite. The silver sulphides afford cerargerite. The rarer metals alter to corresponding compounds of less frequency. These upper portions are also more cellular and porous, being at times even earthy. The rusty color from the presence of limonite often marks the outcrop and is of great aid to the prospector. It has been called the iron hat, or gossan. This feature has important economic bearings. The character of ores may entirely change at a definite point in depth, and the later products, if not lower in grade, as is often the case, may demand different, perhaps more difficult, modes of treatment. Oxidized ores are the easiest to smelt, and the benefit of careful exploration before indulging in too confident expectations may be emphasized. As examples, the Ducktown copper deposits (see Example 16), the Leadville silver mines (Example 30), the southwest Virginia zinc deposits at Bonsacks (Example 27), the copper and silver veins at Butte, Mont. (Example 17), and others in Llano County, Texas (Example 17c), may be cited. At Ducktown a considerable thickness of chalcocite, melaconite, and carbonates accumulated just at the water line and abruptly changed to low-grade, unworkable pyrite and chalcopyrite below it. At Bonsacks, near Roanoke, Va , very rich, easily treated earthy limonite and smithsonite (30–40 % zinc) passed into a refractory, low-grade (15–20 % zinc), intimate mixture of blende and pyrite. Excavations in dry districts may not reach the water line for great

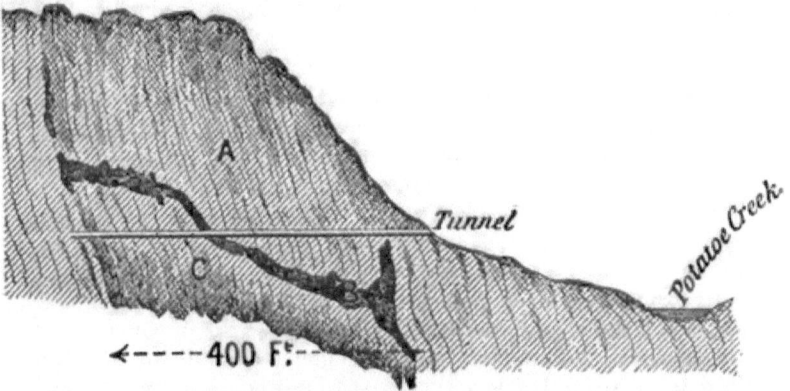

FIG. 7.—*Illustration of the oxidized zone, or gossan, the zone of enrichment, and the unchanged sulphides, at Ducktown, Tenn. After A. F. Wendt, School of Mines Quarterly, Vol. VII., 1886.*

depths. Thus at Eureka, Nev., in the rainless region of the Great Basin, the oxidized ores continue to 900 feet.

It is worthy of remark in this connection that possibly some deposits of oxidized ores may have been formed originally as such. Wendt has argued this for the copper mines of the Bisbee district, Arizona. (See Example 20b.) If oxidized ores are now found below the water line, it may indicate a depression of the rocks from a previous higher position. R. C. Hills has brought out a very interesting instance of the concentration of gold and silver in the lower part of the oxidized zone, or at least at a considerable depth below the outcrop. The upper portion of the vein, in this case with a quartz gangue, was impoverished. The gold is thought to have been carried down in solution with ferrous and ferric sulphates, which were decomposed by feldspar, while the precious metal was thrown down. The ore bodies lie in the Summit district, Rio Grande County, Colorado. (R. C. Hills, *Proc. Colo. Sci. Soc.*, Vol. I., p. 32 ; S. F. Emmons, quoting Hills, *Engineering and Mining Journal*, June 9, 1883.)

1.05.07. The waters of mines which have opened up and exposed sulphides to oxidation are often charged with sulphuric acid and even metallic salts. This is especially true of mines in copper sulphides, and the pumps are much corroded. In instances considerable metallic copper has been removed by passing the mine drainage over scrap-iron, as at Ducktown, Tenn., and as has been lately introduced experimentally at Butte, Mont. Mine timbers have been preserved very long periods by the deposition of copper on them, from their reducing action on the solutions. Pumps and timbers placed by the Romans in the Rio Tinto mines, in Spain, are still in good preservation. Even gold has been detected in Australian mine waters. (See *School of Mines Quarterly*, Vol. XI., p. 364, for review of literature bearing on this subject.)

1.05.08. *Electrical Activity.*—A theoretical agent for the precipitation of ores in veins, which was a great favorite among the writers fifty or sixty years ago, was electrical action, and careful experiments were made in England and Germany to detect it. By connecting the opposite ends of a vein with a wire, in which was a galvanometer, the attempt was made again and again to establish the existence of galvanic action. At times the results gave some grounds for belief ; but at others they were contradictory or uncertain, so that no very definite or reliable conclusions were established. Other experiments were made in Germany about 1844, by Reich, while lately quite elaborate investigations have been carried out by Dr. Carl Barus on the Comstock Lode, and at Eureka, Nev. Great difficulties are met in preserving the necessary insulation throughout the wet and devious underground workings, and amid such surroundings in detecting the currents, which would be necessarily small. With Barus the thesis was not alone to establish a galvanic action, such as might be a precipitating agency, but also to observe what effect, if any, was exerted by the intervention of an ore body on the normal terrestrial currents. Had this latter been proved of sufficient amount, the existence of such bodies might be indicated by plotting electrical observations. While in some respects of interest, the results of Dr. Barus are not very decisive, and this line of investigation is hardly to be considered a promising one. The importance attached to it in former years may be illustrated by these words of De la Beche, one of the ablest of English writers, in 1839. Speaking of veins in general, after discussing those of Cornwall in particular, he says: "Mineral veins result from the filling of fissures in rocks by chemical deposits, from substances in solution in the fissures, such deposits being greatly due to electro-chemical agency."

CHAPTER VI.

THE CLASSIFICATION OF ORE DEPOSITS, A REVIEW AND A SCHEME BASED ON ORIGIN.

1.06.01. In the classification of ore deposits the same systematic arrangement is not to be expected as in the grouping of plants, animals, or minerals. Ore deposits have not the underlying affinities and relationships of living organisms or of definite chemical compounds. The series of objects is too diverse, and, in the nature of the case, the standards of appeal must be different. The subject is, however, one of great practical importance as well as of great scientific interest. A vocabulary of intelligible terms is indispensable for description and comparison, and, under our mining laws, often for valid titles, while as a vehicle for the spread of knowledge and reasonable conceptions regarding these phenomena, its importance cannot be overestimated.

1.06.02. All schemes of classification rest on these principles: form, origin,—or the genetic principle (including method, relative time of origin as contrasted with the walls, etc.),—state of aggregation, and mineral contents. Of these, the principle of form is usually esteemed the weightiest, and is given the greatest prominence, partly because it has been thought to be the one most closely affecting exploitation, and partly because it involves less that is or has been, up to very recent times, more or less hypothetical. Yet form is largely fortuitous, and it has, of course, no law, while, with sufficient knowledge, the genetic principle is the one giving a far more thoroughly scientific basis. Every one, in opening up or searching for an ore body, must be influenced by some hypothesis, either of shape or of origin. It is the conviction of the writer that, with all our deficiencies of knowledge, the genetic principle is also the best guide, even in practical development.

1.06.03. Very early in the development of mining literature the distinction was made between those ore bodies which are parallel to the stratification and those which break unconformably across it. This took place long before the epoch-making time of Werner, and even before the conception of the relative ages of strata had been at all generally grasped. Thus among the Germans we find the terms "Lager" and "Flötze"[1] on the one side, being set off in contrast to "Gang" (vein) on the other. Werner, writing in 1791, quotes Von Oppel's distinctions between Flötze (strata, beds) and Gänge (veins), which were published in 1749; but without doubt, as mining terms, they go much further back. Beyond this simple indication of the views of the older writers, no attempt will be made here to quote authorities earlier than 1850. This is justifiable because the important works, like De la Beche's *Geology of Cornwall and Devon*, and Henwood's *Metalliferous Deposits of Cornwall and Devon*, are rather discussions of veins than systematic attempts at classification.

In the following pages the principal schemes of classification are grouped according to certain relationships and similarities that run through them. It would be interesting to arrange them in chronological order, but points of likeness and unlikeness would not be thus brought out, nor can the influence of one writer on another be so clearly manifested. The underlying object, aside from showing in a bird's-eye view what has been done, is to lead up to an attempt at a purely genetic classification from which

---

[1] Lager and Flötze are difficult to render into English while retaining their native shades of meaning. The later writers in Germany (Serlo, Gätzschmann, Von Groddeck, Köhler) define them as being interbedded bodies, each later than the foot wall in formation, and older than the hanging; and that Lager are much more limited in horizontal extent than Flötze. R. Wabner shows, however, in the *Berg. u. Huet. Zeitung*, Jan. 2, 1891, p. 1, that writers in the earlier part of the century did not entirely restrict the term Lager as regards age relative to the foot and hanging, but applied it to ore bodies, which follow the general bedding, although they may have been introduced much later than the formation of the walls. Thus the frequent occurrence of lead ores in limestone along certain beds (southeast Missouri, for example) would be called Lager. We would apply the terms impregnation, or dissemination, or bed-vein, to such. Flötz we would call stratum, and Lager, as defined by the later authors, "bed" or "seam." Werner, for instance, in his classification of the rock formations of the globe, made: I. Urgebirge (Primitive, Primary, etc., having no fossils). II. Secondary, subdivided into A. Uebergangsgebirge (Transition, more or less metamorphosed sediments, but fossiliferous). B. Flötzgebirge (Unaltered strata). From this the meaning of Flötz can be grasped. By contrast, a magnetite lense is a good illustration of Lager.

mere form is eliminated to the last degree, and well-recognized geological phenomena are brought to the foreground. It has indeed been said with force that the origin of ore deposits is a subject which is very largely a matter of hypothesis, and that it involves profound subterranean causes, of which we know but little. Still, it is held that an acquaintance with what has been accomplished in recent years by our best workers, and a rigid adherence to well-recognized principles in geology and mineralogy, especially as developed in rock study (*i.e.*, in that department of geology that of late years we have grown to call petrography), will establish much that cannot be questioned, and will aid in differentiating the cases which are still objects of reasonable doubt. It is, however, true that among the subjects on which human imagination, often superstitious, has run to wild extremes, and on which cranky dreamers have exercised their wits, the origin of ore deposits stands out in particularly strong relief.

1.06.04. A. *Schemes Involving only the Classification of Veins.*

(1)

G. A. von Weissenbach [1] (*Gangstudien*, 1850, p. 12).
  (*a*) Sedimentärgänge (Sedimentary Veins).
  (*b*) Kontritionsgänge (Attrition Veins).
  (*c*) Stalactitische oder Infiltrationsgänge (Stalactitic or Infiltration Veins).
  (*d*) Plutonische oder Gebirgsmassengänge (Masses, dikes, knobs, bosses, etc., not necessarily with ores).
  (*e*) Ausscheidungsgänge (Segregated Veins).
  (*f*) Erzgänge (True or Fissure Veins).

(2)

B. von Cotta, in comments on Von Weissenbach's Scheme. *Gangstudien*, 1850, p. 79. According to the vein filling.
  1. Gesteinsgänge (Dikes).
    (*a*) Not crystalline (Sandstone).
    (*b*) Crystalline (Granite).
  2. Mineralgänge (Veins).
    (*a*) Of one non-metallic mineral.
    (*b*) Of several non-metallic minerals.
  3. Erzgänge. Ore veins.

---

[1] See also Whitney's *Metallic Wealth of the U. S.*, 1854, p. 44.

CLASSIFICATION OF ORE DEPOSITS.

(3)

B. von Cotta, *idem*, p. 80.   According to Shape and Position.
   I. Wahre, einfache Spaltengänge (Fissures).
      (*a*) Querdurchsetzende, Cross fissures.
      (*b*) Lagergänge, Bed veins.
      (*c*) Klüfte (Cracks), Adern (Veinlets).
   II. Gangzüge (Linked Veins).[1]
   III. Netzgänge (Reticulated Veins).
   IV. Contaktgänge (Contact Veins).
   V. Lenticulargänge (Lenses).
   VI. Stockförmige Gänge (Stocks, Masses).

(4)

B. von Cotta, *idem*, p. 80.   According to the texture of the vein filling.
   I. Dichte Gänge (Compact Veins).
   II. Krystallinische Gänge.
   III. Krystallinisch, körnige (granular) Gänge.
   IV. Krystallinisch, massige (massive) Gänge.
   V. Gänge mit Lagentextur (Banded veins).
      (*a*) Ohne Symmetrie der Lagen (unsymmetrical).
      (*b*) Mit Symmetrie der Lagen (symmetrical).
   VI. Gänge mit Breccien oder Conglomerattextur.

(5)

J. Leconte, *Amer. Jour. Sci.*, July, 1883, p. 17.
   1. Fissure Veins.
   2. Incipient Fissures, or Irregular Veins.
   3. Brecciated Veins.
   4. Substitution Veins.
   5. Contact Veins.
   6. Irregular Ore Deposits.

1.06.05.   In Von Weissenbach's table the sedimentary veins are much the same as the "sandstone dikes" which J. S. Diller has recently described. (*Bull. Geol. Soc. Amer.*, I. 411.) They and the stalactitic veins have small practical value, although of great scientific interest. Under (*d*), the stockworks with tin ores

---

[1] Gangzüge is happily translated "linked veins," by Mr. G. F. Becker (*Quicksilver Deposits*, p. 410). Any attempt to render the original by preserving the figure of a flock of birds or of a school of fish, etc., is, as Mr. Becker remarks, infelicitous, if not impossible.

are the principal illustration of economic prominence. The attrition veins are an important class, and increasing study has widened the application of this or synonymous terms. Segregated veins and true veins are well-known forms. In the comments of Von Cotta, which follow Von Weissenbach's paper, veins are grouped from every possible standpoint, Von Weissenbach's scheme being taken as the one based on origin. Nos. 2 and 4 have small claims to attention. No. 3 foreshadows the drift of many subsequent writers. The meanings of the terms are self-evident, except perhaps *Gangzüge* (linked veins). This refers to a group of parallel and more or less overlapping veins, deposited along a series of openings, evidently of common origin. It is a convenient term.

The terms used by Leconte may be passed without comment as being self-evident in their meaning, except (4) and (6). The scheme was devised, as a perusal of the citation will show, after the author had set forth some original views of the causes which lead to the precipitation of ores, and had forcibly stated others very generally accepted. In the explanatory text some quite curious associations are found, which are cited by way of illustration. Thus under group (4), stalactites, caves, gash veins, and the Leadville ore bodies are considered examples, and under group (6) the grouping together of beds, igneous masses, and all other forms of so-called irregular deposit is decidedly open to criticism. This is the more emphatic because the concluding sentences of the paper (of whose general value and excellence there can be no question) give the impression that the author felt he had cleared up all the points in the origin of ore bodies which would be of interest or importance to a purely scientific investigator as contrasted with a practical miner.

1.06.06. B. *General Schemes Based on Form.*

(6)

Von Cotta and Prime. *Ore Deposits*, New York, 1870.
   I. Regular Deposits.
      A. Beds.
      B. Veins.
         (*a*) True (Fissure) Veins.
         (*b*) Bedded Veins.
         (*c*) Contact Veins.
         (*d*) Lenticular Veins.
   II. Irregular Deposits.

C. Segregations.
    (*a*) Recumbent.
    (*b*) Vertical.
D. Impregnations (Disseminations).

(7)

Lottner-Serlo, *Leitfaden zur Bergbaukunde*, 1869.
  I. Eingelagerte Lagerstätten (Inclosed Deposits).
    A. Plattenförmige (Tabular).
      (*a*) Gänge (Veins).
      (*b*) Flötze und Lager (Strata, beds, seams).
    B. Massige Lagerstätten (Massive Deposits).
      (*a*) Stöcke   } Masses.
      (*b*) Stockwerke }
    C. Andere unregelmässige Lagerstätten (other irregular deposits).
      (*a*) Nester (Pockets).
      (*b*) Putzen.
      (*c*) Nieren (Kidneys).
  II. Aufgelagerte Lagerstätten (Superficial Deposits).
    D. Trümmerlagerstätten (Placers).
    E. Oberflächliche Lager (Surface beds).

(8)

Koehler, *Lehrbuch der Bergbaukunde*, 1887.
  I. Plattenförmige Lagerstätten (**Tabular** Deposits).
    (*a*) Gänge (Veins).
    (*b*) Flötze und Lager (Strata, beds, seams).
  II. Lagerstätten von unregelmässige Form (Deposits of irregular Form).
    (*a*) Stöcke und Stockwerke (Masses).
    (*b*) Butzen, Nester, und Nieren (Pockets, concretions, etc.).

(9)

Callon, *Lectures on Mining*, 1866 (Foster and Galloway's translation).
  I. Veins.
  II. Beds.
  III. Masses (*i.e.*, not relatively long, broad, and thin).

1.06.07. The scheme of Von Cotta and Prime carries out the principle of form to its logical and somewhat trivial conclusion.

Thus under irregular deposits it is a matter of extremely small classificatory moment whether an ore body is recumbent or vertical. Otherwise the scheme is excellent, and its influence can be traced through most of those that are of later date. The original draft came out in the German in 1859. All the others are from treatises on mining, in which this subject plays a minor rôle, and indicates the tendency, referred to above, of mining engineers, when writing theoretically, to imagine certain fairly definite forms, which are to be exploited. As previously remarked, however, considering the uncertainty of ore bodies and their variability in shape, it is here argued that the genetic principle might better take precedence. Several of the German terms are difficult to render into English mining idioms, as for example, Stock, Butzen (Putzen), Nester, and Nieren.

1.06.08. C. *Schemes, Partly Based on Form, Partly on Origin.*

(10)

J. D. Whitney, *Metallic Wealth*, 1854.
- I. Superficial.
- II. Stratified.
  - (*a*) Constituting the mass of a bed or stratified deposit.
  - (*b*) Disseminated through sedimentary beds.
  - (*c*) Originally deposited from aqueous solution, but since metamorphosed.
- III. Unstratified.
  - A. Irregular.
    - (*a*) Masses of eruptive origin.
    - (*b*) Disseminated in eruptive rocks.
    - (*c*) **Stockwork** deposits.
    - (*d*) **Contact** deposits.
    - (*e*) **Fahlbands.**
  - B. Regular.
    - (*f*) Segregated veins.
    - (*g*) **Gash** veins.
    - (*h*) **True** or fissure veins.

(11)

J. S. Newberry, *School of Mines Quarterly*, March, 1880, May, 1884
- I. Superficial.

CLASSIFICATION OF ORE DEPOSITS. 49

II. Stratified.
    (a) Forming entire strata.
    (b) Disseminated through strata.
    (c) Segregated from strata.

III. Unstratified.
    (a) Eruptive masses.
    (b) Disseminated through eruptive **rock**.
    (c) Contact deposits.
    (d) Stockworks.
    (e) Fahlbands.
    (f) **Chambers.**
    (g) Mineral veins.
        1. Gash veins.
        2. Segregated veins.
        3. Bedded veins.
        4. Fissure veins.

(12)

J. A. Phillips, *Ore Deposits*, 1884.

I. Superficial.
    (a) By mechanical action **of water.**
    (b) By chemical **action.**

II. Stratified.
    (a) **Deposits constituting the bulk of metalliferous beds formed by precipitation from aqueous solution.**
    (b) **Beds** originally deposited **from solution but subsequently** altered by metamorphism.
    (c) **Ores disseminated** through sedimentary beds, **in which they have been** chemically deposited.

III. Unstratified.
    (a) True veins.
    (b) Segregated **veins.**
    (c) Gash veins.
    (d) **Impregnations.**
    (e) **Stockworks.**
    (f) Fahlbands.
    (g) Contact **deposits.**
    (h) **Chambers or pockets.**

1.06.09. It is at once apparent that Whitney's scheme contains the essentials of the others, which are merely slight modifications. Newberry introduces impregnations, **chambers, and** bedded veins.

The first named is a useful term, although it is not always easy to distinguish impregnations from others earlier given. Thus, they may be very like the division, disseminated through strata, or disseminated through eruptive rock, or, if in metamorphic rock, fahlbands. The word is also used to indicate places along a vein where the ore has spread into the walls. The term "chambers," or "caves," has found application in the West, and is a useful addition. Bedded veins appear also in Von Cotta above (No. 6). Phillips seeks to explain the methods of origin in his use of Whitney's scheme and clearly feels the importance of emphasizing the genetic principle more strongly. Much of it is implied in the simpler phraseology, however, and the extended sentences lack the incisiveness of the earlier schemes. The arrangement as set forth by Whitney is worthy of high praise, and the scheme is one of the many valuable things in a book that has played a large part in the economic history of the United States.

1.06.10. D. *Schemes Largely Based on Origin.*

(13)

J. Grimm, *Lagerstätten*, 1869.

    I. Gemengtheile oder grössere Einschlüsse in den Gebirgsgesteinen. Einsprengung, Imprägnation. (Essential component minerals and inclusions in country rock. Impregnations.)

        (*a*) Ursprüngliche Einsprengung. (Original with the inclosing rock.)

        (*b*) Von anderen Lagerstätten weggeführte Bruchtheile, etc. (Fragments brought from a distance. Placers, ore-bearing boulders. Breccias.)

    II. Untergeordnete Gebirgsglieder oder besondere Lagerstätten. (Subordinate terranes or special forms of Deposits.)

        (*a*) Plattenförmige Massen. (Tabular masses.)

            1. Lager oder Flötze. Bodensatzbildung. (Beds, strata.)

            2. Gänge, Klüfte, Gangtrümmer, etc. (Veins of varying sizes.)

            3. Plattenförmige Erz-ausscheidungen und Anhäufungen. (Segregated veins.)

        (*b*) Stöcke und regellos gestaltete Massen. (Stocks and irregular masses.)

  1. Lagerstöcke Linsenstöcke, Linsen. (Lenticular deposits, etc.)
  2. Stöcke, Butzen, Nester, etc. (Masses, pockets, etc.)
  3. Stockwerke. (Stockworks.)

(14)

A. von Groddeck, *Lehre von den Lagerstätten, etc.*, 1879, p. 84.
 I. Ursprüngliche Lagerstätten (Primary deposits).
  A. Gleichzeitig mit dem Nebengestein gebildet.
   (Formed at the same time with the walls.)
   1. Geschichtete. (Stratified.)
    (*a*) Derbe Erzflötze. (Entire beds in fossiliferous strata.)
    (*b*) Ausscheidungsflötze. (Disseminated in beds.)
    (*c*) Erzlager. (Lenticular beds, mostly in schists.)
   2. Massige. (Massive; the word is nearly synonymous with igneous.)
  B. Später wie das Nebengestein gebildet. (Formed later than the walls.)
   3. Hohlräumsfüllungen. (Cavity fillings.)
    (*a*) Spaltenfüllungen oder Gänge. (Fissure fillings or veins.)
     (1) In massigen Gesteinen. (In igneous rocks.)
     (2) In geschichteten Gesteinen. (In stratified rocks.)
    (*b*) Höhlenfüllungen. (Chambers.)
   4. Metamorphische Lagerstätten. (Alterations, replacements, etc.)
 II. Trümmer-lagerstätten. (Secondary or detrital deposits.)

(15)

R. Pumpelly, *Johnson's Encyclopædia*, 1886, VI. 22.
 I. Disseminated concentration.
  (*a*) Impregnations, Fahlbands.
 II. Aggregated Concentration.
  (*a*) Lenticular aggregations and beds.
  (*b*) Irregular masses.
  (*c*) Reticulated veins.
  (*d*) Contact deposits.

} Forms due to the texture of the inclosing rock, or to its mineral constitution, or to both causes.

III. Cave deposits. ⎫ Forms chiefly due to pre-
IV. Gash veins.   ⎬ existing open cavities
V. Fissure veins.   ⎭ or fissures.
VI. Surface deposits.
    (*a*) Residuary deposits.
    (*b*) Stream deposits.
    (*c*) Lake or bog deposits.

   1.06.11. These three are all excellent, and give some interesting variations in the several points of view from which each writer regarded his subject. There are instances in the two German schemes where it is difficult to render the original into a corresponding English term and recourse has been had to the explanatory text. Grimm especially writes an obscure style. He divides accordingly as the ore forms an essential and integral part of the walls or a distinct body. Von Groddeck has in view the relative time of formation as contrasted with the walls. Grimm afterward emphasizes geometrical shape, but this Von Groddeck practically does away with, and continues more consistently genetic. His scheme might perhaps come more appropriately in the next section.

   Pumpelly's conception varies considerably from the others. He writes, as his full paper states, in the belief that the metals have all been derived primarily from the ocean, whence they have passed into sedimentary, and, by fusion of sediments, into igneous rocks. The group of residuary surface deposits, carrying out as it does a favorite idea of Professor Pumpelly, as set forth in his papers on the secular decay of rocks, is an important distinction.

   1.06.12. E. *Schemes Entirely Based on Origin.*

(16)

H. S. Munroe. Used in the Lectures on Mining in the School of Mines, Columbia College.
  I. Of surface origin, beds.
    (*a*) Mechanical (action of moving water).
      1. Placers and beach deposits.
    (*b*) Chemical (deposited in still water).
      1. By evaporation (salt, gypsum, etc.).
      2. By precipitation (bog ores).
      3. Residual deposits from solution of limestone, etc. (hematites).
    (*c*) Organic.
      1. Vegetable (coal, etc.).

2. Animal (limestone, etc.).
 (d) Complex (cannel coal, bog ores, etc.).
II. Of subterranean origin.
 (a) Filling fissures and cavities formed mechanically.
  1. Fissure veins, lodes.
  2. Cave deposits—lead, silver, iron ores.
  3. Gash veins. The cavities of 2 and 3 are enlarged by solution of limestone.
 (b) Filling interstitial spaces and replacing the walls.
  1. Impregnated beds.
  2. Fahlbands.
  3. Stockworks.
  4. Bonanzas.
  5. Masses.

1.06.13. This scheme covers all forms of mineral deposits, whether metalliferous or not, while most of those previously given, as well as the one that follows, concern only metalliferous bodies. The scheme is consistently genetic and was elaborated because such a one filled its place in lectures on mining better than one based on form. The general principle on which the main subdivision is made differs materially from any hitherto given. Deposits formed on the surface are kept distinct from those originating below, even though the first class may afterward be buried.

(17)

1.06.14. J. F. Kemp, 1892. Revised from the *School of Mines Quarterly*, November, 1892.
 I. Of Igneous Origin. Excessively basic developments of fused and cooling magmas. Peridotite, forming iron ore at Cumberland Hill, R. I.[1] Magnetite, Jacupiranga, Brazil.[2] Titaniferous magnetite in Minnesota gabbros[3]; in Adirondack gabbros[4]; in Swedish and Norwegian gabbros.[5]

---

[1] M. E. Wadsworth, *Bull. Mus. Comp. Zoöl.*, 1880, VII.

[2] O. A. Derby, *Amer. Jour. Sci.*, April, 1891.

[3] N. H. Winchell, *Tenth Ann. Rep. Minn. Geol. Survey*, pp. 80–83. *Bull. VI.* of same Survey, p. 135.

[4] Forthcoming paper by J. F. Kemp.

[5] J. H. L. Vogt, *Geol. Fören. i. Stockholm-Förhand*, XIII. 476, May, 1891. English abstract and review by J. J. H. Teall. *Geol. Mag.*, February, 1892. See also *Zeitschr. für Praktische Geologie*, I. 4.

II. Deposited from Solution.
1. Surface precipitations, often forming beds, and caused by—
   (a) Oxidation. Bog ores. Ferruginous oölites, as in some Clinton ores.[1]
   (b) Sulphurous exhalations from decaying organic matter etc. (Pyrite.)
   (c) Reduction, chiefly by carbonaceous, organic matter. (Pyrite from ferrous sulphate.)
   (d) Evaporation, cooling, loss of pressure, etc. (Hot spring deposits, as at Steamboat Springs, Nev.[2])
   (e) Secretions of living organisms. (Iron ores by algæ.[3])

NOTE.—These same causes of precipitation operate in subterranean cavities, although not again specially referred to.

2. Disseminations (impregnations) in particular beds or sheets, because of—
   (a) Selective porosity. (Silver Cliff, Colo., silver ore in porous rhyolite.[4] Amygdaloidal fillings, as in copper-bearing amygdaloids, Keweenaw Point, Santa Rita, N. M.[5] Impregnations of porous sandstone, as at Silver Reef, Utah.[6]
   (b) Selective precipitation by limestone. (Lateral enlargements at Newman Hill, near Rico, Colo.[7])

3. Filling joints, caused by cooling or drying. (Mississippi Valley gash veins in part.)

4. Occupying chambers (caves) in limestone. (Cave Mine, Utah.[8])

---

[1] C. H. Smyth, Jr., *Amer. Jour. Sci.*, June, 1892, p. 487.

[2] G. F. Becker, *Monograph XIII.*, *U. S. Geol. Survey*, pp. 331, 468; Laur, *Ann. des Mines*, 1863, p. 421; J. Leconte, *Amer. Jour. Sci.*, June, 1883, p. 424, July, p. 1; W. H. Weed, *idem*, August, 1891, p. 166.

[3] Sjogrun, *Berg.- und Hütt. Zeit.*, 1865, p. 116.

[4] Clark, *Engineering and Mining Journal*, Nov. 2, 1878, p. 314.

[5] A. F. Wendt, *Trans. Amer. Instit. Min. Eng.*, XV. 27.

[6] C. M. Rolker, *idem*, IX. 21.

[7] J. B. Farish, *Proc. Colo. Sci. Soc.*, April 4, 1892.

[8] J. S. Newberry, *School of Mines Quarterly*, March, 1880. See also J. P. Kimball, on Santa Eulalia, Chihuahua, *Amer. Jour. Sci.*, II., xlix. 161.

CLASSIFICATION OF ORE DEPOSITS. 55

    5. Occupying collapsed (brecciated) beds, caused by solution and removal of support, or from dolomitization of limestone. (Southwest Missouri zinc deposits.[1])

    6. Occupying cracks at Monoclinal bends, Anticlinal summits, Synclinal troughs, often with replacement of walls. (Gash veins in part ; galena deposits at Mine la Motte, Missouri ; zincblende deposits in the Saucon Valley, Pennsylvania.[3]

    7. Occupying Shear-zones, or dynamically crushed strips along faults, whose displacement may be slight, closely related to No. 8. (Butte, Mont.[4])

    8. True veins filling an extended fissure, often with lateral enlargements. See also under 5.

    9. Occupying Volcanic necks, in agglomerates. (Bassick and Bull Domingo Mines, near Rosita, Colo.[5])

    10. Contact deposits. Igneous rocks almost always form one wall. Fumaroles. (Marquette hematites, Michigan.[6] Greisen.)

    11. Segregations formed in the alteration of igneous rock. (Chromite in serpentine.)

III. Deposited from Suspension.

    1. Metalliferous Sands and Gravels, whether now on the surface (placers, magnetite beach-sands), or subsequently buried. (Deep gravels, magnetite lenses ?)

    2. Residual Concentrations, left by the weathering of the matrix. (Iron Mountain, Mo., hematite in part.[7])

---

[1] F. L. Clerc, *Lead and Zinc Ores in Southwest Missouri Mines*, Carthage, Mo., 1887; A. Schmidt, *Missouri Geol. Survey*, 1874, p. 384.

[2] T. C. Chamberlain, *Wis. Geol. Survey*, IV., 1882, 367.

[3] F. L. Clerc, *Mineral Resources*, 1882, p. 361; H. S. Drinker, *Trans. Amer. Inst. Min. Eng.*, I. 367.

[4] S. F. Emmons, *Trans. Amer. Inst. Min. Eng.*, XVI. 49; W. P. Blake, idem, XVI. 65.

[5] C. W. Cross, *Proc. Colo. Sci. Soc.*, 1890, p. 269.

[6] C. R. Van Hise, *Amer. Jour. Sci*, February, 1892, p. 116.

[7] R. Pumpelly, *Bull. Geol. Soc. Amer.*, II., p. 230.

1.06.15. It is believed that under the above heads are included all the forms of ore bodies which constitute well-recognized and fairly well understood geological phenomena. To these categories year by year we are enabled, by the results of extended and careful investigation, to refer many that have been obscure. A number of familiar terms for ore bodies in mining literature fail to appear, but are mentioned in the classifications quoted from others. Many of these refer only to form, and geologically considered are only convenient admissions of ignorance as to origin. Some other ore bodies whose methods of origin are involved in the processes of regional metamorphism are placed by themselves farther on. The explanations of them are as yet hypothetical. A few comments on the scheme may now be added, although in the main it explains itself.

1.06.16. I. Much attention has been given of late years to processes of rock formation from igneous magmas. Of these the excessively basic are the only ones primarily concerned with ores. It is well known that in the series of igneous rocks we have successively those with less and less silica. It is quite conceivable that local developments might bring about such a decrease of the silica and such an increase of one of the commonest of the bases, iron oxide, that the limits of an ore might be reached. Such bodies are almost always highly titaniferous, so much so that in this country they are not available. Not a little attention was directed in earlier years to the Cumberland Hill outcrop in Rhode Island, but it was found to be too high in this element to be suited to furnace practice. No analyses are at hand of the Brazilian example, but a considerable percentage of titanium might be expected. The presence of these ores in Canada, the Adirondacks, and Minnesota is very familiar. It is possible, as indicated by Vogt, that when magnetite crystals had formed in the still fused magma, they became aggregated by magnetic currents in the earth. And it is also conceivable that these early and heavy crystallizations may have settled to the lower portions of the magma and have become concentrated. Much that is more or less speculative is involved in these explanations.

1.06.17. Under II. 1, the precipitating agencies are mentioned, which are the chief causes in the chemical reactions of deposition, and these run through all the subterranean cavities as well. The general application is esteemed self-evident. The large part played by organic matter, both when living and when dead and decaying, is notable. Its office even in precipitating the gangue minerals in surface reactions we are just beginning to appreciate. Siliceous sinters have been shown by W. H. Weed to be formed around the hot springs of the Yellowstone Park through the agency of algæ, and A. Rothpletz has recently proved that the calcareous oölites around the Great Salt Lake are referable to minute organisms. Many accumulations of iron ores have, with reason, been attributed to the same agency; but for this metal ordinary and common chemical reactions are oftenest applicable. When organic matter decays, sulphurous gases are one of the commonest products, and likewise one of the most vigorous of precipitants. Thus under Example 24, Paragraph 2.06.03, when speaking of the Wisconsin zinc and lead mines, it will be seen that such an agency from decaying seaweeds has been cited by both Whitney and Chamberlain. When the products of such decomposition become imprisoned in the rocks as oils and gases, their action is unmistakably important and is especially available in limestones. Organic matter is a powerful reducing agent as well, and in this way is capable of bringing down metallic compounds. The silver-bearing sandstones of southern Utah are cases in point, as they afford plant impressions now coated with argentite. The purely physical agencies cited under (d) have also an important rôle.

1.06.18. Under 2 (a) the uprising solutions may be diverted by porous strata so as to soak through them and become subject to precipitating agents of one kind or another. They furnish the simplest kind of cavities, and starting with these the scheme is developed in a crescendo to the most complicated. The purely chemical action of limestone beds, however, seems at times to come into play and to cause precipitation along them. Of all rocks they are the most active chemical reagent. It may be questioned with reason as to whether caves or caverns (4), properly so called, have ever formed a resting-place for ores. So many which have been cited as such may with greater reason be referred to shrunken replacements that a doubt hangs over their character.

1.06.19. Under (5) brecciated beds whose fragments are coated and whose interstices are filled with ore are, with great reason, referred to the collapse from the removal of a supporting layer. In addition to the illustration cited, the red hematite deposits of Dade and Crawford counties, Missouri, have been thought to have had a similar origin. Such phenomena are only to be expected in regions that have long been land. Cracks at the bends of folds may, in cases, have occasioned impregnations and disseminations, even when their character is obscured. The **cracks need** be but small and numerous to have occasioned far-reaching results. **If** a fault fissure, **as a** possible conduit of supply, crosses the axis of the fold, the necessary conditions are afforded **for extended horizontal** enrichment. Recent explorations with **the diamond drill at Mine** la Motte **seem to corroborate such an hypothesis.** Should the anticline or roll afterward sink toward **the** horizontal, a very puzzling deposit might **originate.** Shear zones have been already discussed at length (1.02.03), as have true veins and volcanic necks (see also 2.09.20). **As regards** contact deposits, the igneous rock, which **usually forms one** wall, may serve two different purposes. It may act merely as an impervious barrier which directs solutions along its course, or serves to hold them, **either because it is itself** bent into a basin-like fold, or because it **forms** a trough **with a** dense bed dipping in an opposite direction. **Such** relations occur in the Marquette and Gogebic ranges of the Lake Superior iron region. It is not apparent that in these cases the igneous rock has in any degree stimulated circulations. In the more characteristic "contact deposits" the igneous rock has apparently been **a strong** stimulator of ore-bearing circulations, and often **the source of the** metals themselves. This form of deposit becomes, then, an attendant phenomenon, or even **a variety, of contact** metamorphism. Under 11 chromite is the chief illustration. The mineral is practically limited to serpentinous **rocks, and is distributed** through them in irregular masses. **It appears to be an auxiliary** product of alteration.

1.06.20. III. **The débris that results** from the weathering of rock masses under the action of frost, wind, rain, heat, and cold is **washed along by the** drainage system of a district, and the well-known **sorting action transpires,** which is so important in connection **with the formation of the** sedimentary rocks. Minerals of **great specific gravity tend to concentrate by** themselves, while **lighter materials are washed farther from the starting** point, and **settle only in still water.** Stream bottoms supply the most favorable situations, and in their bars are found accumulations of the **heavier minerals which are in** the surrounding rocks. The commonest of these are magnetite, garnet, ilmenite, wolframite, zircon, topaz, spinel, etc., and with these, in some regions, native gold, platinum, iridosmine, etc.; in other places cassiterite, or stream tin, **as described under tin.** Even an **extremely rare** mineral such as **monazite may make a sandbar of** considerable size. (See O. A.

Derby, *Amer. Jour. Sci.*, III. xxxvii., p. 109.) The action of surf or smaller shore waves is also a favorable agent. The throw of the breaker tends to cast the heavier material on the beach, where its greater specific gravity may hold it stationary. The heavier minerals may be sorted out of a great amount of beach sand. Magnetite sands, which have accumulated in this way, are of quite wide distribution, and at present are of some though not great importance. (Example 15.) With the magnetite are found grains of garnet, hornblende, augite, etc., and often ilmenite. Gold is concentrated in the same way along the Pacific by the wash of surf against gravel cliffs. In abandoned beaches of Lake Bonneville, near Fish Springs, Tooele County, Utah, placers of rolled boulders of argentiferous galena have been worked.

A superficial deposit of somewhat different origin is the bed of hematite fragments that mantles the flanks of Iron Mountain, Missouri, and runs underneath the Cambro-Silurian sandstones and limestones. This seems to have been formed by the subaërial decay of the inclosing porphyry. The heavier specular ore has thus been concentrated by its greater specific gravity and resistant powers. (See R. Pumpelly, "The Secular Disintegration of Rocks," *Proc. Geol. Soc. Amer.*, Vol. II., December, 1890.)

1.06.21. There remain a few of great importance, but whose geological history is less clearly understood. They are nearly all involved in processes of regional metamorphism, and therefore in some of the most difficult problems of the science. Lenticular beds or veins of magnetite and pyrite that are interbedded with schists, slates, or gneisses are the principal group. Such magnetite bodies have been regarded as intruded dikes, as original bodies of bog ore in sediments which have later become metamorphosed, and as concentrated delta, river, or beach magnetite sands. It is possible that in instances they may be replaced bodies of limestone, afterward metamorphosed. The lenticular shape and the frequent overlapping arrangement of the feathering edges in the foot wall are striking phenomena.

The overlap was referred by H. S. Munroe in the *School of Mines Quarterly*, Vol. III., p. 34, to stream action during mechanical deposition, and a figure of some hematite lenses in the Marquette region was given in illustration. The arrangement in instances also suggests the shearing and buckling processes of dynamic metamorphism and disturbance. The individual lenses, now in linear series, were thus all one original bed. The crumpling of the schistose rocks has pinched them by small buckling folds and shoved the ends slightly past each other in the process so familiar in the production of reversed faults from monoclines. Sheared granitic veins on a small scale are a not uncommon thing in areas of schistose rocks, such as Manhattan Island, in the city limits of New York, and suggest strongly this explanation. Should the compression not go so far as to bring rupture of the bed, but only a thickening by the formation of a sigmoid fold, it would occasion an enlarged cross section, as has been suggested by B. T. Putnam (*Tenth Census*, Vol. XV. 110) for the great magnetite ore body at Mine 21, Mineville, in the Lake Champlain region.

1.06.22. Quartz veins, often auriferous and of a lenticular character, furnish another puzzling ore body. They are commonly called segregated veins, and lie interbedded in slates or schistose rocks. If in a pre-existing cavity, it must have been formed, either by the opening of beds under compression, or by displacement along the bedding, so that depressions came opposite each other. Replaced lenses of limestone which had been squeezed into this shape from an original, connected bed should also be instanced as a possibility. The name "segregated" would imply a filling by lateral secretion, but it is by no means impossible that solutions have come from below. The veins are another attendant feature on regional metamorphism, and as such deserve more investigation.

1.06.23. The veins that contain cassiterite in many parts of the world, and that yet have the mineralogical composition of granite, are another product of metamorphic action, both contact and regional. The gangue minerals, feldspar, quartz, and mica, are quite characteristic of acid, igneous rocks, but the coarseness of the crystallization in the comparatively narrow veins bars out a true igneous form of origin. All our artificial methods of reproducing these minerals lead us to infer that the veins were filled at a high temperature and pressure, therefore at considerable depths, and through the aid of steam. Cassiterite has also been detected in a few rare cases, under such circumstances that it seemed to be an original mineral in igneous granite. It is probable, therefore, that it may be an original and early crystallization from an igneous magma, much as is magnetite. More observed cases would be welcome as evidence.

1.06.24. The great iron ore bodies of Vermilion Lake have been referred by N. H. and H. V. Winchell, in the *American*

Geologist, November, 1889, p. 291, to a precipitation from oceanic waters in the vicinity of submarine volcanic eruptions from whose ejectamenta they derived their iron and silica. The hypothesis is regarded as sufficient to account as well for the siliceous deposits associated with the ore. In the words of the authors: "Chemical precipitation in hot oceanic waters, united with simultaneous sedimentary distribution, might produce the Keewatin ores in a manner consistent not only with the physical conditions that prevailed at the time of their formation, and with the structural peculiarities which they exhibit, but also in accordance with the known reactions of heated alkaline waters, and with the chemical character which the ores are known to possess." This hypothesis introduces new conditions and relations which are necessarily somewhat speculative; and while it has claims to attention, it may best be tested by the general consideration of the geological public before being placed with the more simple and certain reactions grouped in the scheme.

1.06.25. Fahlbands should be mentioned here. The term refers to belts of schists, which are impregnated with sulphides, but not in sufficient amount in the locality where the name was first applied (Kongsberg, Norway) to be available for ores. The decomposition of the sulphides gave the schists a rusty or rotten appearance that suggested the name. Whether the ores are an introduction into the schist, subsequent to metamorphism, or a deposit in and with the original sediment, is a doubtful point. The practical importance of these fahlbands lies in the enriching influence that they exert on the small fissure veins that cross them.

1.06.26. The phraseology of the above schemes will be employed in the subsequent descriptions. In addition, much emphasis will be placed on the character of the rocks containing the deposits, whether unaltered sedimentary, igneous, or metamorphic, and whether in the first and last cases igneous rocks are near, for these considerations enter most largely into questions of origin. The ore deposits are illustrated by examples, somewhat as has been done by the best of modern writers abroad, Von Groddeck. The word "example" is preferred to "type," which was employed by Von Groddeck, because it implies less of an individual character. We may cite deposits under different metals thus which all might belong to one type. Under each metal will be given, first, a list of general treatises and papers. These will be marked "Hist." when especially valuable as history, and "Rec." when recommended for ordinary examination. If not marked by either, they are more adapted for special investigations.

GENERAL REFERENCES ON ORE DEPOSITS.

Ansted, D. T. "On Some Remarkable Quartz Veins," *Quar. Jour. Geol. Sci.*, XIII. 246.

Barus, Carl. "The Electrical Activity of Ore Bodies," *M. E.*, XIII. 417. (See also Becker's *Monograph on the Comstock Lode*, p. 310, for references to other papers.)

Becker, G. F. "The Relations of the Mineral Belts of the Pacific Slope to the Great Upheavals," *Amer. Jour. Sci.*, III. 28, 209. 1884.

Belt, Th. Mineral Veins: an Inquiry into their Origin, founded on a Study of the Auriferous Quartz Veins of Australia. London, 1861.

Bischof, G. "On the Origin of Quartz and Metalliferous Veins," *Jameson's Journal*, April, 1845, p. 344. Abstract, *Amer. Jour. Sci.*, I. 49, 396. Advocates aqueous deposition.

Brown, A. J. "Formation of Fissures and the Origin of their Mineral Contents," *M. E.*, II. 215.

Bulkley, F. G. "The Separation of Strata in Folding," *M. E.*, XIII. 384.

Campbell, A. C. "Ore Deposits," *Engineering and Mining Journal*, July 17, 1880, p. 39.

Von Cotta-Prime. Ore Deposits. German, 1859; English, 1870. Rec.

Emmons, E. *American Geology*, 134, 1853. General discussion.

Emmons, S. F. "The Structural Relations of Ore Deposits," *M. E.*, XVI. 304. Rec.

"Notes on Some Colorado Ore Deposits," *Proc. Colo. Sci. Soc.*, II., Part II., p. 35.

"On the Origin of Fissure Veins," *Proc. Colo. Sci. Soc.*, II., p. 189. Rec. (See also R. C. Hills, *idem*, III., p. 177.)

"The Genesis of Certain Ore Deposits," *M. E.*, XV. 125. Rec.

Endlich, F. M. *Hayden's Survey*, 1873, p. 276. General description of veins.

Foster, C. L. "What is a Mineral Vein?" Abstract in *Geol. Mag.*, Vol. I., 513.

Fox, R. W. "Formation of Metallic Veins by Galvanic Agency," *Amer. Jour. Sci.*, I. 37, 199. Abstract from *London and Edinburgh Phil. Mag.*, January, 1839.

"On the Electro-Magnetic Properties of Metalliferous Veins in the Mines of Cornwall," *Amer. Jour. Sci.*, I. 20, 136. Abstract of paper before the Royal Society.

Grimm, J. "Die Lagerstätten der Nutzbaren Mineralien," 1869.

Von Groddeck, A. "The Classification of Ore Deposits," *Engineering and Mining Journal*, June 27, 1885, p. 437.

"Die Lehre von den Lagerstätten der Erze," 1879. Rec. (See also *Engineering and Mining Journal*, Jan. 3, 1880, p. 2, for a review of same.)

Hague, A. D. Mining Industries, Paris Exposition, 1878.

Henrich, C. "On Faults," *Engineering and Mining Journal*, Aug. 24, 1889, p. 158.

Hunt, T. S. "The Geognostical History of the Metals," *M. E.*, I. 331.

"The Origin of Metalliferous Deposits," in "Chemical and Geological Essays."

"Contributions to the Chemistry of Natural Waters," *Amer. Jour. Sci.*, II. 39, 176.

Julien, A. A. "On the Part played by Humus Acids in Ore Deposit, Wall Rock, Gossan," etc., *Proc. A. A. A. S.*, 1879, pp. 382, 385.

Keck, R. "The Genesis of Ore Deposits," *Engineering and Mining Journal*, Jan. 6, 1883, p. 3.

Review of Ore Deposits in Various Countries. Denver, 1892. 31 pages.

Kemp, J. F. "A Brief Review of the Literature on Ore Deposits," *School of Mines Quarterly*, X. 54, 116, 326; XI. 359; XII. 219.

"On the Filling of Mineral Veins," *School of Mines Quarterly*, October, 1891.

"The Classification of Ore Deposits," *School of Mines Quarterly*, November, 1892.

"On the Precipitation of Metallic Sulphides by Natural Gas," *Engineering and Mining Journal*, December, 1890.

Kleinschmidt, J. L. "Gedanken ueber Erzvorkommen," *B. and H. Zeit.*, 1887, p. 413.

Koehler, G. "Die Störungen der Gänge, Flötze, u. Lager." Leipzig, 1886. Translated by W. B. Phillips under title of "Irregularities of Lodes, Veins, and Beds," *Engineering and Mining Journal*, June 25, 1887, p. 454; also July 2, p. 4.

Leconte, J. "Mineral Vein Formation in Progress at Steamboat Springs and Sulphur Bank," *Amer. Jour. Sci.*, III. 25, 424.

"Genesis of Metalliferous Veins," *Amer. Jour. Sci.*, July, 1883. See other references under "Mercury."

Miller, A. Erzgänge. Basel, 1880.

Necker. "On the Sublimation Theory," *Proc. Geol. Soc. of London*, Vol. I., p. 392; also Ansted's *Treatise on Geology*, Vol. II., p. 271. Hist.

Newberry, J. S. "The Origin and Classification of Ore Deposits," *School of Mines Quarterly*, I., 1887, 1880. *Engineering and Mining Journal*, June 19 and July 23, 1880. *A. A. A. S.*, Vol. XXXII., p. 243, 1883. Rec.

"The Deposition of Ores," *School of Mines Quarterly*, V. 329, 1884; *Engineering and Mining Journal*, July 19, 1884.

"Genesis of Our Iron Ores," *School of Mines Quarterly*, II. 1, 1880; *Engineering and Mining Journal*, April 23, 1881. See also under "Iron."

"Genesis and Distribution of Gold," *School of Mines Quarterly*, III., 1881; *Engineering and Mining Journal*, Dec. 24 and 31, 1881. Rec.

Ochsenius, Carl. "Metalliferous Ore Deposits," *Geol. Mag.*, I. 310. Hist.

Pearce, Rich. "On Replacement of Walls," *Chem. News*, March 3, 1865.

Phillips, J. A. "The Rocks of the Mining District of Cornwall and Their Relations to Metalliferous Deposits," *Quar. Jour. Geol. Soc.*, XXXI. 319.

"A Contribution to the History of Mineral Veins," *Quar. Jour. Geol. Soc.*, XXXV. 390.

"Treatise on Ore Deposits," London, 1884.

Pumpelly, R. *Johnson's Encycl.*, Vol. VI., p. 22. Rec.

Raymond, R. W. "What is a Pipe Vein?" *Engineering and Mining Journal*, Nov. 23, 1878, p. 361.

Translation of Lottner, and general remarks on classification. *Min. Stat. West of Rocky Mountains*, 1870, p. 447.

Indicative Plants, *M. E.*, XV. 645.

"Geographical Distribution of Mining Districts in the United States," *M. E.*, I., p. 33.

Sandberger, F. "Untersuchungen über Erzgänge," 1882; "Theories of the Formation of Mineral Veins," *Engineering and Mining Journal*, March 15, 22, 29, 1884, pp. 197, 212, 232.

Wabner, R. "Ueber die Eintheilung der Minerallagerstätten nach ihrer Gestalt, sowie die Anwendung und die Benützung der Wörte, Lager und Flötz," *B. and H. Z.*, Jan. 2, 1891, p. 1.

Wadsworth, M. E. "The Theories of Ore Deposits," *Proc. Boston Soc. Nat. Hist.*, 1884, p. 197. Rec.

"The Lateral Secretion Theory of Ore Deposits," *Engineering and Mining Journal*, May 17, 1884, p. 364.

"Classification of Ore Deposits," *Rep. of Mich. State Geologist*, 1891-92, p. 144. Rec. (See Addenda.)

Whitney, J. D. "Remarks on the Changes which take place in the Structure and Composition of Mineral Veins near the Surface," *Amer. Jour. Sci.*, ii. XX. 53.

"Metallic Wealth of the United States," 1854. Rec.

Whittlesey, C. "On the Origin of Mineral Veins," *A. A. A. S.*, XXV. 213.

Williams, Albert. "Popular Fallacies Regarding the Precious Metal Ore Deposits," *Fourth Ann. Rep Director U. S. Geol. Survey*, pp. 257-278.

PART II.

THE ORE DEPOSITS.

CHAPTER I.

THE IRON SERIES (IN PART).—INTRODUCTORY REMARKS ON IRON ORES.—LIMONITE.—SIDERITE.

GENERAL LITERATURE.

Birkinbine, J. "Prominent Sources of Iron Ore Supply," *M. E.*, XVII. 715. Statistical; Rec.

Various statistical papers in the volumes on *Mineral Resources, U. S. Geol. Survey*, especially 1886, p. 39; 1887, p. 30.

Chester, A. H. "On the Percentage of Iron in Certain Ores," *M. E.*, IV. 219.

Dunnington, F. P. "On the Formation of the Deposits of Oxides of Manganese," *Amer. Jour. Sci.*, iii., XXXVI. 175. The paper treats of Iron also.

Hewitt, A. S. "Iron and Labor," *M. E.*, September, 1889. The paper contains valuable statistics.

"A Century of Metallurgy," *M. E.*, V. 164.

Hunt, T. S. "The Iron Ores of the United States," *M. E.*, October, 1890.

Kimball, J. P. "Genesis of Iron Ores by Isomorphous and Pseudomorphous Replacement of Limestone," *Amer. Jour. Sci.*, September, 1891, p. 231. Continued in *Amer. Geologist*, December, 1891.

Julien, A. A. "The Genesis of the Crystalline Iron Ores," *Trans. Phil. Acad. Nat. Sci.*, 1882, p. 335; *Engineering and Mining Journal*, Feb. 2, 1894.

"Origin of the Crystalline Iron Ores," *Trans. N. Y. Acad. Sci.*, II., p. 6; *Amer. Jour. Sci.*, iii., XXV. 476.

Lesley, J. P. "The Iron Manufacturer's Guide," 1886. Hist. Rec.

Newberry, J. S. *International Review*, November and December, 1874.

"Genesis of the Ores of Iron," *School of Mines Quarterly*, November, 1880. Rec. *Amer. Jour. Sci.*, iii., XXI. 80.

"Genesis of the Crystalline Iron Ores," *Trans. N. Y. Acad. Sci.*, II., October, 1882. Rec.

Newton, H. "The Ores of Iron: Their Distribution with Reference to Industrial Centers," *M. E.*, III. 360.

Pumpelly, R., and Others. *Tenth Census*, Vol. XV., 1886, especially pp. 3–17. Rec.

Reyer, E. "Geologie des Eisens," *Oest. Zeit. f. B. und H.*, 1882, Vol. XXX., pp. 89, 109.

Rogers, W. B. "On the Origin and Accumulation of the Protocarbonate of Iron in the Coal Measures," *Proc. Bost. Soc. Nat. Hist.*, 1856.

Smock, J. C. "On the Geological Distribution of the Ores of Iron," *M. E.*, XII. 130.

"Iron Mines and Iron Ore Districts in New York," *Bull. N. Y. State Mus.*, June, 1889. Rec.

Swank, J. M. Chapters on iron in *Mineral Resources, U. S. Geol. Survey*, since 1883.

"History of the Manufacture of Iron in All Ages." 1891.

Whitney, J. D. "Metallic Wealth of the United States," 1854, p. 425. Hist.

"On the Occurrence of Iron in the Azoic System," *A. A. A. S.*, 1855, 209; *Amer. Jour. Sci.*, ii., XXII. 38.

Winchell, N. H. and H. V. "The Iron Ores of Minnesota," *Bull. No. 6, Minn. Geol. Survey*. Part IV. contains an exhaustive review of methods of origin, and Part V. a very complete annotated bibliography.

Table of the Iron Ores, Limonite, Siderite, Hematite, Magnetite, Pyrite.

	Fe.	H_2O.	CO_2.	S.
Limonite (brown hematite, bog ore), $2Fe_2O_3 \cdot 3H_2O$	59.89	14.4		
Siderite (Spathic ore, clay ironstone, blackband), $FeCO_3$	48.27		37.92	
Hematite (red and specular), Fe_2O_3	70.0			
Magnetite, $FeO.Fe_2O_3$ or Fe_3O_4	72.4			
Pyrite, FeS_2	46.7			53.3

2.01.01. No one of the iron ores ever occurs pure in large amounts. Only a few closely approach this condition. The largest quantity of rich ore as yet mined in the United States was doubtless obtained from the Lovers' Pit opening, operated by Witherbee, Sherman & Co., on Barton Hill, near Mineville, N. Y. The record shows that 40,000 tons of magnetite averaged 68.6% Fe. In general the ores run much less. The richest are the magnetites and specular hematites. In numerous instances the mines of the Lake Champlain district have produced the former, and Lake Superior mines the latter, at 63 to 65%, or even more. The separated ores in the Lake Champlain district run about 65%. The unseparated ores have much less, and indeed all percentages from 50 to 65. Thus the lump ore (shipped as mined) from Chateaugay, N. Y., has about 50%. The Cornwall (Penn.) magnetite holds even less. The Clinton red hematites from New York afford about 44% in the furnace, as the result of long experience. The limonites, as usually mined, produce from 40 to 50%. The crude spathic ores are the lowest of all, and in the variety black-band may even be about 30%. They are easily calcined, however, and on losing their carbonic acid, moisture, and bituminous matter the percentage of iron rises a third or more. A. H. Chester found in 1875, as the result of an endeavor to determine the average yield of certain standard ores in the furnace, Lake Superior specular, 62.5%; Lake Superior limonite, 49.5%; Rossie (N. Y.) red hematite, 54.5%; Wayne County (N. Y.) Clinton ore, 40%.

2.01.02. The common impurities in iron ores are the common elements or oxides that enter most largely into rocks, and those which make up the walls of the deposit are usually the ones that appear most abundantly in the ore. Silica (SiO_2), alumina (Al_2O_3), lime (CaO), magnesia (MgO), titanium oxide (TiO_2), carbonic acid (CO_2), and water (H_2O) appear in large amounts and determine to a great extent the character, fluxing properties, etc., of the ore. With these, and of more far-reaching influence, are smaller amounts of sulphur and phosphorus. The last two and titanium chiefly determine the character of the iron which is yielded in the furnace and are the first foreign ingredients sought. The sulphur is present in pyrite, the phosphorus in apatite. As is well known, 0.1% of phosphorus is set as the extreme limit for Bessemer pig irons, and as ores for these command the best market, they are eagerly sought. To obtain the allowable limit of phosphorus in the ore, its percentage in iron is divided by 1000. Thus a 65.3% ore should not have over 0.065% phosphorus to be ranked as Bessemer. If at the same time, with sufficiently low phosphorus, the gangue is highly siliceous, a composition desirable for Bessemer practice, ores may be of value, although of comparatively low grade and remotely situated.

Thus the lump magnetite of the Chateaugay mines, in the northern Adirondacks, affords but 50% iron, yet it is mined and shipped forty miles to Plattsburg, and thence four hundred miles to the furnaces. It has 18.44 SiO_2 and only 0.029 P and 0.052 S. Late operations in the siliceous specular hematites of Pilot Knob, Mo., have utilized low-grade ore rejected in former years. Other instances could be cited. On the other hand, a moderate amount of phosphorus is not only no drawback for ordinary foundry irons and such as are to be subjected to tool treatment, but it is a prime necessity. Excessive amounts are desired only for weak but very fluid irons. Considerations like these, which are rather metallurgical than geological, largely determine the availabilty of a deposit, and to some extent the present locations of the mining districts.

2.01.03. Iron itself is one of the most abundant and widely disseminated elements entering into the composition of the earth. A careful estimate of the probable average composition of the outer crust of the earth afforded Alexander Winchell (*Geological Studies*, pp. 19, 20) the results in Column I. Prestwich, the English geologist, gives the figures in Column II. (*Geology*, Vol. I., p. 10.)

I.		II.	
O	45	O	50
Si	25	Si	25
Al	10	Al	10
Fe	8	Ca	4.5
Ca	6	Mg	3.5
Na	2.5	Na	2
C, H, S, N, Cl, Mg, etc.	1.5	C, Fe, S, Cl	2.4
K	2	K	1.6
	100.		99.0

The figures were obtained by taking the general prevalence of the common rocks and estimating on their known compositions. In Winchell's estimate iron is fourth in abundance, while in Prestwich's it is ninth. Either case illustrates its great abundance and wide distribution, although the former with much greater emphasis.

2.01.04. A general comparison of the tabulated analyses of igneous rocks (See Roth's *Gesteinsanalysen* and *Allgemeine Chem. und Physikal. Geologie*) shows that the granites contain 0.0–7% iron oxides, the porphyries 0.0–14%, the rhyolites 0.0–8%, the diorites and diabases 4–16%, the andesites 3–15%, the basalts 12–20%.

Limestones invariably have at least small amounts, and at times very considerable percentages. Sandstones are often low, but not seldom are stained through and through. The metamorphic rocks offer close analogies to the igneous. In general distribution and in quantity, iron leads the list of the distinctively metallic elements. Its peculiar property of possessing two oxides, of different chemical quantivalence, assists greatly in the formation of ores and the general circulation of the metal. This is set forth under the following examples.

LIMONITE.

2.01.05. Example 1. *Bog Ore.*—Beds of limonite, superficially formed in marshes, swamps, and pools of standing water. The general circulation of water through the rocks enables it very frequently to take up iron in solution. Ferruginous minerals are among the first and easiest that fall a prey to alteration. Carbonic acid in the water aids in dissolving the iron, which thus, in waters containing an excess of CO_2, passes into solution as the protocarbonate $FeCO_3$. Organic acids may also play a part. The alteration of pyrite affords sulphuric acid and ferrous sulphate, and the latter enters readily into solution. On meeting calcium carbonate, both ferric and ferrous sulphate are decomposed, yielding in the first case calcium sulphate, ferric hydrate, and carbonic acid; in the second, if air is absent, ferrous carbonate and calcium sulphate, but on the admission of air ferric hydrate soon forms. (See F. P. Dunnington, *Amer. Jour. Sci.*, iii., XXXVI. 176. Experiments 10 and 11. See also Addenda.)

2.01.06. Bodies of limonite that become exposed to a reducing action from the favorable presence of decaying organic matter likewise furnish the protocarbonate. In general it may be stated that free oxygen must be absent or only in small quantity where solution takes place. Sooner or later the ferruginous (or chalybeate) waters come to rest, especially in swamps. The protosalt is exposed to the evaporation of the excess of CO_2, that held it in solution, and also to the action of oxygen. Two molecules of carbonate, together with one atom of oxygen and some water, break up into CO_2 and $Fe_2O_3, x H_2O$. The latter forms as a scum and then sinks to the bottom and accumulates in cellular masses. The sesquioxide is insoluble, and as against ordinary waters free from reducing agents it remains intact. Deposits of mud and peat forming above may cover the beds with a protecting layer.

Hardly a bog exists which does not show, when cut in cross section, the bog ore beneath. Frequent associates of the ore are diatomaceous earth and shell marl, formed by the remains of organisms which once inhabited the waters. At times excellent impressions of leaves and shells are preserved in the ore. Such ore bodies are not often practically available on account of the low percentage in iron, due to the abundance of sand and silt washed in, and to the frequent large amounts of sulphur and phosphorus which they contain. The sulphur is present in pyrite and the phosphorus in vivianite, sometimes in sufficient quantity to be visible (Mullica Hill, N. J., var. *Mullicite*). In certain parts of the country bog deposits have been and others may yet be utilized.

2.01.07. In eastern North Carolina bog-ore beds are frequent and are found lying just below the grass roots. Scattered nodules occur in the overlying soil, which are succeeded by a bed three feet or less in thickness, resting on sand.[1]

In Hall's Valley and Handcart Gulch, Park County, Colorado, interesting and extensive deposits of limonite are in active process of formation. The iron comes from neighboring great beds of pyrite.[2]

Bog ore of good quality has recently been reported from the vicinity of Great Falls, Mont.[3]

At Port Townsend Bay, in the vicinity of Puget Sound, and at the Patton mines, near Portland, Ore., the ores are of such quality as to be available.[4] (For bog ore in Quebec, see Addenda.)

2.01.08. A somewhat different variety of Type 1 is formed when the ferruginous waters come to rest in the superficial hollows of the rock which has furnished the iron. Depressions in the serpentines of Staten Island, N. Y., carry such deposits, and the iron is referred by N. L. Britton to the leaching of the underlying rock. The ore contains a notable percentage of chromium, which is known to occur in the serpentine. The mines have been in former years quite large producers. Similar limonites occur at Rye, N. Y.[5]

---

[1] W. C. Kerr, *Geology of North Carolina*, 1875, p. 218. B. Willis, *Tenth Census*, Vol. XV., p. 302.

[2] R. Chauvenet, "The Iron Resources of Colorado." *M. E.*, June, 1889. "Notes on Iron Prospects in Northern Colorado," *Ann. Rep. Colo. School of Mines*, 1886.

[3] *Mineral Resources, U. S. Geol. Survey*, 1888, p. 34.

[4] B. T. Putnam, *Tenth Census*, Vol. XV., p. 496.

[5] N. L. Britton, *School of Mines Quarterly*, May, 1881. Compare also *Amer. Jour. Sci.*, iii., XX. 32, and XXII. 488.

At the Prosser mines, near Portland, Ore., deposits of limonite are found in the superficial hollows of a Tertiary basalt of the Cascade range. The ore contains roots and trunks of trees, and is covered by a later flow of basalt. Similar bodies of limonite resulting from basalt are known in the German province of Hesse and in Ireland.[1]

2.01.09. The limonite sand, or oölite, that forms in the Swedish lakes about ten meters from the banks and in water up to ten meters in depth is another variety of this type. A layer half a meter and less in thickness accumulates every fifteen to thirty years and is periodically dredged out. The ore precipitates first as a slime that breaks up afterward into small concretions. It has been thought that the formation of these and similar bodies of limonite has been aided by small algæ and other plants or microscopic organisms.[2]

2.01.10. Example 2. Bodies of limonite in cavities of ferruginous rocks, on the outcrop, or below the surface, which have resulted either from the alteration of the rock *in situ* or from its partial replacement by limonite. Residual clay, quartz, and other remains of alteration usually occur with the ore. Ferruginous limestones are the commonest sources of such deposits, but other rocks may afford them. The deposits are not limited to any one geological series, but in different parts of the country occur whenever the conditions have been favorable. Some of the ore may have been brought in by subterranean circulations which have leached the neighboring rocks. Considerable limonite has also resulted from the weathering of clay ironstone nodules and blackband beds in the Carboniferous system (to be mentioned later), and not infrequently from the alteration of nodular masses of pyrites.

---

[1] B. T. Putnam, *Tenth Census*, Vol. XV., p. 16, on the Oregon ore. Tasche, *Berg.- und Hütt. Zeit.*, 1886, p. 209; also Wurtemberger, *Neues Jahrb.*, 1867, p. 685, on the Hessian ores. Tate and Holden, "On the Iron Ore Associated with the Basalt of Northeastern Ireland," *Quar. Jour. Geol. Sci*, XXVI. 151.

[2] F. M. Stapff, *Zeitschr. d. d. geolog. Gesellsch.*, 1866, Vol. XVIII., p. 8, on the geology of the ores. Sjogrun, *Berg.- und Huett. Zeit.*, 1865, p. 116, on the agency of algæ. On the general formation of bog ores the following papers are of interest: G. J. Brush and C. S. Rodman, "Observation on the Native Hydrates of Iron," *Amer. Jour. Sci.*, ii., XLIV. 219; J. S. Newberry, *School of Mines Quarterly*, November, 1880; J. Roth, *Chem. und Phys. Geologie*, I., pp. 58, 97, 221; F. Senft, *Humus, Marsch, Torf- und Limonit-bildungen.*

The limonite is in cellular lumps, in pipes, pots, and various imitative forms which often have a beautiful luster. The hollow masses have in general resulted form the filling of reticulated cracks in shattered rock. The ore thus deposits around the cores of country rock, which afterward are removed, leaving a hollow shell or geode. (See *Tenth Census*, Vol. XV., pp. 275, 369, 370.)

2.01.11. Reserving the Siluro-Cambrian limonites for a subtype the ore bodies are described in order from east to west, taking up first the Alleghany region, then the Mississippi Valley, and lastly

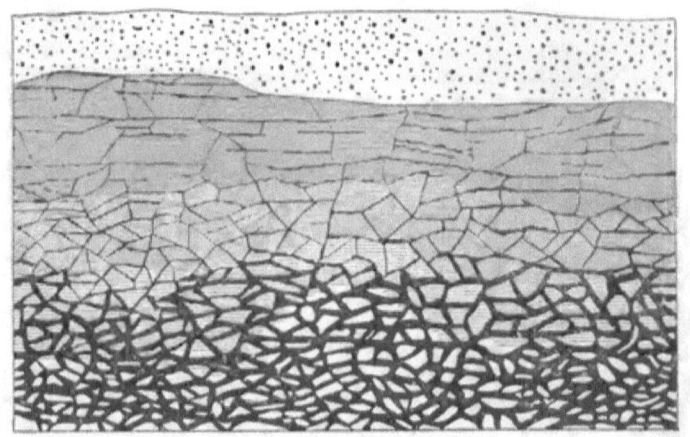

Fig. 8.—*Section of the Hurst limonite bank, Wythe County, Virginia, illustrating the replacement of shattered limestone with limonite and the formation of geodes of ore. After E. R. Benton, Tenth Census, Vol. XV., p. 275.*

the Rocky Mountains. The limonites of New England and New York belong in the subsequent subtype, as do those of eastern Pennsylvania and the more important ones in Virginia, Tennessee, Georgia, and Alabama. In central and western Pennsylvania, however, not a small amount is obtained from the higher lying terranes. The Hudson River slates furnish small amounts in Franklin County, which are thought by McCreath to have resulted from the alteration of nodules of pyrites. (*Second Penn. Geol. Survey*, M3, p. x.) The Medina sandstones contain highly ferruginous portions in Huntingdon County. (McCreath, *Second Penn. Geol. Survey*, MM, p. 108.) The Lower Helderburg and Oriskany are locally quite productive in Blair County, affording several great banks of ore. (Report MM, 196, M3, p. 33.) The Oriskany is of greater importance in Virginia than in Pennsylvania. East of these last mentioned exposures, and in southern Carbon County, is a bed of paint ore between the Oriskany and the Marcellus. (C. E. Hesse, "The Paint-Ore Mines at Lehigh Gap," *M. E.*, New York meeting, 1890.) The Marcellus is the most productive of the Devonian stages. It affords considerable ore in Perry County (Report MM, p. 193; M3, p. 29), Juniata County, Mifflin County, Huntingdon County (Report M, p. 66; MM, p. 194; M3, p. 140), Fulton County (Report M3, p. 42), and Franklin County (Report M3, p. 1). All these are in southern Pennsylvania. Lesley states (*Iron Manufacturers' Guide*, p. 650) that the ore is weathered carbonate. As shown under Example 4, beds of carbonate ore occur in Ulster County, New York, in the Marcellus. (Additional details on the above Pennsylvania deposits will be found in the geological reports on the particular counties.)

2.01.12. As already remarked, the greater part of the limonites in Virginia belong under the Siluro-Cambrian division and are there described, but in the James River basin, on Purgatory and May's Mountains, there are deposits in sandstones of the Clinton. (J. L. Campbell, *The Virginias*, July, 1880.) Other limonite beds occur in the Oriskany on Brushy Mountain (Longdale mines), on Rich Patch Mountain (Low Moor mines, involving also the Medina), on Warm Spring Mountain, and on Peter Mountain. In the Shenandoah Valley, on Massanutton Mountain, the limonite is referred by Prime to the Clinton stage. (*The Virginias*, March, 1880, p. 35.) On North Mountain it lies in the Oriskany, according to Campbell (*The Virginias*, January, 1880, p. 6), and on the Great North Mountain in the Upper Silurian. Considerable oxide of zinc collects in the tunnel heads of the furnaces running on Low Moor ores, indicating the presence of this metal in the limonite.[1]

The iron ores in Kentucky are found in three widely separated districts, one near Greenup, in the northeastern corner of the State, known as the Hanging Rock region; the second near the central part along the Red and Kentucky rivers, known as the Kentucky and Red River region; and the third in the southwestern part near

---

[1] E. C. Means, "Flue Dust at Low Moor, Va.," *M. E.*, 1888; E. C. Pechin, "Virginia Oriskany Iron Ores," *Engineering and Mining Journal*, Aug. 13, 1892, p. 150.

Lyon and Trigg Counties, known as the Cumberland River region. Although the first two contain much limonite, it has altered from nodules of carbonate, and the ores are therefore described under Example 5. One locality near Owingsville, in the second region, has limonites altered from the Clinton hematite. (See Example 6.) The Cumberland region affords limonites in the Subcarboniferous. They are in rounded masses, either solid or hollow, and are distributed through a red clay along with angular fragments of chert. The limonite pots are themselves filled with clay or water.[1]

2.01.13. In Tennessee the limonites of the eastern portion come mostly under Example 2a. In the west they are a southern extension of the pot-ore deposits of Kentucky, and show the same associated chert and clay. Safford has called the rocks containing them the Siliceous Group. The west Tennessee district projects into Alabama to a small extent.[2]

2.01.14. The principal limonite deposits of Alabama come under Example 2a, as do those of western North Carolina and Georgia. Some limonite is produced in Ohio, but it is all weathered carbonate and is mentioned under Example 5. Limonites form an abundant ore in the Marquette district of Michigan, but are mentioned with the vastly greater deposits of hematite under Example 9a. Deposits of brown hematite are worked in a small way in the southeastern part of Missouri, where they rest upon Cambrian strata and have a marked stalactitic character. (P. N. Moore, *Geol. Survey of Missouri*, Report for 1874; F. L. Nason, *Mo. Geol. Survey*, 1892, II., p. 158.) Limonites referred to the Cretaceous by N. H. Winchell occur in western Minnesota. (*Bull. VI., Minn. Geol. Survey*, p. 151.) (For Arkansas limonite, see Addenda.)

2.01.15. In eastern Texas, along the latitude of the northern boundary of Louisiana, extended beds of limonite are found capping the mesas or near their tops, and associated with glauconitic sands of Tertiary age. They are described by Penrose (*First Ann. Rep. Texas Geol. Survey*, p. 66; also *G. S. A.*, III. 44) as (1) Brown laminated ores, (2) Nodular or geode ores, (3) Conglomerate ores. The first form extended beds whose firmness has prevented the erosion of the hills, and which are thought to have originated by the weathering of the pyrites in the greensands and

---

[1] W. B. Caldwell, "Report on the Limonite Ores of Trigg, Lyon, and Caldwell Counties," *Kentucky Geol. Survey*, New Series, Vol. V., p. 251.

[2] W. M. Chauvenet, *Tenth Census*, Vol. XV., p. 357; T. H. Safford, *Geology of Tennessee*, p. 350.

from the iron of the glauconite itself. The second group occur just north of the last, and have probably resulted from the alteration of clay ironstone nodules (Cf. Example 5), while the third has formed in the streams by the erosion of the first two and from the smaller ore-streaks and segregations. Limonite also occurs in northwestern Louisiana. (*Mineral Resources*, 1887, p. 51.) Limonite is known in a number of localities of Colorado. The chief productive mines lie in Saguache County, near Hot Springs. They furnish a most excellent ore from cavities in limestones, which are generally, but with no great certainty, considered Lower Silurian. R. Chauvenet states that the ores yield about 43% Fe in the furnace.[1]

A great body of limonite nodules, bedded in red, residual clay, has been reported from the Clinton series of Allamakee County, Iowa. (E. Orr, *Amer. Geologist*, Vol. I., p. 129.)

Much limonite occurs at Leadville in connection with the lead-silver ores, and is used as a flux by the lead smelters. Some grades low in silver and rich in manganese have even been used for spiegel at Pueblo. For the geological relations, see Example 30.

2.01.16. Limonites in supposed Carboniferous limestone occur in the East Tintec mining district in Utah, and seem to be associated with a decomposed eruptive rock, somewhat as at Leadville. The limonite is chiefly used as a flux by lead-silver smelters.[2]

2.01.17. Example 2a. *Siluro-Cambrian Limonites.*—Beds of limonite in so-called hydromica (talcose, damourite), slates and schists, often also with limestones of the Cambrian and Lower Silurian systems of the Appalachians. The great extent, the geological relations, and the importance of these deposits warrant their grouping in a subtype by themselves. They extend along the Appalachians from Vermont to Alabama, and are in the "Great Valley," as it was early termed, which marks the trough between the Archæan on the east and the first corrugations of the Paleozoic rocks, often metamorphosed, on the west. The masses of limonite are buried in ocherous clay, and the whole often preserves the general structure of the schistose rocks which they have replaced. The

---

[1] R. Chauvenet, "Preliminary Notes on the Iron Resources of Colorado," *Ann. Rep. Colo. State School of Mines*, 1885, p. 21; "Iron Resources of Colorado," *M. E.*, 1889. F. M. Endlich, *Hayden's Reports*, 1873, p. 333. B. T. Putnam, *Tenth Census*, Vol. XV., p. 482. C. M. Rolker, "Notes on Certain Iron Ore Deposits in Colorado," *M. E.*, XIV. 266. Rec.

[2] B. T. Putnam, *Tenth Census*, Vol. XV., p. 490.

original stringers of quartz remain following the original folds. Dolomitic limestone often forms one of the walls, and still less often (but especially in New England) masses of siderite are found inclosed. Manganese is at times present, and in Vermont is of some importance of itself.

2.01.18. The deposits begin in Vermont, where in the vicinity of Brandon they have long been ground for paint. A curious pocket of lignite occurs with them and affords Tertiary fossils. This prompted President Edward Hitchcock, about 1850, to refer all the limonites to the Tertiary, making an instructive example of the occasional hasty generalizations of the early days. Lignite has also been found at Mont Alto, Penn. In northeastern Massachusetts, at Richmond and West Stockbridge; and just across the State line, in Columbia and Dutchess counties, New York, and at Salisbury, Conn., the mines are large, and were among the first worked in the

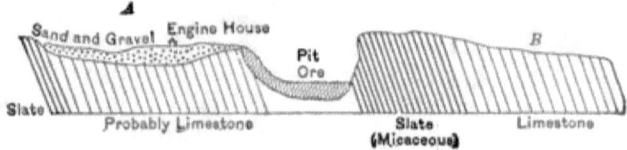

FIG. 9.—*Geological section of the Amenia Mine, Dutchess County, New York, illustrating a Siluro-Cambrian limonite deposit. After B. T. Putnam, Tenth Census, Vol. XV., p. 133.*

United States. The limonite forms geodes, or "pots," pipes, stalactitic masses, cellular aggregates, and smaller lumps from which the barren clays and ochers are removed by washing. The ore is but a fraction of the material mined and occurs in irregular streaks through the clays, etc. It is mostly obtained by stripping and open cuts, and only rarely by underground mining, which would present difficulties with such poor material for walls.[1]

---

[1] J. D. Dana, "Occurrence and Origin of the New York and New England Limonites," *Amer. Jour. Sci.*, iii., XIV. 132, and XXVIII. 398. Rec. E. Hitchcock, "Description of a Brown Coal Deposit at Brandon, Vt., with an Attempt to determine the Geological Age of the Principal Ore Beds of the United States," *Amer. Jour. Sci.*, ii., XV. 95; *Hist. Geol. Survey of Vermont*, I. 233. See also Lesley, below. A. L. Holley, "Notes on the Salisbury (Conn.) Iron Mines and Works," *M. E.*, VI. 220. J. P. Lesley, "Mont Alto (Penn.) Lignites," *Proc. Amer. Acad. Sci.*, 1864, 463-482; *Amer. Jour. Sci.*, ii., XL. 119. L. Lesquereux, "On the Fossil Fruits found in Connection with the Lignite at Brandon, Vt.," *Amer. Jour. Sci.*,

A gap occurs in the succession of the deposits across southern New York and New Jersey, although a few minor ones are known in the western part of the latter State, in the magnesian limestone of the valleys between the hills of gneiss.[1]

2.01.19. In Lehigh County and to the southwest through York County, in eastern Pennsylvania, the limonites are again developed in great amount, and run southwesterly, with few gaps, to Alabama. It is in this portion that the "Great Valley" (called also the Cumberland Valley, or Valley of Virginia) is especially marked. Wherever the great limestone formation, No. II. of Rogers, is developed the ores are found. This corresponds to the Calciferous, Chazy, and Trenton of New York. Limonites also occur still lower in the Cambrian at about the horizon of the Potsdam sandstone or in the overlying slates. According to McCreath, they are distinguishable in Pennsylvania as ores at the top, ores in the middle, and ores at the bottom of the great limestone No. II. Those at the top form the belt along the central part of the valley where the Trenton limestone underlies the Utica or Hudson River slates. Those in the middle are connected with various horizons of ferruginous limestones in the Chazy and Calciferous. Those at the bottom along the north or west part of the South Mountain-Blue Ridge range are geologically connected with the Potsdam sandstone, or the slates which intervene between it and the base of the Calciferous. (*Second Penn. Survey*, Rep. MM, p. 199.) Cobalt has been detected on those of Chester Ridge by Boye, but it is a rare and unique discovery.[2]

---

ii., XXXII. 355. H. Carvill Lewis, "The Iron Ores of the Brandon Period," *Amer. A. A. Sci.*, XXIX. 427, 1880. J. F. Lewis, "The Hematite (Brown) Ore Mines, etc., East of the Hudson River," *M. E.*, V. 216. J. G. Percival, *Rep. on the Geol. of Conn.*, p. 132; also, *Amer. Jour. Sci.*, ii., II. 268. R. A. F. Penrose, "Report on Manganese Ores," *Geol. Survey Ark.*, 1890, Vol. I. (Contains many valuable descriptions of Vermont limonites.) B. T. Putnam, *Tenth Census*, Vol. XV. C. N. Shepard, "Notice, etc., of the Iron Works of Salisbury, Conn.," *Amer. Jour. Sci.*, i., XIX. 311. J. C. Smock, *Bull. VII. New York State Museum*, pp. 12, 52. N. H. and H. V. Winchell, "Taconic Ores of Minnesota and Western New England," *Amer. Geol.*, VI. 263. 1890.

[1] B. T. Putnam, *Tenth Census*, Vol. XX., p. 176. See also *Geol. Survey New Jersey*, 1880.

[2] Dr. Boye, "Oxyd of Cobalt with the Brown Hematite of Chester Ridge, Penn., *Amer. Phil. Soc.*, January, 1846. P. Fraser, *Second Geol. Survey Penn.*, Reps. C and CC; "Origin of the Lower Silurian Limonites

2.01.20. The Siluro-Cambrian limonites run across Maryland in Carroll and Frederick counties, and are mined to small extent. (E. R. Benton, *Tenth Census*, Vol. XV., p. 254.)

These limonites are again strongly developed in the Shenandoah Valley along the western base of the Blue Ridge, and in southwestern Virginia in the Cripple Creek and New River belt. The ores occur in connection with calcareous shales, calcareous sandstones, and impure limestones, but have not justified the expectations formed of them. In Carroll County, Virginia, the gossan of the great deposit of pyrite is dug for iron ore. The walls, however, are older than the Cambrian.[1]

2.01.21. The limonites of eastern Tennessee are the southern prolongation of the area of southwest Virginia. They lie between the Archæan of the Unaka range on the east, and the Upper Silurian strata in the foot of the Cumberland tableland on the west. The ores outcrop in the longitudinal valleys or "coves." The bottoms of these valleys, according to Safford (p. 449), are formed by the shales, slates, and magnesian limestones of the Knox group, and in

---

of York and Adams Counties," *Proc. Amer. Phil. Soc.*, March, 1875. J. W. Harden, "The Brown Hematite Ore Deposits of South Mountain between Carlisle, Waynesborough, and the Southeast Edge of the Cumberland Valley," *M. E.*, I. 136. J. P. Lesley, Summary, Final Report, Vol. I., 1892, pp. 205, 341. Rec. A. S. McCreath, *Second Geol. Survey Penn.*, Vol. MM, 199. F. Prime, *Second Geol. Survey Penn.*, Reps. D and DD; "On the Occurrence of the Brown Hematite Deposits of the Great Valley," *M. E.*, III. 410; *Amer. Jour. Sci.*, ii., IX. 433; also, XI. 62, and XV. 261. Rec. B. T. Putnam, *Tenth Census*, Vol. XV., p. 181.

[1] E. R. Benton, *Tenth Census*, Vol. XV., p. 261. J. L. Campbell, "Report on the Mineral Prospects of the St. Mary Iron Property," etc., *The Virginias*, February, 1883, p. 19. See also *The Virginias*, January, 1880, p. 4; March, p. 43. F. P. Dewey, "The Rich Hill Iron Ores," *M. E.*, X. 77. W. M. Fontaine, "Notes on the Mineral Deposits of Certain Localities in the Western Part of the Blue Ridge," *The Virginias*, March, 1883, p. 44; April, p. 55; May, p. 73; June, p. 92. B. S. Lyman, "On the Lower Silurian Brown Hematite Beds of America," *A. A. A. S.*, XVII. 114. A. S. McCreath, "The Iron Ores of the Valley of Virginia," *M. E.*, XII. 103; *Engineering and Mining Journal*, June, 1883, p. 334. E. C. Moxham, "The Great Gossan Lead of Virginia," *M. E.*, February, 1892. E. C. Pechin, "The Iron Ores at Buena Vista, Rockbridge County, Virginia," *Engineering and Mining Journal*, Aug. 3, 1889, p. 92; "Mining of Potsdam Brown Ores in Virginia," *Engineering and Mining Journal*, Sept. 19, 1891, p. 337; "Iron Ores of Virginia and Their Developments," *M. E.*, XIX. 101; "Ore Supply for Virginia Furnaces," *Engineering and Mining Journal*, Vol. LI., 1891, pp. 322, 349. Rec.

FIG. 10.—View of the Siluro-Cambrian brown hematite bank at Baker Hill, Ala. From the Engineering and Mining Journal, Jan. 28, 1893, p. 77.

the residual clay left by their alteration the ore is found. The gossan of the neighboring veins of copper pyrites, best known at Ducktown (see Example 16), were originally exploited for iron.[1]

The Tennessee limonite extends across northwestern Georgia, and still farther east the Huronian limestones of North Carolina also enter the State. But as even these Huronian schists and associated marbles have been considered by F. P. Bradley to be metamorphosed Silurian (Cambrian), the ores may also belong under Example 2a. The well-determined Siluro-Cambrian rocks form but a narrow belt of no great importance in North Carolina.[2]

The limonites are again strongly developed in Alabama and furnish a goodly proportion of the ore used in the State. They form a belt lying east of the Clinton ores (Example 6), later described. As in Tennessee, they are associated with strata of the Knox group.[3]

2.01.22. Extensive deposits of limonite also occur in the Lake Superior district, near Negaunee, Mich., as stated above, but they are mentioned again under Example 9a. They are in Huronian strata.

2.01.23. *Origin of the Siluro-Cambrian Limonites.*—Dr. Jackson of the First Pennsylvania Survey argued in 1839[4] that they originated *in situ;* that is, by the alteration of the rocks in and with which they occur. Percival, in his report on the Geology of Connecticut, in 1842 (p. 132) attributed them to the alteration of pyrite in the neighboring mica-slate. Prime, in Pennsylvania, in 1875 and 1878 (Reports D and DD), considers that the iron has been obtained by the leaching of the neighboring dolomites and slates, it being in them either as silicate, carbonate, or sulphide; that the ore has reached its position associated with the slates, because, being impervious, they retained the ferruginous solutions; and that the potash abundantly present in the slates probably assisted in precipitating it.[5] Frazer, in 1876,[6] in studying the beds of

---

[1] J. M. Safford, *Geol. of Tenn*, p. 448, 1869. B. Willis, *Tenth Census*, Vol. XV., p. 331.

[2] F. P. Bradley, "The Age of the Cherokee County Rocks, North Carolina," *Amer. Jour. Sci.*, iii., IX. 279 and 320; B. Willis, *Tenth Census*, Vol. XV., p. 367.

[3] W. M. Chauvenet, *Tenth Census*, Vol. XV., p. 383. For other references to Alabama iron ore deposits, see under Example 6.

[4] *Ann. Rep. First Penn. Survey*, 1839.

[5] *M. E.*, II. 410.

[6] *Second Penn. Survey*, Rep. C, p. 136.

York and Adams counties, Pennsylvania, found the hydromica slates filled with the casts of pyrite crystals, and held these to have been the sources of the iron, by affording ferrous sulphate and sulphuric acid. The latter reacted on the alkali of the slates, producing sodium sulphate. This, meeting calcium carbonate, afforded calcium sulphate and sodium carbonate, which later precipitated the iron. Calcium carbonate alone is, however, abundantly able to precipitate iron carbonate and oxide from both ferrous and ferric sulphate solutions (even when natural) without the introduction of the alkali, although this might account for the alteration of the slates.[1]

2.01.24. J. D. Dana has written at length on the New England and New York deposits, and finds them always at or near the junction of a stratum of limestone, proved in many cases to be ferriferous, and sometimes entirely siderite, and one of hydromica slate or mica schist. In several mines bodies of unchanged spathic ore are embedded in the limonite. Hence Professor Dana explains the limonite as derived by the weathering of a highly ferruginous limestone, from which the limonite has been left behind by the removal of the more soluble elements, so as practically to replace the limestone in connection with other less soluble matter. The limonite has also at times replaced the schists, probably deriving its substance in part from iron-bearing minerals in them, and changing these rocks to the ochers and clays now found with the ores. These views are undoubtedly very near the truth for the region studied. (Cf. also Example 4.) Weathering limestones do furnish residual clay ocher, etc., as is shown by the deposits of western Kentucky and Tennessee under Example 2.

2.0.25. Another hypothesis early formulated and advocated by many is that the limonites have been derived by the surface drainage of the old Appalachian highlands and then precipitated in still water where they are now found. A precipitation around the shores of a ferruginous sea has also been urged on the analogy of certain explanations of the Clinton ore. (Example 6.) Their supposed Tertiary age has already been remarked. All these views are essentially hypothetical.[2]

---

[1] See F. P. Dunnington, "On the Formation of Deposits of Manganese," *Amer. Jour. Sci.*, iii., XXXVI., p. 175. (Experiments 10 and 11.)

[2] See H. D. Rogers, *Trans. Asso. Amer. Geol. and Nat.*, 1842, p. 345; E. Hitchcock, *Geol. Vt.*, Vol. I., p. 233; J. P. Lesley, *Iron Man.*, p. 501; Rep. A, *Second Penn. Survey*, p. 83; J. S. Newberry, *International Review*, November and December, 1874.

ANALYSES OF LIMONITES.

2.01.26. All published analyses, except when forming a sufficiently large and continuous series from the output of any one mine, are to be taken with caution. Ores necessarily vary much, and a single analysis or a selected set may give a very wrong impression. The percentage in iron is different for different parts of the same ore body. The few that follow have been selected to show the range and the average. The highest are exceptionally good, the lowest less than the average, and the medium values indicate approximately the general run. Limonites afford from 40 to 50% Fe as actually exploited, but it is not difficult to find individual analyses that run higher. They are not, generally speaking, Bessemer ores.

Analyses of Limonites.

	Fe.	P.	S.	SiO_2	Al_2O_3	H_2O.
Berkshire County, Mass.	47.52	0.187				
Connecticut	50.48	0.353				
Dutchess County, New York	46.45	0.370		14.10	3.053	
Staten Island	39.72	0.059	0.391	14.19	3.59	12.41
Pennsylvania	56.3	0.125	0.02	5.165		
Virginia (Low Moor)	43.34	0.636				
Tennessee (Lagrange Furnace)	50.91	0.237				
Alabama	50.89	0.225				
Colorado	53.37	0.034	0.20	7.90	0.70	
Colorado, average	48.00	0.030		20.		13.
Prosser mine, Oregon	44.71	0.666				
Pure mineral	59.92					14.4

SIDERITE, OR SPATHIC ORE.

2.01.27. Siderite is the **protocarbonate of iron**. As a mineral it often contains more or less calcium, magnesium, and manganese. When of concretionary structure, embedded in shales and containing much clay, the ore is called clay ironstone. When the concretions enlarge and coalesce, so as to form beds of limited extent, generally containing much bituminous matter, they are called black-band, and are chiefly developed in connection with coal seams.

2.01.28. Example 3. *Clay Ironstone.*—The name is applied to isolated masses of concretionary origin (kidneys, balls, etc.) which may at time coalesce to form beds of considerable extent. They are usually distributed through shales, and on the weathering of the matrix are exposed and concentrated. They are especially characteristic of Carboniferous strata and differ from black-band only in the absence of bituminous matter and in the consequent drab color. They weather to limonite, generally in concentric shells with a core of unchanged carbonate within. Fossil leaves or shells often furnish the nucleus for the original concretion, and are thus, as at Mazon Creek, Ill., beautifully preserved. When in beds the ore is sometimes called flagstone ore; when broken into rectangular masses by joints, it is called block ore.

2.01.29. Example 3a. *Black-band.*—The name is applied to beds consisting chiefly of carbonate of iron with more or less earthy and bituminous matter. They are of varying thickness, though rarely more than six feet, and are almost invariably associated with coal seams. They are thus especially found in the Carboniferous system, and to a far less degree in the eastern Jura-Trias. They are also recorded with the Cretaceous coals of the West. It is not possible to separate the two varieties in discussing their distribution. The various productive areas are taken up geographically, beginning with the Appalachian region.

2.01.30. The carbonate ores are of great importance in the Carboniferous of western Pennsylvania and in the adjacent parts of Ohio, West Virginia, and Kentucky. In these States the system is subdivided in connection with the coal, from above downward, as follows: I. The Upper Barren Measures, Permo-Carboniferous, or Dunkard's Creek Series; II. The Upper Productive Coal Measures, or Monongahela River Series; III. The Lower Barren Measures, or Elk River Series; IV. The Lower Productive Coal Measures, or Alleghany River Series; V. The Great or Pottsville Conglomerate. In the Upper Barren Measures of Pennsylvania, according to McCreath, there is hardly a stratum of shale or sandstone without clay-ironstone nodules, but no continuous beds are known.[1] The deposits are not of great actual importance, and are worthy of only passing mention. In the Upper Productive Coal Measures some ore occurs associated with the Waynesburg Coal seam, and again, just under the Pittsburg seam, there is considerable known as the Pittsburg Iron Ore Group. This latter ore becomes of great importance in Fayette County and extends through several beds.[2] The Lower Barren Measures in Pennsylvania also contain carbonate ore in a number of localities. The most per-

---

[1] *Second Penn. Survey,* Rep. K, p. 386; MM, p. 159.
[2] Rep. MM, p. 162; KK, p. 111; L, p. 98.

sistent is the Johnstown ore bed, near the base of the series. There are two additional beds just over the Mahoning sandstone.

The Lower Coal Measures are the chief ore producers in all the States. They furnish balls of clay ironstone in very many localities in western Pennsylvania, which will be found recorded with many additional references in Report MM, p. 174, *Penn. Geol. Survey*. The nodules are scattered through clay and shales. The so-called Ferriferous Limestone, which lies a few feet below the Lower Kittaning Coal Seam, affords in its upper portion varying thicknesses of carbonate ore, known as "buhrstone ore," which is altered in large part to limonite. Some little carbonate ore was found in the early days in the anthracite measures of eastern Pennsylvania. Several beds of the same occur in the Great Conglomerate and its underlying (Mauch Chunk) shales. They are chiefly developed in southwestern Pennsylvania (Report KK), and may form either entire beds or disseminated nodules. The limonites of the Marcellus stage that pass into carbonate in depth in Perry and the neighboring counties have already been mentioned under Example 2. In West Virginia both Upper and Lower Measures afford the ore. From the latter black-band is extensively mined on Davis Creek, near Charleston.[1]

2.01.31. In Ohio a number of nodular deposits are known, but practically no ore is produced above the Mahoning sandstone of the Lower Coal Measures. Below this sandstone the ores are extensively developed. They extend up and down the eastern part of the State and are both black-band and clay ironstone. Orton identifies twelve different and well-marked horizons distributed through the Lower Measures. He distinguishes the stratified ores mostly black-band, and the concretionary ores, including kidney ores, block ores, and limestone ores.[2]

2.01.32. The general distribution of the iron ores of Kentucky has already been outlined under Example 2. The Hanging Rock region is a southern prolongation of the Ohio district of the same geological horizon. P. N. Moore has classified the local ores as limestone ores which are associated with limestone, block ores, and kidney ores. The last two names refer to the fracture or shape of

---

[1] M. F. Maury and W. M. Fontaine, *Resources of West Virginia*, 1876, p. 247.

[2] *Geol. of Ohio*, V., p. 378, and supplemental report on the Hanging Rock region in Vol. III.

the masses. They occur associated with the usual clay and shale. Farther west, between the Kentucky and Red rivers, are the other deposits, the principal one of which comes low in the series, just over the Subcarboniferous limestone.[1]

2.01.33. Small quantities of black-band have been found in the Deep River coal beds, in North Carolina, associated with the Triassic coals.[2]

A large bed, or series of beds, has recently been reported from Enterprise, Miss., in strata of the Claiborne stage. They run from ten to eighteen feet in thickness and extend for miles.[3] Scattered nodules have been noted at Gay Head, Martha's Vineyard.[4] Carbonate ores are as yet of no importance in the coal measures of the Mississippi Valley. They have been found associated with the Cretaceous coals of Wyoming and Colorado,—and indeed the first pig iron of the latter State was made from them in Boulder County,—but they are not an important source of ore.[5] An extended bed of very excellent carbonate has recently been discovered with coal near Great Falls, in the Sand Coulée region of Montana. Being near coal, limestone, and other iron ores, it promises to be of considerable importance.[6]

2.01.34. Example 4. Burden Mines, near Hudson, N. Y. Elongated lenticular beds of clay ironstone, passing into subcrystalline siderite, inclosed conformably between underlying slates, and overlying calcareous sandstone, of the Hudson River stage. The ore occurs in four "basins," which outcrop along the western slope of a series of moderate hills, just east of the Hudson River. The hills have been shown by Kimball to be the eastern halves of anticlinal folds now reduced by erosion to easterly dipping monoclines. The western half of the ore bodies has been eroded away, leaving an outcrop forty-four feet thick as a maximum, which pinches out along the strike and dip. The basins ex-

---

[1] P. N. Moore, "On the Hanging Rock District in Kentucky," *Kentucky Geol. Survey*, Vol. I., Part 3.

[2] B. Willis, *Tenth Census*, Vol. XV., p. 306; W. C. Kerr, *Geology of North Carolina*, 1875, p. 225.

[3] A. F. Brainard, "Spathic Ore at Enterprise, Miss.," *M. E.*, XIV. 146.

[4] W. P. Blake, "Notes on the Occurrence of Siderite at Gay Head, Mass.," *M. E.*, IV. 112.

[5] R. Chauvenet, "Notes on the Iron Resources of Colorado," *Ann. Rep. Colo. School of Mines*, 1885, 1886; *Trans. Amer. Inst. Min. Eng.*, Colorado meeting, 1889.

[6] O. C. Mortson, *Mineral Resources U. S.*, 1888, p. 34.

tend from southwest to northeast, parallel to the trend of the hills. The beds are more or less faulted. The southern part of the second basin affords Bessemer ores, but the others are too high in phosphorus. At this point the principal mining has been done. Ac-

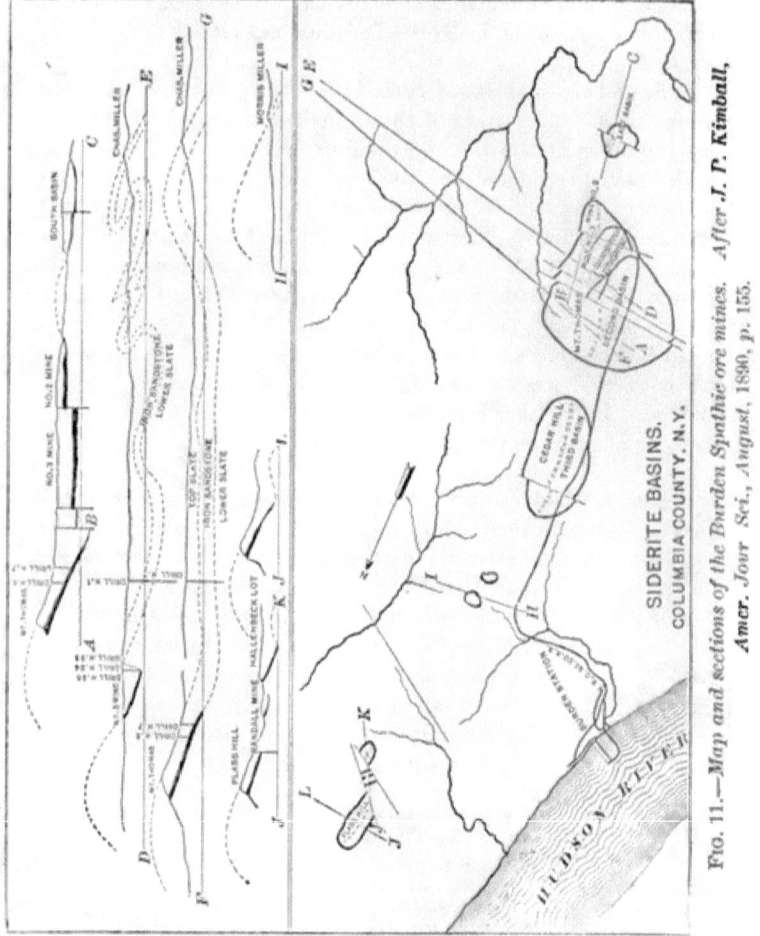

Fig. 11.—Map and sections of the Burden Spathic ore mines. After J. P. Kimball. Amer. Jour. Sci., August, 1890, p. 155.

cording to Olmstead, some varieties are richer in phosphorus than others, but they are so intimately mixed as not to be practicably separated. Up to 1889 the mines had produced 450,000 tons of roasted Bessemer ores.

2.01.35. In their geological relations the ores are of the greatest interest, as they occur on the western limit of the metamorphic belt, which forms the basis of the Taconic controversy, yet in strata which have been identified by fossils. Beds of limonite hitherto regarded as Siluro-Cambrian occur to the east; and should further study, on the lines developed chiefly by J. D. Dana, W. B. Dwight, and C. D. Walcott, clear up their stratigraphical relations, the work done in developing the structure of the siderite basins, as pointed out by Kimball, may be of great aid in explaining them. Very similar bodies of siderite occur with these limonites. (Example 2a.) The Burden ores are relatively high in magnesia, and this leads Kimball to suggest their original deposition from the off-shore drainage of the basic rocks of the Archæan highlands. Further, it may be added that the ores in their lenticular shape are highly suggestive of a possible origin for magnetite deposits, and they are again referred to under "Magnetite." Other deposits of siderite in the shales of the Marcellus stage are known and were formerly worked at Wawarsing, Ulster County, across the Hudson River.[1]

2.01.36. Example 5. Roxbury, Conn. A fissure vein in gneiss, six to eight feet wide, of crystalline siderite, with which are associated quartz and a variety of metallic sulphides, galena, chalcopyrite, zincblende, etc. Although productive in former years, it is no longer worked, and is of scientific more than economic interest, being a unique deposit. It has furnished many fine cabinet specimens.[2]

---

[1] J. P. Kimball, "Siderite Basins of the Hudson River Epoch," *Amer. Jour. Sci.*, III., xl. 155. I. Olmstead, "Distribution of Phosphorus in the Hudson River Carbonate," *M. E.*, 1889. R. W. Raymond, "The Spathic Ores of the Hudson River," *M. E.*, IV. 309. J. C. Smock, *Bulletin of New York State Museum on Iron Ores*, p. 62.

[2] J. P. Lesley, *Iron Manufacturers' Guide*, p. 649. C. U. Shepherd, "Report on the Geology of Connecticut," 1837, p. 30, *Amer. Jour. Sci.*, I., xix. 311.

CHAPTER II.

THE IRON SERIES CONTINUED.—HEMATITE, RED AND SPECULAR.

2.02.01. The sesquioxide of iron, F_2O_3, is always of a red color when in powder. If it is of earthy texture, this color shows in the mass, and the ore is called red hematite; if the ore is crystallized, the red color is not apparent, and the brilliant luster of the mineral gives it the name specular hematite. The red hematites are first treated.

2.02.02. Example 6. *Clinton Ore.*—Wherever the Clinton stage of the Upper Silurian outcrops, it almost invariably contains one or more beds of red hematite, interstratified with the shales and limestones. These ores are of extraordinary persistence, as they outcrop in Wisconsin, Ohio, and Kentucky in the interior, and then beginning in New York, south of Lake Ontario, they run easterly across the State. Again in Pennsylvania they follow the waves of the Appalachian folds, and extend south into West Virginia and Virginia in great strength. They are found in eastern Tennessee and northwestern Georgia, and finally in Alabama are of exceptional size and importance. The structure of the ore varies somewhat. At times it is a replacement of fossils, such as crinoid stems, molluscan remains, etc. (fossil ore); again as small oölitic concretions, like flaxseed (flaxseed ore, oölitic ore, lenticular ore); while elsewhere it is known as dyestone ore. The ore in many places is really a highly ferruginous limestone, and below the water level in the unaltered portion it often passes into limestone, while along the outcrop it is quite rich.

2.02.03. In Dodge County, southeastern Wisconsin, the ore is 14 to 26 feet thick and consists of an aggregate of small lenticular grains.[1] In Ohio it outcrops in Clinton, Highland, and

---

[1] T. C. Chamberlain, *Geol. Survey Wis.*, Vol. I., p. 179. R. D. Irving, "Mineral Resources of Wisconsin," *M. E.*, VIII. 478; *Geol. Survey Wis.*, Vol. I., p. 625.

Adams counties, in the southwestern portion of the State along the flanks of the Cincinnati Arch, but it is thin and poor in iron, although rich in fossils.[1] A small area of the Clinton has furnished considerable ore in Bath County, Kentucky, where it is altered to limonite.[2]

2.02.04. Coming eastward, the limestones and the shales of the Clinton outcrop in the Niagara River gorge in New York, but show no ore. This appears first in quantity in Wayne County, a hundred miles east and just south of Lake Ontario. One bed reaches 20 to 22 inches. Farther east are the Sterling mines, in Cayuga County; and again near Utica, in the town of Clinton, which first gave the ore its name, it is of great economic importance. There are two

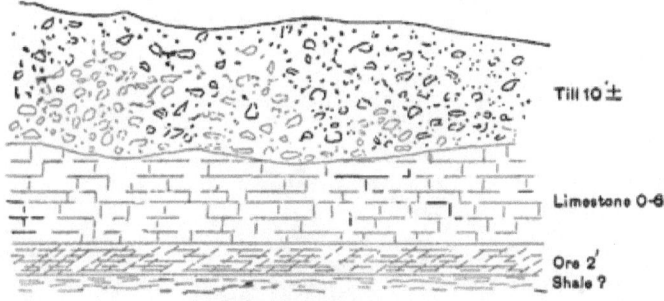

FIG. 12.—*Clinton Ore, Ontario, Wayne County, New York. After C. H. Smyth, Jr.*

workable beds, the upper of which, with a thickness of about two feet, is the only one now exploited. Beneath this are 12 or 15 inches of shale, and then the second bed of 8 inches of ore.[3] Some 25 feet over the upper bed is still a third, which is too low grade for mining. It is four to six feet thick, and is locally called red flux. It consists of pebbles and irregular fragments of fossils, which are coated with hematite and cemented with calcite.

2.02.05. The rocks of the Clinton thicken greatly in Pennsylvania and run southwestward through the central part of the State.

---

[1] J. S. Newberry, *Geol. of Ohio*, Vol. III., p. 7. E. Orton, *Geol. of Ohio*, Vol. V., p. 371.

[2] N. S. Shaler, *Geol. of Ky.*, Vol. III., 163.

[3] A. H. Chester, "The Iron Region of Central New York;" address before the Utica Merchants and Manufacturers' Association, Utica, 1881. J. C. Smock, *Bull. of N. Y. State Museum.* C. H. Smyth, Jr., "On the Clinton Iron Ore," *Amer. Jour. Sci.*, June, 1892, p. 487.

Six different ore beds have been recognized, of which the lower are probably equivalent to the **southern dyestone ores.**[1]

The ores are of chief importance in the Juniata district. The belt extends southwestward across Maryland and **eastern West Virginia,** where the beds are quite thick, although as yet not much developed, and appears in the extreme southwest corner of Virginia. Thence it runs across eastern **Tennessee, and is of very great im-**

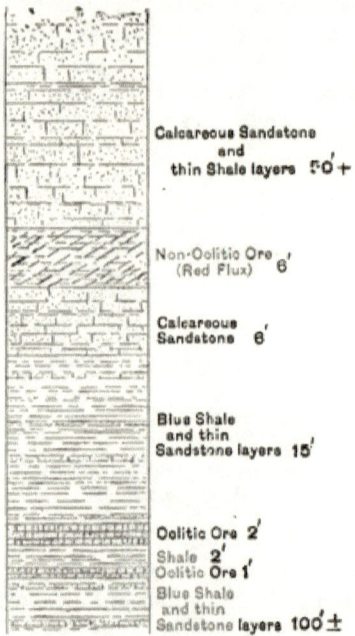

FIG. 13.—*Clinton Ore, Clinton, New York. After C. H. Smyth, Jr.*

portance. The lines of outcrop are known as "dyestone ranges." They lie west of the Siluro-Cambrian limestones (Example 2a) and in the edges of the Cumberland tableland. Four or five are known, of which the largest extends across the State. This ore is

---

[1] J. H. Dewees, "**Fossil Ores of the Juniata Valley,**" *Penn. Geol. Survey*, **Rep. F.** E. d'Invilliers, *Ibid.*, Rep. F3 (Union, Snyder, Mifflin, and Juniata counties). A. S. McCreath, *Ibid.*, Rep. MM, p. 231. J. J. Stevenson, *Ibid.*, Reps. MM and T2 (Bedford and Fulton counties). I. C. White, *Ibid.*, Reps. MM and T3 (Huntington County). H. H. Stock, "Ores at **Danville,** Montour County," *M. E.*, October, **1891.**

more fossiliferous toward the south and more oölitic toward the north. It is very productive in the Chattanooga region.[1]

2.02.06. The Clinton just appears in northwestern Georgia, and continues thence into Alabama, where it is again of great importance,

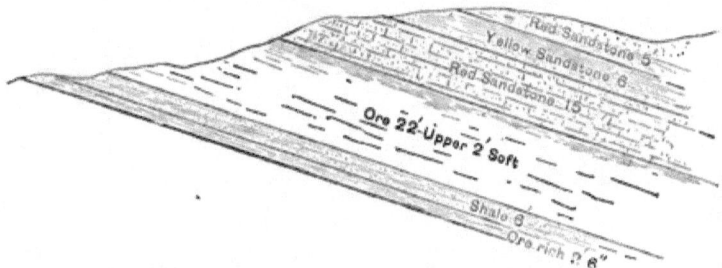

Fig. 14.—*Clinton Ore, Eureka Mine, Oxmoor, Ala. After C. H. Smyth, Jr.*

and, with the less productive Siluro-Cambrian limonites, furnishes practically all the ore of the State. The outcrop can be traced almost continuously for 130 miles. The ore is rich in fossils and occurs in several beds, which, although averaging much less, may

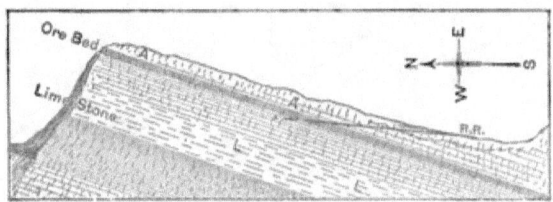

Fig. 15.—*Cross-section of the Sloss Mine, Red Mountain, Ala.*

aggregate, as at the Eureka furnace, as much as 34 to 37 feet. The chief mines are in Red Mountain, a northeast and southwest ridge, east and south of Birmingham. Folds and faults have brought the beds into close proximity with the coal and limestone of the region, and thus into a position very favorable for economic working.[2]

---

[1] Killebrew and Safford, *Resources of Tennessee*. E. C. Pechin, "The Iron Ores of Virginia," etc., *M. E.*, XIX. 1016. J. B. Porter, "Iron Ores, Coal, etc., in Alabama, Georgia, and Tennessee," *M. E.*, XV. 170. J. M. Safford, *Geol. of Tenn.*

[2] A. F. Brainerd, "On the Iron Ores, Fuels, etc., of Birmingham,

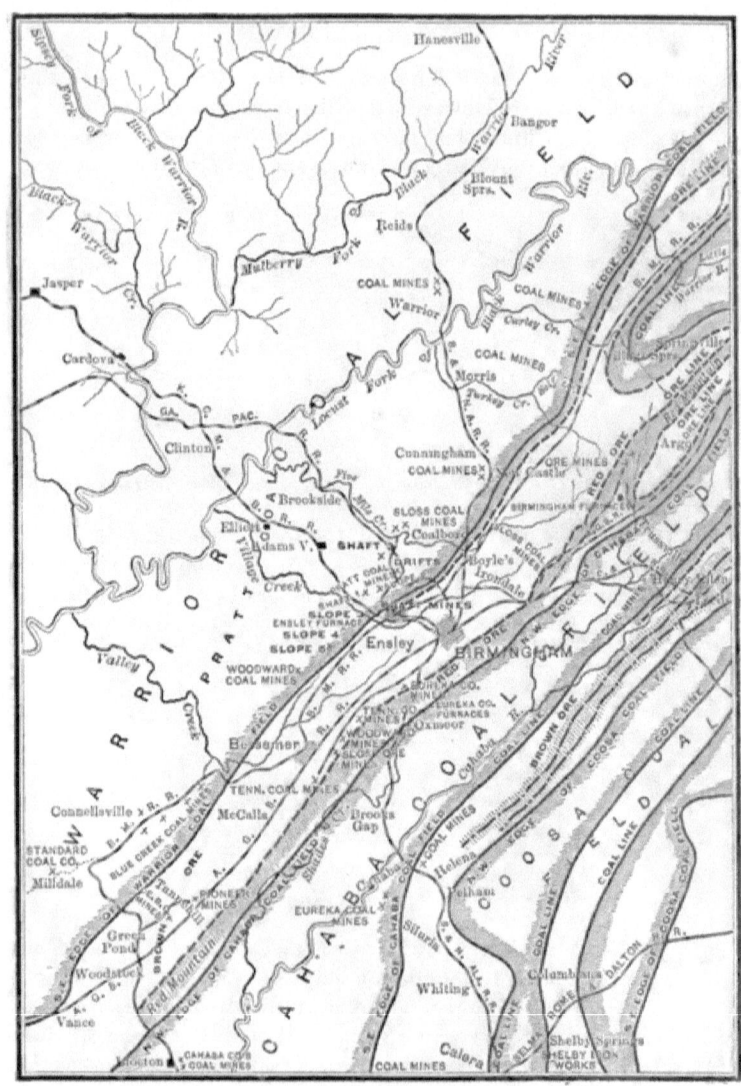

FIG. 16.—*Map of the Vicinity of Birmingham, Ala. From the Transactions of the American Institute of Mining Engineers, Vol. XIX., Plate IV.*

2.02.07. Red hematite, supposed to be of the Clinton stage, occurs in Nova Scotia in very considerable amount, in Pictou and Antigonish counties.[1]

2.02.08. In general the Clinton ore is characterized by a high percentage of phosphorus, and is seldom, if ever, available for Bessemer pig. It is chiefly employed for ordinary foundry irons. The percentage in iron varies much. Experience at Clinton, N. Y., shows that it averages about 44% Fe in the furnace. These hematites have undoubtedly originated in some cases by the weathering of ferruginous limestones above the water level. I. C. Russell has shown that the unaltered limestones at the bottom of a mine in Atalla, Ala., 250 feet from the surface, contained but 7.75% Fe, while the outcrop afforded 57.52%. J. B. Porter has recorded the gradual increase of lime also in another Alabama mine, from a trace at the outcrop to 30.55% at 135 feet. Other writers have explained these beds as due to the bringing of iron in solution into the sea of the Clinton age and to its deposition as small nodules, etc., or as ferruginous mud. (Roger, Lesley, Newberry.) In this way an oölitic mass has originated, as in the modern Swedish lakes (Newberry). (See Example 1.) N. S. Shaler has argued, on the basis of the Kentucky beds, that the iron has been derived from the overlying shales, and descending in solution has been precipitated by the lower lying limestones. As the shales are themselves calcareous, this seems improbable. A. F. Foerste has shown that the ore is very often deposited either in the interstices of fragments of bryozoans or as replacing their substance. The rounded, water-worn character of the original fragments is regarded as occasioning the apparent concretionary character. Admirable work upon the origin of the ore has also been done by C. H. Smyth, Jr. He finds that the small oölites, or concretions, as they occur at Clinton, N. Y., and many other localities, have a water-worn grain of quartz as a nucleus. The character of the grain is such that it has evidently been derived from granitoid or schistose rocks. The hematite comes off at times in concentric layers, when

---

Ala.," *M. E.*, XVII. 151. "The Sloss Iron Ore Mines," *Engineering and Mining Journal*, Oct. 1, 1892, p. 318. T. S. Hunt, "Coal and Iron in Alabama," *M. E.*, XI. 236. J. B. Porter, "Iron Ores, Coal, etc., in Alabama, Georgia, and Tennessee," *M. E.*, XV. 170. E. A. Smith, *Alabama Geol. Survey*, 1876; also *A. A. A. S.*, XXVII. 246.

[1] Sir J. W. Dawson, *Acadian Geology*, p. 591. Fletcher, *Can. Geol. Survey*, 1886.

tapped gently. It may also be dissolved away so as to leave a siliceous cast or skeleton of the spherule. Dr. Smyth thus makes a strong argument that the ores in such cases are concretionary, and that they were formed in shallow waters around the nuclei of sand. But he also admits, as others quoted above have indicated, that the replacement of bryozoa and the weathering of ferruginous limestone have in many localities played their part. The iron ore is in the latter case a residual product, but now the mine waters are depositing calcium carbonate rather than removing it.[1]

2.02.09. Glenmore Estate, Greenbrier County, West Virginia. A bed of red hematite in Oriskany sandstones. Limonites are abundant in the Oriskany of Virginia, and the hematite may have been derived from such [2] or *vice versa*.

2.02.10. Mansfield Ores, Tioga County, Pennsylvania. Three beds of ore are found in the strata of the Chemung stage of Tioga County, Pennsylvania. They are known as the (1) Upper or Spirifer Bed, (2) the Middle or Fish Bed, and (3) the Lower Ore Bed. No. 1 is full of shells and is about 200 feet below the Catskill red sandstones, and at Mansfield is two to three feet thick. No. 2 is oölitic, resembles the Clinton ore, and affords fish remains. It lies about 200 feet below No. 1 and varies up to six or seven feet thick. No. 3 is 100 to 200 feet lower, and contains small quartz pebbles.[3] The ore is not rich, and but little has been mined. It is a brownish red hematite.[4]

2.02.11. Beds of red hematite are reported by Schmidt in the Lower Carboniferous of western central Missouri.[5]

---

[1] A. F. Foerste, "Clinton Group Fossils, with Special Reference to Collections from Indiana, Tennessee, and Georgia," *Amer. Jour. Sci.*, iii., XL. 252. (Abstract; original not cited.) "Clinton Oölitic I on Ores," *Amer. Jour. Sci.*, iii., XLI. 28. Rec. "Notes on Clinton Group Fossils, with Special Reference to Collections from Maryland, Tennessee, and Georgia," *Proc. Bost. Soc. Nat. Hist.*, XXIV. 263. J. P. Lesley, *Iron Manufacturers' Guide*, p. 611. J. S. Newberry, "Genesis of the Ores of Iron," *School of Mines Quarterly*, November, 1880, p. 13. Rec. H. D. Rogers, *Geol. of Penn.*, Vol. II., p. 127. N. S. Shaler, *Geol. of Ky.*, Vol. III., p. 163. C. H. Smyth, Jr., "On the Clinton Iron Ore," *Amer. Jour. Sci.*, June, 1892, p. 487. Rec.

[2] W. N. Page, "The Glenmore Iron Estate, Greenbrier County, West Virginia," *M. E.*, XVII. 115.

[3] A. S. McCreath, Rep. MM, *Second Penn. Survey*, p. 231.

[4] J. P. Lesley, *Geol. of Penn.*, 1888, Vol. I., p. 311. A. Sherwood, Rep. G, *Second Penn. Survey*, pp. 33, 37, 41, 42, 67. A. S. McCreath, Rep. MM, *Second Penn. Survey*, p. 251.

[5] A. Schmidt, "Iron Ores and Coal Fields," *Missouri Geol. Survey*, 1872, p. 169.

2.02.12. **Example 7.** Crawford County, Missouri. Bodies of **specular** and hard and soft red hematite, associated with **clay and chert**, and filling **cone-shaped** (base down) or rude cylindrical depressions in the Second Sandstone of the **Missouri Cambrian.** The hard and soft hematites have resulted from the alteration of the specular. Clay and chert always accompany the ore, and with it fill the cavities in the broken and faulted Second Sandstone. The deposits are distributed over several counties in central Missouri, of which Crawford, Dent, and Phelps are the most productive. The ore, etc., was thought by A. Schmidt of the Missouri Survey (Rep. 1878, p. 66) to have either replaced the pre-existing rock or to have been deposited in the hollows at the then existing surface. Pumpelly, however, regards the iron as having been derived from the weathering of the overlying **First and Second Limestones**, to whose decay he likewise attributed the clay and chert. The latter, it may be remarked, very generally mantle southern and central Missouri, and it is probable that the rocks have not been submerged since Paleozoic times. Much of the drainage, it is thought, passed off through subterranean channels forming caves in the Third Limestone. To the collapse of these is attributed the formation of the cavities, in which the ores were laid down with the residual clay, etc.[1] The region has afforded from 100,000 to 200,000 tons of ore annually.[2]

2.02.13. **Example 8.** Jefferson County, New York. Beds of red hematite, with more or less specular, lying **beneath** sandstones of the **Potsdam stage** and associated with **serpentine**, crystalline limestone, and other **sandstone beds** of uncertain relations. Not far away the **Laurentian** gneiss outcrops, although nowhere associated with the ores. The beds occur along a northeast belt from Philadelphia, Jefferson County, to Gouverneur, St. Lawrence County. They range up to 30 feet in thickness, and consist mostly of red, earthy hematite with included masses of specular. Many interesting **minerals** (siderite, millerite, chalcodite, quartz, etc.) are found in cavities. Brooks gives the following geological section. 1. Potsdam sandstone, 40 feet. 2. Hematites, 40 feet. 3. Soft schistose, slaty, green magnesian rock with pyrite and graphite (thought by Emmons to be igneous), 90 feet and more. 4. Granular,

---

[1] R. Pumpelly, *Tenth Census*, Vol. XV., p. 12.

[2] W. M. Chauvenet, *Tenth Census*, Vol. XV., p. 403; see also Pumpelly's paper, p. 12. A. Schmidt, "Iron Ores and Coal Fields," *Missouri Geol. Survey*, 1874, p. 124. F. L. Nason, *Idem.*, Vol. II., 1892, p. 116. Rec.

crystalline limestone with phlogopite and graphite. 5. Sandstone like (1), 15 feet. 6. Crystalline limestone with beds and veins of granite. E. Emmons, in the early New York Survey (*Geology Second District*), attributed an eruptive origin to these ore bodies and to the associated serpentine and limestone. Such an origin is controverted by Brooks, who first recorded the lower lying sandstone. The deposits need further study.[1]

2.02.14. **Example 9. Lake Superior Hematites.** Bodies of hematite, both red and specular, soft and hard, in metamorphic rocks. They vary widely in shape, although at times quite perfectly lenticular. They are usually associated with jasper and chert, and have for a footwall a relatively impervious rock of some sort. Magnetite is at times present. Although of varying physical structure and associations, all the Lake Superior hematites are here grouped under one general example, in order to avoid unnecessary subdivisons, and to emphasize their related characters. There are five principal ore-producing belts or districts, which are also called in instances "ranges," as they follow ranges of low hills. They are, in the order of their chronological exploitation, the Marquette, just south of Lake Superior, in Michigan; the Menominee, on the southern border of the Upper Peninsula and partly in Wisconsin; the Gogebic or Penokee-Gogebic, on the northwestern border between Michigan and Wisconsin; the Vermilion Lake, in Minnesota, northwest of Lake Superior; and the Mesabi (Mesaba), in the same general region as the last.

2.02.15. The geology of these districts has been a subject of much controversy, not alone in the relations of the separate areas, but in the subdivisions of a single one. The ever-present difficulty of classifying and correlating metamorphic rocks has here been very great. Moreover, there are other separate districts, of related geological structure, which ought also to be brought into harmony, and only at a very recent date has this been even partially attained.

2.02.16. The ores and their inclosing rocks have usually been called Huronian, as this is the name formerly applied to the

---

[1] T. B. Brooks, "On Certain Lower Silurian Rocks in St. Lawrence County, New York," *Amer. Jour. Sci.*, iii. IV., p. 22. G. S. Colby, *Jour. U. S. Asso. Charcoal Iron Workers*, Vol. XI., p. 263. E. Emmons, *N. Y. Geol. Survey, Second District*, p. 93. T. S. Hunt, "Mineralogy of the Laurentian Limestones of North America," 21*st Ann. Report Regents of N. Y. State Univ.*, 1871, p. 88. J. C. Smock, *Bulletin of N. Y. State Museum*.

schistose and metamorphic rocks overlying what was conceived to be the basal, gneissic Laurentian. The later and more careful work has essentially modified such grouping. The reorganization has been brought about by the brothers N. H. and Alexander Winchell, by R. D. Irving, C. R. Van Hise, and the Canadian geologists, especially A. C. Lawson, who has worked in the Rainy Lake region. The definite introduction of Huronian in the classification is especially due to Logan (1857). Previously Foster and Whitney had merely called all the metamorphic rocks concerned with the iron ores in the Lake Superior regions "Azoic." T. B. Brooks in the Marquette district distinguished twenty members (1873), but, as Major Brooks frankly states, the classification was chiefly intended to aid explorations for ores. Rominger made the classification much simpler (1884), and many others have since written on the subject.[1]

2.02.17. As now viewed, the Laurentian is regarded as consisting of granites and gneisses and a higher series of gneisses and schists. They are grouped under the name of "Fundamental Complex" by Irving and Van Hise (Cascade Formation of Wadsworth, 1892), but the upper series is called Coutchiching in the Rainy Lake Region by Lawson. The unconformity is an eruptive one. Above these, after an unconformity not always clearly marked, comes the succession of schistose rocks, which are grouped together under the name Algonkian. They consist of a lower series, called by various names in the different regions, but which in the Marquette, Menominee, and Vermilion Lake districts contains some of the most important mines. It is variously denominated Lower Huronian, Lower Marquette, Keewatin, Lower Vermilion, and Menominee proper in the different exposures, and probably the great cherty limestone of the Penokee-Gogebic series is its local equivalent. In the Marquette district Wadsworth has recently divided it still further into the Republic and Mesnard formations. The upper part follows an unconformity and is called in the different regions Upper Huronian, Animikie, Upper Vermilion, Upper Marquette, Western Menominee, and Penokee-Gogebic proper.

---

[1] See M. E. Wadsworth, *Notes on the Geology of the Iron and Copper Districts*, 1880; N. H. Winchell, "A Last Word with the Huronian," Geol. Soc. Amer., Vol. II., p. 85; C. R. Van Hise, "An Attempt to Harmonize some apparently conflicting Views of Lake Superior Stratigraphy." Amer. Jour. Sci., ii., XLI. 117; and *Tenth Ann. Rep. Director U. S. Geol. Survey*. The papers give many references.

For the Marquette region this has also been further divided by Wadsworth into two, the Holyoke and the Negaunee formations. It is much less metamorphosed than the lower member, and in the Marquette district contains some ore. In the Menominee region of Wisconsin it affords the deposits there wrought and carries the ore in the Gogebic range. Still higher, after another unconformity follows the Keweenawan (Keweenian) or Nipigon. This closes the Algonkian. Still above is the Potsdam sandstone.

2.02.18. **Example 9a.** Marquette District. The Marquette district was earliest known and has been most thoroughly studied; but owing to the confused geological structure, there has been, as already remarked, much discordance of interpretation. In the Marquette district the Huronian Algonkian rocks form a broad synclinal trough with many subordinate folds and several tongues or projections running out from the main body. They rest on and are bounded by Laurentian gneiss. They consist of green schists, quartzites, banded jaspers, slates, ore bodies, and dikes altered to "soapstone" or "soaprock." Brooks divided them into twenty members, of which Beds VI., X., XIII., and a horizon below V. afford the ore. Bed XIII. contains the magnetite, which increases in amount toward the western portion of the field. Rominger in Vol. IV. of the *Michigan Survey*, 1884, reduced the number to seven. Irving and Van Hise have contributed much in late years toward a solution of the geology. Irving regarded the series as separable into a lower division of greenstone schists and more acidic rocks, both of which are dynamically metamorphosed eruptive rocks, and an unconformable, overlying, iron-bearing division of sedimentary origin. (See papers cited below.) Van Hise, however, in his latest paper places the break above the most important ore bodies.

2.02.19. The ores were classed by Brooks under five heads—(*a*) Red specular, (*b*) Magnetic, (*c*) Mixed, (*d*) Soft Hematites, and (*e*) Flag ores; and the grouping illustrates very well their general characters. Class (*a*) includes the slaty hematites that break into irregular tapering plates, and the massive so-called granular ores. Class (*b*) includes granular and more or less friable magnetites, which are related to those described under Example 13. Class (*c*) includes the highly siliceous ores, consisting of hematite, closely interlaminated with red or white jasper. When containing over 50% iron they are at present valuable, but the advance in concentration, especially magnetic, promises to make the low-grade

FIG. 17.—*Open cut in the Republic mine, Marquette range, showing a horse of jasper. From a photograph by H. A. Wheeler.*

ores available. Class (*d*) embraces limonites closely related to those of Example 2*a*, where they are referred to. Class (*e*) has a flaggy structure and its ores are related to Class (*c*), but are less distinctly banded and are mere local varieties of ferruginous schists.

2.02.20. The ore bodies have been in earlier years generally regarded as true beds of greater or less extent and often of great irregularity. They approximate a lenticular shape in the simplest development, as is better shown in the other less disturbed districts. In the Marquette region this is at times obscured by the excessive disturbances. They often follow the foldings of the walls, particularly in synclinal troughs. Later developments have brought out the fact that the ore bodies are associated with some underlying rock that is relatively impervious. The favorite one is the so-called soaprock, an altered igneous intrusion that is chiefly in dikes. Beds of jasper seem to play the same rôle. Van Hise, in his paper of February, 1892, notes four varieties—(1) Deposits on the contact of a quartzite conglomerate (the base of the Upper Marquette) and the ore-bearing formation ; (2) deposits resting upon soaprock, which grades into massive diorite ; (3) deposits resting upon dikes of soaprock, which follow along or cut across the ore-bearing formations ; (4) deposits interbedded in the jasper or chert. In Figure 18 a generalized section taken from Van Hise exhibits these different varieties. On the east the soft hematites (limonites) are first met ; then in going west the red and specular hematites ; and then the magnetic character increases, until at the western end of the district the magnetites are most abundant. South of the Marquette region and between it and the Menominee is found the Felch Mountain area. It consists of three small basins now cut off by erosion from the main exposures, with which it was doubtless at one time connected.

2.02.21. The origin of these ore bodies has been a subject of much controversy. Detailed descriptions of the various hypotheses will be found in Wadsworth's monograph.[1] Only the important attempts at explanation are instanced here. The early survey of Foster and Whitney (1851) attributed an eruptive origin, and the same difficult thesis has since been attempted by Wadsworth (1880), who bases his argument chiefly on the analogy of the banded jaspers to laminated felsites, and to the fact that they and the ore curve around masses of inclosing schist or break across

---

[1] M. E. Wadsworth, "Notes on the Iron and Copper Districts of Lake Superior," *Bull. Mus. Comp. Zoöl.*, Vol. VII., No. 1, July, 1880.

THE IRON SERIES, CONTINUED.   105

them, like intruded dikes. Wadsworth mentions Dr. Selwyn of
the Canadian Survey, and the late C. E. Wright, as supporters of
this view. While it may not as yet be possible to demonstrate
beyond question the true origin, it is quite inconceivable that

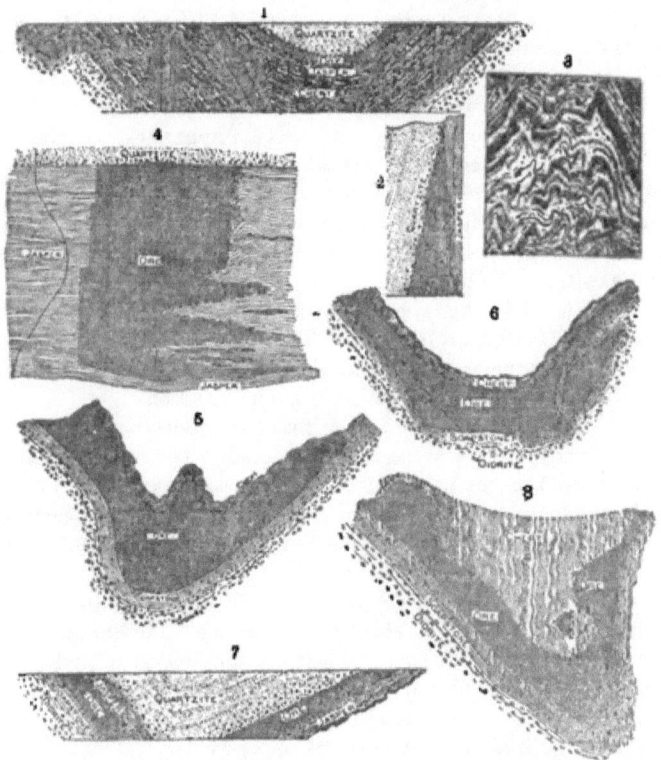

Fig. 18.—*Cross sections to illustrate the occurrence and associations of
iron ore in the Marquette district, Michigan. After C. R. Van Hise,
Amer. Jour. Sci., February, 1892; Engineering and Mining
Journal, July 9, 1892.*

a nearly pure siliceous rock and an equally pure basic oxide should
be side by side and intimately associated as intrusive masses. They
would combine. Quite large masses of iron oxide may and do occur as segregations of basic magmas ; not, however, in any amount
in acidic. All other geologists who have given the matter attention concur in some form of sedimentary origin, or in origin by re-

placement. The beds thus formed may have been afterward metamorphosed. Credner, Brooks, Wright in his published work, and at first Irving, described them as having formed as beds of limonite. These were conceived to have been metamorphosed in the general metamorphism of the region. Brooks thought it possible that the hematites were altered magnetites, an idea confirmed by the presence of martite, but he considered all to have been limonite originally. Irving's views are set forth under Example 9c. Van Hise's latest work traces much the same relations as in the less disturbed Gogebic district. He emphasizes the almost invariable occurrence of the ore along the contact of chert and intrusive dikes, which are now altered to so-called soapstone. This relation is shown in Fig. 18. The ores are thought to have replaced these walls, synclinal troughs having been favorite points of deposition. E. Reyer considered that the iron had been leached from the neighboring, basic eruptive rocks (green schists), and had been precipitated as hydrate. The eruptives are, in this view, regarded as submarine, and the similar association of basic eruptives with the iron ores of Elba is commented on.

2.02.22. It was in the forties that the importance and extent of the ore bodies were first vaguely suspected. The trouble that they made with the compasses of the early land surveyors indicated their existence. Important mining began in 1854. Somewhat over 100,000 tons were produced in 1860, over 800,000 in 1870, nearly 1,500,000 in 1880. In 1877 the Menominee region was opened, and in 1885 the Penokee-Gogebic and Vermilion districts began to ship. The total shipments from the Lake Superior region in 1890 were 8,982,531 tons. The total production to 1891 of the Marquette district was 32,700,000 tons. A quite complete citation of the literature is to be found in Wadsworth's monograph, already referred to, and in Irving's "Copper-bearing Rocks of Lake Superior," *Monograph No. V., U. S. Geol. Survey*. See also under Examples 9b, 9c, and 9d. Only the most important or most recent papers are mentioned here.[1]

---

[1] J. Birkinbine, "Resources of the Lake Superior District," *M. E.*, July, 1887. T. B. Brooks, *Geol. Survey of Michigan*, Vol. I., 1873; *Geol. Survey of Wisconsin*, Vol. III., p. 450. H. Credner, "Die vorsilurischen Gebilde der oberen Halbinsel von Michigan in Nord Amerika," *Zeitsch. d. d. Geol. Ges.*, 1869, XXI. 516; also *Berg.- und Huett. Zeit.*, 1871, p. 369. Foster and Whitney, *Geol. of the Lake Superior District*, Vol. I., "Iron Lands," 1851. R. D. Irving, "On the Origin of the Ferruginous Schists and

2.02.23. Example 9b. Menominee District. The Menominee River, which gives the district its name, forms the southeasterly boundary between the Upper Peninsula of Michigan and Wisconsin. The mines are situated about forty miles south of the Marquette group, and the same distance west of Lake Michigan. The larger number are in Michigan, but the productive belt extends also into Wisconsin. They lie along the south side of an east and west range of hills, which rise from 200 to 300 feet above the surrounding swampy land. The geological section immediately associated with the ore involves the following, all of which corresponds to the Lower Marquette as outlined in the introduction. The cherty limestone is local. (1) Norway limestone (named from the Norway mine), a belt of siliceous limestone, 1200 feet thick; (2) Quinnesec ore group, consisting of limestone, siliceous or jaspery slates, black and flesh-colored, hydromica schists and slates and ore beds, 1000 feet; (3) Lake Hanbury slate group, slates and schists with quartzose bands. Unconformably on these lies the horizontal Potsdam sandstone. The ore occurs along two or three

---

Iron Ores of the Lake Superior Region," *Amer. Jour. Sci.*, iii., XXXII. 263; "Preliminary Paper on an Investigation of the Archæan of the Northwestern States," *Fifth Ann. Rep. Director U. S. Geol. Survey*, p. 131; *Seventh Ann. Rep.*, p. 431; also, Administrative Reports, in subsequent volumes. J. P. Kimball, "The Iron Ore of the Marquette District," *Amer. Jour. Sci.*, ii., XXXIX. 290. H. S. Munroe, *School of Mines Quarterly*, III., p. 43. E. Reyer, "Geologie der Amerikanischen Eisenerzlagerstätten (insbesondere Michigan)," *Oest. Zeit. f. Berg.- u. Hütt.*, Vol. XXXV., pp. 120, 131, 1887. C. Rominger, *Geol. Survey of Michigan*, Vol. IV., 1884. C. R. Van Hise, "An Attempt to Harmonize Some Apparently Conflicting Views of Lake Superior Stratigraphy," *Amer. Jour. Sci.*, iii., XLI., p. 117, February, 1891; *Tenth Ann. Rep. Director U. S. Geol. Survey;* "The Iron Ores of the Marquette District of Michigan," *Amer. Jour. Sci.*, February, 1892, p. 115. M. E. Wadsworth, "Notes on the Iron and Copper Districts of Lake Superior," *Bull. Mus. Comp. Zoöl.*, VII. 1, 1880; "On the Origin of the Iron Ores of the Marquette District, Lake Superior," *Proc. Bost. Soc. Nat. Hist.*, Vol. XX., p. 470; *Engineering and Mining Journal*, Oct. 29, 1881, p. 286; *Ann. Rep. Mich. State Geologist*, 1891-92. "The Geology of the Lake Superior Region," in a pamphlet issued by the Duluth, South Shore and Atlantic R. R., 1892. Dr. Wadsworth announces a new subdivision of Formations in this and in *Amer. Jour. Sci.*, January, 1893, p. 73. H. Wedding. *Zeitsch. f. Berg.-, Hütt.-, und Salinenwesen in Preus. Staat.*, XXIV., p. 339. C. E. Wright and C. D. Lawton, *Reps. of the Commissioners of Mineral Statistics of Michigan*, 1880, and annually to date. G. H. Williams, "Greenstone Schist Areas of the Menominee and Marquette Regions of Michigan," introduction by R. D. Irving, *Bull.* 62, *U. S. Geol. Survey.*

planes of deposition in (2) and not far from the contact with (1), while minor bodies have been found in the Potsdam, which seem to have resulted by the erosion of the older lenses in the Potsdam times. (See paper by J. Fulton, cited below.) Especially instructive exposures of green schists are found which have furnished some of the best evidence that they are metamorphosed, igneous, intrusive rocks. The ores are generally soft, blue, earthy hematites, which give a red powder and consist of very finely divided particles of specular. The brown hematites are of very limited occurrence, being known only in the Emmet mine. The lenticular shape of the ore bodies is better shown than in the Marquette district, and even the large masses clearly exhibit this cross section. They strike about N. 75° W., and dip 70° to 80° N. They also pitch to the

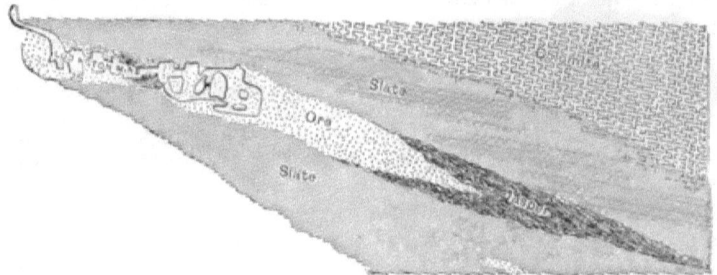

FIG. 19.—*Plan of Ludington ore body, Menominee district, Michigan. After P. Larsson, M. E., July, 1887.*

west; *i.e.*, run down diagonally on the dip. (Cf. New Jersey Magnetite, Example 13*d*). There were produced up to 1891 a grand total of 12,800,000 tons since mining began.[1]

2.02.24. Example 9*c*. Penokee-Gogebic District. This lies in an east and west range of hills, which crosses the westerly boundary of the Upper Peninsula and Wisconsin, and is from ten to twenty miles south of Lake Superior, and eighty to one hundred

---

[1] T. B. Brooks, *Geol. Survey of Wisconsin*, Vol. III., 430-663. D. H. Brown, "Distribution of Phosphorus in the Ludington Mine," *M. E.*, XVI. 525. J. Fulton, "Mode of Deposition of the Iron Ores of the Menominee Range, Michigan," *M. E.*, XVI. 525. Per. Larsson, "The Chapin Mine," *M. E.*, XVI. 119. C. E. Wright, *Geol. Survey of Wisconsin*, III. 666, 734. G. H. Williams, "Greenstone-Schist Areas of the Menominee and Marquette Regions of Michigan, with an Introduction by R. D. Irving," *Bull.* 62, *U. S. Geol. Survey*.

miles west of the Marquette mines. The rocks are less metamorphosed than in the previous two districts. The strata run east and west with a northerly dip of 60° to 80°, and with no subordinate folds. They consist of cherty limestone at the base, followed by quartz, slates, quartzite, iron ore, and ferruginous cherts, and finally slate and schists. The strata are traversed by dikes. The ore is a soft, red, somewhat hydrated hematite, with more or less manganese, which is often considerable and is most abundant in the southern mines. Hard specular is rare. Irving first showed that these ore bodies had originated from the replacement of dolomitic or

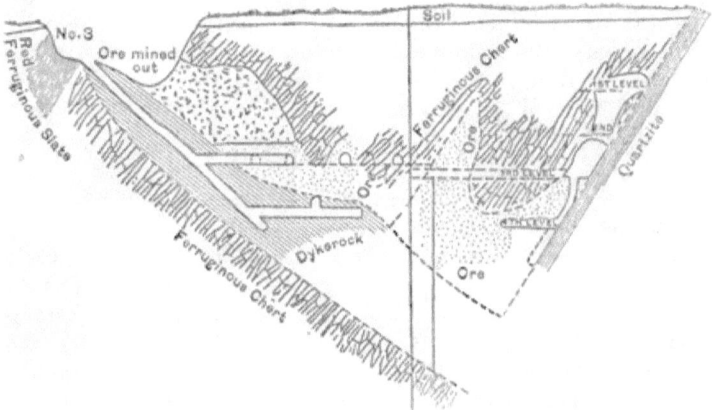

FIG. 20.—*Cross section of the Colby mine, Penokee-Gogebic district, Michigan, to illustrate occurrence and origin of the ore. After C. R. Van Hise, Amer. Jour. Sci., January, 1891.*

calcitic beds with iron oxide. Since then Van Hise has proved them to be in the troughs formed by the intersection of northerly dipping compact quartzites and southerly dipping trap dikes. He has traced the iron to a source in the layers of cherty carbonates, parallel with the quartzites and above them. From this it has been leached out by the percolating water and has been deposited in the apices of the troughs, where it has replaced the original carbonate rocks. Somewhat the same process is outlined for the Marquette ores in his latest paper. Up to 1891, there were produced a grand total of 8,300,000 tons of ore.[1]

---

[1] R. D. Irving, *Geol. Survey of Wisconsin*, III.; pp. 100–167, 1880. "Origin of the Ferruginous Schists and Iron Ores of the Lake Superior Re-

2.02.25. **Example 9*d*.** Vermilion Lake, Minnesota. Beds of hard specular with but little soft, intimately associated with jasper (or jaspilyte, as locally called), and both contained in green schists. The district is situated in the northeastern corner of Minnesota, and northwest of Lake Superior. Two Harbors, the shipping point, is twenty-six miles east of Duluth, and Tower, the chief mining town, is sixty-seven miles from the docks. Leaving the lake, the railroad first crosses the north flank of the Lake Superior synclinal, consisting of southerly dipping igneous rocks belonging to the Keweenawan. Underlying these are a series of gabbros and augite-syenites that contain titaniferous magnetite and may be a parallel to the Adirondack norites. Next follow the black slates of the Animikie, and then a heavy quartzite called the Pewabic quartzite. N. H. Winchell applies to these collectively the name Taconic, a term which the best work in the East rejects in its home. They form the Mesabi range of hills. Sedimentary, gneissic, and eruptive exposures, referred to the Laurentian, succeed in the north. Next come the Vermilion mica and hornblende schists, and after these the Keewatin sericitic schists, jaspilyte, etc., containing the ore bodies at Tower. The Laurentian rocks appear again on the north, and beyond to the northwest is the Rainy Lake region, studied by A. C. Lawson. All the formations referred to above run in belts, having a general direction north and east. The ore bodies which are important as yet are all in the Keewatin. They vary in size from small bodies up to masses, which extend as much as a mile on the strike. They approximate the lenticular shape so characteristic of crystalline iron ore deposits. The jasper, or jaspilyte, is everywhere associated, often very intimately, in parallel bands with the ore, while the containing rock is a green schist which is regarded as an altered igneous rock or tuff. The principal mines are located at Tower, on Vermilion Lake, and at Ely, which is twenty-three miles farther northeast. N. H. and H. V. Winchell, in the Bulletin referred to be-

gion," *Amer. Jour. Sci.*, iii., XXXII. 263, 265; see also under Van Hise. C. D. Lawton, "Gogebic Iron Mines," *Engineering and Mining Journal*, Jan. 15, 1887, p. 42. C. R. Van Hise, "On the Origin of the Mica Schists and Black Mica Slates of the Penokee-Gogebic Iron bearing Series," *Amer. Jour. Sci.*, iii., XXXI. 453–459. "The Iron Ores of the Penokee-Gogebic Series in Michigan and Wisconsin," *Amer. Jour. Sci.*, iii., XXXVII. 32. C. E. Wright, *Geol. Survey of Wisconsin*, III., pp. 239–301.

low, draw a parallel between the more crystalline rocks and the Vermilion schists on the one hand, and the more sericitic and hydrated rocks of the Keewatin on the other. The latter consist chiefly of a chloritic mineral, a sericitic mineral, a feldspathic mineral, mostly plagioclase, and of hematite. The former exhibit a hornblendic mineral, a micaceous mineral, a feldspathic mineral, mostly orthoclase, and magnetite. A passage of one series of rocks into the other is not at all an inconceivable metamorphic process. These last mentioned magnetites are not as yet productive, although regarded as promising. There are titaniferous magnetites in the gabbros and also other undeveloped hematites in the Animikie. Still other magnetites occur in the Pewabic quartzite, recently shown to be of value. The ores from Tower and Ely are high-grade Bessemer, and are produced in great quantity. Nearly 400,000 tons of all kinds were shipped in 1887 and 891,910 in 1890, making a grand total of 3,200,000.

2.02.26. N. H. and H. V. Winchell have argued that these ores originated as marine chemical precipitates. The great extent of igneous rocks associated with them leads to the suggestion that the inclosing rocks have been formed by submarine volcanoes, whose lapilli, etc., have largely contributed their materials. Deposits of iron and silica are thought to have formed from the heated overlying waters. Somewhat similar views of the extensive ore bodies of Elba are held abroad.[1]

The general geological relations are also discussed in many of the papers cited under the other districts, especially those of Irving and Van Hise.

2.02.27. Example 9e. Mesabi Range. Of much more recent development than the other districts is the Mesabi range of Minnesota. The mines are not yet shipping ore (1892), but promise to in 1893. The indications are that the deposits are not less extensive than in any other of the Lake Superior localities, if in-

---

[1] A. H. Chester, *Eleventh Ann. Rep. Minn. Geol. Survey*, 155, 167. T. B. Comstock, "Vermilion Lake District in British America," *M. E.*, July, 1887. H. V. Winchell, "The Diabasic Schists containing the Jaspilyte Beds of Northeastern Minnesota," *Amer. Geol.*, II. 18. N. H. and H. V. Winchell, "On a Possible Chemical Origin of the Iron Ores of the Keewatin in Minnesota," *Amer. Geol.*, IV. 291, 389; "The Taconic Iron Ores of Minnesota and of Western New England," *Amer. Geol.*, VI. 263; "The Iron Ores of Minnesota," *Bull. VI., Minn. Geol. Survey*. Rec. Ann. Reports *of the Minn. Geol. Survey*.

deed they are not even larger and of a form to be more easily mined. The present developments are situated southwest of Vermilion Lake, and nearer Duluth and Lake Superior. The ore lies under the black slates called Animikie in the section given in Paragraph 2.02.25, and over the quartzite, there called the Pewabic; but they are situated twenty miles or so west of the line of that section. The ore bodies are all south of the granite ridge called the Giants' range. Upon the southern slopes of this range lie the green schists of the Keewatin, which are unconformably overlain by the Pewabic quartzite. On this rests the ore-bearing

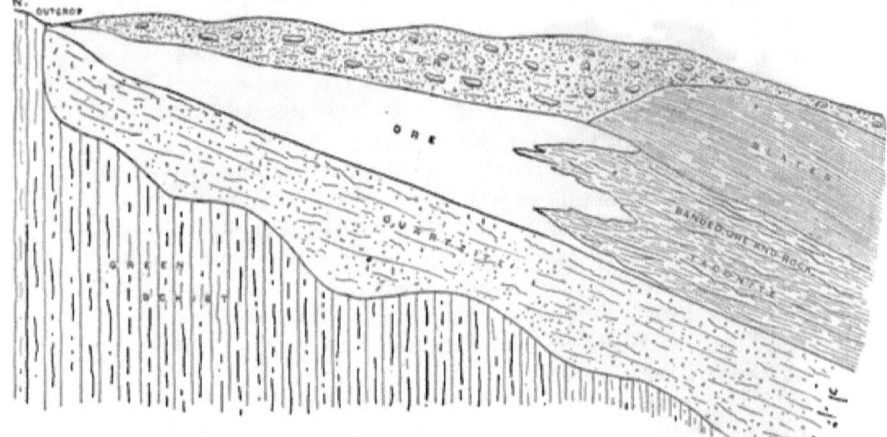

Fig. 21.—*General cross section of ore body at Biwabik, Mesabi Range, Minn. After H. V. Winchell, 20th Ann. Rep. Minn. State Geologist.*

rock, which is a jaspery quartzite, called taconyte by H. V. Winchell. Over this, in order, come greenish siliceous slates and cherts, black slates (referred to the Animikie), and great masses of gabbro. On the flanks of the Giants' range the dip is steep, but it flattens out nearly to horizontality away from the granite. All the formations above the Keewatin are called Taconic by the Winchells.

2.02.28. The ore bodies lie on the southerly slopes of low hills, and are found immediately below the mantle of glacial drift, which varies up to 100 feet in thickness. Ore indications have long been known on the range, and various reports have been made in previous years, although always unfavorably. The indications then available showed only siliceous limonites of low grade. Deep test pits, which penetrated these caps and the drift, have, however, rewarded persistent prospecting. The ore is both soft, blue, earthy, and sandy hematite, and hard specular. With these are limonites and paint ores. The sections show at times fifty feet and more of excellent hematite, which may be of exceptional purity and far below the Bessemer limit of phosphorus, or which may slightly exceed it, but in general the published analyses would show them to be quite high-grade, siliceous, low phosphorus ores.

The ore bodies lie in the jaspery quartzite (taconyte) along its outcrop. They may be directly on the Pewabic quartzite, as seems usual, or else entirely in the taconyte. They fade out into the latter along the dip. They are regarded by H. V. Winchell, to whose description the above is chiefly due, as having originated by replacement of the taconyte. This rock sometimes contains calcareous streaks, which have perhaps aided in furnishing the carbonic acid, which, it is thought, has dissolved the silica of the quartzite (taconyte) in the replacement process. The greater part of it, however, has doubtless been atmospheric, and has by its solvent action concentrated the iron already disseminated in the taconyte. Mesabi is also written Mesaba and Missabe.[1]

2.02.29. Example 10. James River, Virginia. Specular hematite in narrow beds (lenses), interstratified with quartzites and slates of metamorphic character and Archæan age. They run four to six feet, or less, in thickness, with prevailingly vertical dip, but they also pitch diagonally down on the dip like the lenses of magnetite, later described. They furnish a very excellent grade of ore. The ore bodies are found along both sides of the James River, a few miles above Lynchburg. Some magnetite also occurs in the region, and some limonite. More or less clay accompanies the ore.[2]

2.02.30. Similar lenses of specular ore and magnetite are found

---

[1] H. V. **Winchell**, *Twentieth Ann. Rep. Minn. State Geologist*, p. 112, 1892; reprinted *M. E.*, **1892**. **Rec.** The *New York Times* of Dec. 14, 1892, has a quite extended account. **H. V. Winchell** and **J T. Jones**, "The Biwabik Mine," *M. E.*, February, 1893.

[2] **E. B. Benton**, *Tenth Census*, Vol. XV., p. 263 (on Virginia). J. L. Campbell, *Geology and Resources of the James River Valley*, p. 49, New York, 1882. B. Willis, *Tenth Census*, Vol. XV., p. 301. **The** *Virginias*, a monthly formerly published by Jed. Hotchkiss, at Staunton, contains much information on Virginia in general.

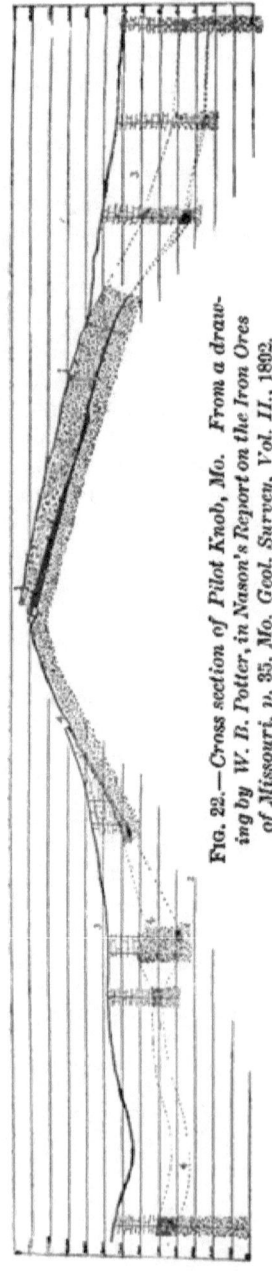

FIG. 22.—Cross section of Pilot Knob, Mo. From a drawing by W. B. Potter, in Nason's Report on the Iron Ores of Missouri, p. 35, Mo. Geol. Survey, Vol. II., 1892.

in central North Carolina, in schistose rocks, which have been referred to the Huronian.

2.02.31. Lenses of specular hematite of very excellent quality are found also in metamorphic rocks, north of Fort Laramie, Wyoming, which may prove productive in time. But little is as yet known about them.

2.02.32. Example 11. Pilot Knob, Mo. Two beds of hard specular hematite separated by a thin seam of so-called slate (probably volcanic tuff), and interstratified with breccias and sheets of porphyry. Along the eastern limit of the Ozark uplift of Missouri and Arkansas a series of knobs of granite and porphyritic rocks project through the Cambrian limestones and sandstones. They are older than the limestones, and clearly were not intruded through them. The limestones and sandstones lie up against the porphyry and in the valleys between. The underlying porphyry has been found in the valley near Pilot Knob, after penetrating four hundred feet of sedimentary rocks. The porphyry and ores have often been called Huronian, but in view of the recent reorganization of the Huronian (see Example 9), this is not done, nor ever has been, on any accurate grounds. Pilot Knob is formed by one of these eruptive knobs. It consists of sheets of porphyries that are capped by porphyry breccia, the two ore beds, and the intervening tuff. The beds strike and dip 13° S. S. W. The hill is over 600 feet high. The lower bed has furnished most of the ore, running from 25 to 40 feet thick, and affording a dense bluish, specular hematite of from 50 to 60% Fe, siliceous, and very low in phosphorus. The upper bed is irregular and of lower grade and runs from 6 to 10 feet thick. The Pilot Knob mines in this solid ore are now substantially exhausted.

Recent drill holes on the northerly slope and below the outcropping face of ore have shown that under the Cambrian strata of

FIG. 23.—*View of open cut at Pilot Knob, Mo., showing the bedded character of the iron ore. From a photograph by J. F. Kemp*, 1888.

the valley there is a great bed of ore boulders or breccia in clay, much as is the case at Iron Mountain, later described. Analyses of cores were not, however, sufficiently encouraging for development during the present low prices for iron. Doubtless the bed will afford important reserves.

2.02.33. Near Pilot Knob are two other hills of porphyry,

116   KEMP'S ORE DEPOSITS.

FIG. 24.—View of Iron Mountain, Mo., from the east.   From a photograph by H. A. Wheeler.

Shepherd Mountain and Cedar Mountain, whose ores are structurally more related to Example 11a. The first contains three veins, the Champion, the North, and the South. They are long and narrow (4 to 10 feet), strike north 60° to 70° east, and dip 70° north. The Champion vein contained a little streak of natural lodestone, but the ore is mostly specular.

The North vein shows a good breast of ore five feet wide, but too full of pyrite to be available. Cedar Mountain has a vein of specular ore. Neither hill has been an important producer. Minor veins have been found on neighboring porphyry hills (Buford, Hogan, and Lewis mountains), some of which contain much manganese.

2.02.34. Example 11a. Iron Mountain, Missouri. Veins of hard, specular hematite irregularly seaming a knob of porphyry. Iron Mountain is five or six miles north of Pilot Knob, and is a low hill with a westerly spur called Little Mountain, and has also a northerly spur. It consists of feldspar porphyries, more or less altered. These are seamed with one large, and on the west somewhat dome-shaped, parent mass of ore, and innumerable minor veins that radiate into the surrounding rock. Upon the flanks of the porphyry hill rests a mantling succession of sedimentary rocks, that dip away on all sides. The lowest member is a conglomerate of ore fragments, weathered porphyry, and residual clay left by its alteration. It is regarded by Pumpelly as formed by pre-Silurian surface disintegration and not by shore action, inasmuch as sand does not fill the interstices, while white porphyry clay does. It is, however, overlain by a thin bed of coarse, friable sandstone, which marks the advance of the sea, and whose formation preceded the limestones. This conglomerate is now the principal source of the ore. It is mined underground, hoisted and washed by hydraulic methods, like those employed in the auriferous gravels of California, and then jigged. The apatite has largely weathered out of it. The rock of the mountain itself, in the cuts of the mines, is largely kaolinized, and exhibits everywhere the effects of extreme alteration. The smaller veins that penetrate the porphyry show at times casts or much altered cores of apatite crystals.

2.02.35. The porphyries of Pilot Knob and Iron Mountain, in thin section, are seen to belong to quartz-porphyries, feldspar-porphyries, and porphyrites. Both orthoclase and plagioclase are present in them, and many interesting forms of structure. One significant fact is that they are everywhere filled with dusty particles of iron oxide, probably magnetite. Our knowl-

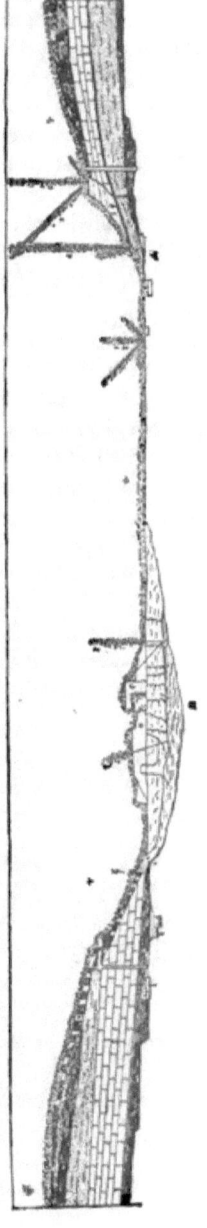

FIG. 25.—Cross section of Iron Mountain, Mo., showing the knob of porphyry, with the veins of ore, and the mantling succession of ore-bearing conglomerate, sandstone, and limestone. From a drawing by W. B. Potter in Nason's Report on the Iron Ores of Missouri, p. 29, Mo. Geol. Survey, Vol. II., 1892.

edge of the chemical composition of the porphyries is, however, as yet very imperfect. An eruptive origin was originally assigned to these ores by J. D. Whitney (*Metallic Wealth of the United States*, p. 479, 1854), just as to the Lake Superior hematites. The later investigations of Adolph Schmidt for the Missouri Survey in 1871 arrived at a different conclusion. Dr. Schmidt considered them, whether occurring in an apparent bed, as at Pilot Knob, or in various more or less irregular veins, as at Iron Mountain, to have been formed either by a replacement of the porphyries with iron oxide deposited from solution, or by a filling in the same way of fissures, probably formed by the contraction of the porphyry in cooling. In the valuable report on iron ores by F. L. Nason in the *Missouri Geological Survey* a sedimentary origin is advocated for the Pilot Knob beds. They are conceived to have been deposited in a body of water in a hollow, between formerly existing porphyry hills, which rose above. In the course of weathering, the hills became the valleys and the early sedimentary beds the hilltop. It is, however, somewhat difficult to understand how the more or less incoherent sediments withstood degradation better than the hard, firm porphyry hills. Some such origin as sedimentation or replacement is, however, the only reasonable one. It is not improbable that the Pilot Knob ores originated in the saturation and more or less complete replacement of tufaceous layers with infiltrating iron oxide.

An extended table of analyses of Iron Mountain ores will be found in *Mineral Resources of the United States*, 1889–90, p. 47.[1]

ANALYSES OF HEMATITES, RED AND SPECULAR.

(The same discrimination must be employed in looking over these analyses that was emphasized under limonite.)

	Fe.	P.	S.	$S.O_2$.	Al_2O_3	H_2O.
Clinton, N. Y. (fossil ore)........	44.10	0 650	0 23	12.63	5.45	2.77
Wisconsin (fossil ore)............	51.75	1.392				
Pennsylvania (Mifflin ore).......	44.4	0.115	0.028			
Tennessee (Meigs County).......	51.63	0.345				
Birmingham, Ala................	56.4	0.34		16.8	0.5	
Antwerp, N. Y..................	46.32	0.883				
Missouri (Crawford County).....	59.41	0.085				
Marquette dist., Mich. (specular).	68.4	0.53		2 07		
" " " "	64.83	0.067		3.60	2.03	
Menominee district, Michigan..	60.47	.009		3.38		
Iron Mountain, Mo..............	65.5	0.040	0.010	5.75		
Pilot Knob, Mo	59.15	0.015		13.27	2.19	
James River (Maud vein)	49.89	0.139				
Elba	61.81	0.02	0.17	5.97	3.47	
Pure mineral....................	70.0					

[1] G. C. Broadhead, "The Geological History of the Ozark Uplift,"

These analyses are mostly taken from State reports and from *Mineral Resources of the United States*. They are intended to illustrate the general run of compositions, but for Birmingham and Marquette are high. Analyses vary widely.

Amer. Geol., III. 6.   J. R. Gage, "On the Occurrence of Iron Ores in Missouri," *Trans. St. Louis Acad. Sci.*, 1873, Vol. III., p. 181.   E. Harrison, "Age of the Porphyry Hills," *Ibid.*, Vol. II., p. 504.   E. Haworth, "A Contribution to the Archæan Geology of Missouri," *Amer. Geol.*, I. 280–363; "Age and Origin of the Crystalline Rocks of Missouri," *Bull. V., Mo. Geol. Survey*, 1891.   A. V. Leonhard, "Notes on the Mineralogy of Missouri," *Trans. St. Louis Acad. Sci.*, Vol. IV., p. 440.   F. L. Nason, "Report on the Iron Ores of Missouri," *Mo. Geol. Survey*, II.   Rec.   R. Pumpelly, "Geology of Pilot Knob and Vicinity," *Mo. Geol. Survey*, 1872, p. 5; see also remarks on Iron Mountain, *Bull. Geol. Soc. Amer.*, Vol. II., p. 220.   Rec.   W. B. Potter, "The Iron Ore Regions of Missouri," *Journal U. S. Asso. Charcoal Iron Workers*, Vol. VI., p. 23.   Rec.   F. A. Sampson, "A Bibliography of the Geology of Missouri," *Bull. II., Mo. Geol. Survey*, 1890. (This is a valuable book of reference.)   F. Shepherd, *Ann. Rep. Mo. Geol. Survey*, 1853–54.   Hist.   A. Schmidt, "Iron Ores of Missouri," *Mo. Geol. Survey*, 1872, p. 45, and especially p. 94.   Rec.   J. D. Whitney, *Metallic Wealth of the U. S.*, p. 479

CHAPTER III.

MAGNETITE AND PYRITE.

2.03.01. Example 12. Magnetite Beds. Beds of magnetite, often of lenticular shape, interstratified with Archæan gneisses and crystalline limestones. They are extensively developed in the Adirondacks, in the New York and New Jersey Highlands, and in western North Carolina. The presence of magnetite in Michigan (Example 9a), in Minnesota (Example 6b), on Shepherd Mountain in Missouri (Example 11), and in Virginia (Example 12) has already been referred to. Other magnetite bodies are known in Colorado, Utah, California, and Wyoming, and will be mentioned subsequently. Titanium is often present in such amounts as to render the ore of no value. The same is true of pyrite and pyrrhotite. Apatite is always found, although it may be in very small quantity. Chlorite, hornblende, augite, epidote, quartz, feldspar, and a little calcite are the common associated minerals. In New Jersey the beds occur in several parallel ranges or belts.

2.03.02. Example 12a. Adirondack Region. The magnetite deposits occur in the foothills of the Adirondacks on all sides, and to a less extent in the mountains themselves. The mountains are very largely knobs of a rock, which is chiefly labradorite, with some hypersthene and other bisilicates, and is variously called labrador-rock, norite, hypersthene-rock, anorthosite, etc. Whether this is igneous or a series of metamorphosed sediments, it is not yet generally admitted, as the geology of the region largely remains to be worked out. It certainly exhibits both metamorphic and igneous facies and is undoubtedly a more or less metamorphosed igneous rock. Associated with it, especially in the foothills, are gneiss and crystalline limestones, in the former of which occur the magnetite deposits now wrought, but field work in the summer of 1892 has convinced the writer that there are large bodies of magnetite in true igneous gabbros in the town of Westport, if not elsewhere. At Lyon Mountain, on the north, the Chateaugay ore body is really a bed of gneiss very rich in magnetite, rich enough in places to afford a merchantable ore. The greater part of it, however, requires concentration. It is known for several miles on the outcrop. On either side the ore shades out into the walls of barren rock. The Little River mines on the west side, in St. Law-

FIG. 26.—*View of open cut and underground work in Mine 21, Mineville, near Port Henry, N. Y. Photographed by J. F. Kemp, 1892.*

rence County, appear to be a similar but larger bed or group of beds of lean ore, said to be from 800 to 1500 feet wide and two miles long. The Barton Hill mines, near Mineville, are openings on a single connected zone and extend, together with Fisher Hill, over a mile on the strike.

2.03.03. Besides these extended beds there are other deposits exhibiting more perfectly the peculiar lenticular shape characteristic of magnetite, and to this class are to be referred the greater number of smaller bodies (Hammondville). They pinch and swell, roll and fold and feather out, and often come off sharply from the walls. They frequently follow and overlap one another like shingles, the second one succeeding the first in the footwall. They are not infrequently cut by trap dikes and are thrown by these and by normal faults. At Hammondville small gulches seem to cut off the ore, and are probably due to faults. Other deposits are of enormous size, as at Mineville (200 to 300 feet clear ore between the walls), and their relations are less clear. They may be large lenses doubled over in a sigmoid fold. The Champlain magnetite is quite notably granular as contrasted with New Jersey, which tends rather more to break in prisms.

2.03.04. C. E. Hall has divided the metamorphic rocks of the Adirondacks into the (*a*) Lower Laurentian Magnetite Iron Ore series, containing the most important ore beds. (*b*) The Laurentian Sulphur Ore Series. (*c*) The Limestones and the Labrador or Upper Laurentian, with Titaniferous Iron Ores. (*c*) is thought to be certainly later than (*a*), but the relations of (*b*) are uncertain. T. S. Hunt also states that the titaniferous ores are associated with the Labradorite series or Norian, which he places as of later age than the gneisses with the good ore.

Commercially the ores are divided into (1) ores high in phosphorus but low in sulphur; (2) ores low in both phosphorus and sulphur; (3) pyritous ores; (4) titaniferous ores (*Tenth Census*, Vol. XV.). Under class (4) come numerous beds which are worthless, but which if the titanium could be neutralized would be very valuable (Lake Henderson; see Addenda.). Mineville is by far the most productive region. It ships 400,000 to 500,000 tons yearly. Chateaugay and Hammondville are next. The Arnold mines produce some, while the Palmer Hill mines with the decline of the bloomaries have gradually ceased. The mines on the west side of the mountains are less important. They afford, so far as developed, a low grade of ore, that, however, with the improvements in magnetic concentration, seems to promise well. The largest openings are at Jayville and Little River.

There are numerous magnetite deposits in Canada of analogous geological relations, but they are often highly titaniferous.[1]

---

[1] L. C. Beck, *Mineralogy of New York*, Part I., pp. 1–38. J. Birkin-

MAGNETITE AND PYRITE. 123

2.03.05. Example 12b. New York and New Jersey Highlands, and the South Mountain of Pennsylvania. Beds of lenticular shape in Archæan gneiss and crystalline limestone. From Putnam County, New York, a ridge of Archæan rocks runs southwest across the Hudson River, traversing Orange County, New York, and northern New Jersey, and running out in Pennsylvania. Lenses of magnetite occur throughout its entire extent. They are not as large as in the Adirondacks, but are more regularly distributed. East of the Hudson, in Putnam County, the Tilly Foster mine is the most important, and the descriptions and figures of it are the best illustrations of the shape of lenses published. West of the Hudson, in Orange County, the Forest of Dean mine affords considerable ore yearly. It is cut by an interesting trap dike. As the results of study of the Archæan of this region, N. L. Britton has divided it into a Lower Massive group, a Middle Iron Bearing, and an Upper Schistose. (*Geol. of N. J.*, 1886, p. 77.) F. L. Nason has also sought to classify it on the basis of rock types, of which he makes four, named, from their typical occurrences, Mount Hope type, Oxford type, Franklin type, and Montville type. They are ar-

---

bine, "Crystalline Magnetite in the Port Henry (N. Y.) Mines," *M. E.*, February, 1890. Rec. H. Credner, *Zeitsch. d. d. g. Gesell.*, 1869, XXI., p. 516; *B. und H. Zeit.*, 1871, 369. J. D. Dana, "On the Theories of Origin," *Amer. Jour. Sci.*, iii., XXII. 152, 402. E. Emmons, *Geology of New York, Second District*, pp. 87, 98, 231, 255, 291, 309, 350. Hist. C. E. Hall, "Laurentian Magnetite Ore Deposits of Northern New York." 32d *Ann. Rep. State Museum*, 1884, p. 133. Rec. J. F. Kemp, "Notes on the Minerals Occurring near Port Henry, N. Y.," *Amer. Jour. Sci.*, iii., XI. 62, and *Zeitsch. f. Kryst.*, XIX. 183. G. W. Maynard, "The Iron Ores of Lake Champlain," *Brit. Iron and Steel Inst.*, Vol. I, 1874. W. C. Redfield, "Some Account of Two Visits to the Mountains of Essex County, N. Y., in 1836-37," *Amer. Jour. Sci.*, i., XXXIII. 301. Hist. B. Silliman, "Remarks on the Magnetites of Clifton, St. Lawrence County, N. Y.," *M. E.*, I. 364. J. C. Smock, "Iron Mines of New York," *Bull. VII., N. Y. State Museum*. Rec. J. Stewart, "Laurentian Low Grade Phosphate Ores," *M. E.*, February, 1892. Wedding, *Zeitschr. f. B., H., und S. im. p. St.*, XXIV. 330, 1876. See also the general works on Iron Ores cited at beginning of Part II. On Canadian magnetites the following papers may be mentioned. F. P. Dewey, "Some Canadian Iron Ores," *M. E.*, XII. 192. B. J. Harrington, "On the Iron Ores of Canada," *Can. Geol. Survey*, 1873-74. T. S. Hunt, *Can. Geol. Survey*, 1866-69, pp. 261, 262. T. D. Ledyard, "Some Ontario Magnetites," *M. E.*, XIX. 28, and July, 1891. W. H. Merritt, "Occurrence of Magnetite Ore Deposits in Victoria County, Ontario," *A. A. A. S.*, XXXI. 413, 1882.

ranged in their order of probable age. They correspond in some respects to Britton's grouping, but differ materially in others. (*Geol. of N. J.*, 1889, p. 30.) Four courses, or mine-belts, have been recognized in New Jersey,—the Ramapo, the Passaic, the Musconetcong, and the Pequest,—in order from east to west. The lenses strike northeast with the gneisses, and usually have, like them, high dips. In addition they have also a so-called "pitch"

FIG. 27a. FIG. 27b.

FIGS. 27a and 27b.—*Model of the Tilly Foster ore body. 27a, Top view, showing faulted shoulder. After F. S. Ruttmann, Trans. Amer. Inst. Min. Eng., XV. 79. 27b, View of bottom of same. Photographed by J. F. Kemp from the model now at the School of Mines, Columbia College.*

along the strike, so that they run diagonally down the dip. They have been observed to pitch northeast with an easterly dip and southwest with a westerly. Either by the overlapping of lenses or by an approximation to an elongated bed, they sometimes, as at Hibernia, extend a mile or more in unbroken series. Again, they may be almost circular in cross section (Hurd mine). At Franklin Furnace one is found in crystalline limestone.[1]

---

[1] E. S. Breidenbaugh, "On the Minerals Found at the Tilly Foster Mine, New York," *Amer. Jour. Sci.*, iii., VI. 207. J. F. Kemp, "Diorite

2.03.06. **South Mountain, Penn.** Small lenses of magnetite occur in Berks, Bucks, and Lehigh counties of **southeastern Pennsylvania**, in the metamorphic rocks of the South Mountain belt. They are very like those to the north in New Jersey, but are lower in both iron and phosphorus. Their product is about **100,000 tons** yearly. The Cornwall magnetite is described under Example 13, for its geological structure is entirely different from the lenses.[1]

2.03.07. **Example 12c.** Western North Carolina and Virginia. Beds of magnetite, of the characters already described, in Archæan gneisses and schists. The ore body at **Cranberry, N. C.,** is the largest and best known. It occurs in **Mitchell County**, and has lately been connected by rail with the lines in **east Tennessee**. According to Kerr, **the principal outcrop is 1500 feet long** and 200 to 800 **feet broad; but, of course, all of this is not ore**. The mines can afford very large quantities of excellent **Bessemer** grade. Pyroxene and epidote are associated with the ore. Kerr has referred the magnetite to the Upper Laurentian. In the **southern** central portions of North Carolina other magnetites occur in the mica and talcose schists, which have been referred to the Huronian. Specular hematite is associated with them. (Example 10.) **Magnetite** has also been lately reported from Franklin and Henry counties, Virginia, and Stokes County, North Carolina, which may be available in the future. Some doubt, however, is cast on its amount and quality.[2]

Dike at the Forest of Dean Mine," *Amer. Jour. Sci.*, iii., XXXV. 331. F. H. McDowell, "The Reopening of the Tilly Foster Mine," *M. E.*, XVII. 758; *Engineering and Mining Journal*. Sept. 7, 1889, 206. F. S. Ruttman, "Notes on the Geology of the Tilly Foster Ore Body, Putnam County, New York," *M. E.*, XV. 79. Rec. J. C. Smock, *Bull. VII., N. Y. State Museum.* Rec. A. F. Wendt, "The Iron Mines of Putnam County," *M. E.*, XIII. 478. "Iron Mines of New Jersey," *School of Mines Quarterly*, iv., III. N. L. Britton, *Ann. Rep. N. J. Survey*, 1886, p. 77. Rec. G. H. Cook and J. C. Smock, *Geol. of N. J.*, 1868. Rec. (See also subsequent annual reports, especially 1873, p. 12.) F. L. Nason, *Ann. Rep. N. J. Survey*, 1889. Rec. J. W. Pullmann, "The Production of the Hibernia Mine, New Jersey," *M. E.*, XIV. 904. J. C. Smock, "The Magnetite Iron Ores of New Jersey," *M. E.*, II. 314; "A Review of the Iron Mining Industry of New Jersey," *M. E.*, **June**, 1891. Rec.

[1] E. D'Invilliers, Rep. D3, *Penn. Survey*, Vol. II. (South Mountain Belt of Berks County). Rec. F. Prime, Rep. D3, Vol. I. *Penn. Survey* (Lehigh County). B. T. Putnam, *Tenth Census*, Vol. XV., p. 179.

[2] W. C. Kerr, *Geol. of N. C.*, 1875, 264. H. B. C. Nitze, "On Some of the **Magnetites** of Southwestern Virginia, etc., and Discussion of Same, by E. C. Pechin," *M. E.*, June, 1891. B. Willis, *Tenth Census*, Vol. XV., p. 325. *Engineering and Mining Journal*, **Jan. 7, 1888**.

2.03.08. Example 12*d*. Colorado Magnetites. Beds of magnetite of a lenticular character in rocks described as syenite (Chaffee County) and diorite (Fremont County). With these a number of others are mentioned which vary from the example, but of which more information is needed before they can be well classified. The last are mere prospects. The mines in Chaffee County have been the only actual producers. There are three principal claims—the Calumet, Hecla, and Smithfield. They extend continuously over 4000 feet. The wall rock is called syenite. Chauvenet describes them as having resulted from the oxidation of pyrites, and as being in rocks of Silurian age. They average 57% Fe, with only 0.009 P, but are comparatively high in S, reaching 0.1 to 2.0%. These mines and those at the Hot Springs, mentioned under Example 2, have furnished the Pueblo furnaces with most of their stock. The deposit in Fremont County is at Iron Mountain, but is too titaniferous to be valuable. It is a lenticular mass in so-called diorite. A large ore body has been reported from Costillo County, in limestone (Census Report) or syenite (Rolker). In Gunnison County, at the Iron King and Cumberland mines, excellent ore occurs in quartzites and limestones, called Silurian. At Ashcroft, near Aspen, high up on the northern side of the Elk Mountains, is a great bed or vein of magnetite, in limestones of Carboniferous age, with abundant eruptive rocks near. It is thought by Devereux to be altered pyrite. Still, pyrite is a common thing with magnetite elsewhere. There are other smaller deposits in Boulder County and elsewhere in the State.[1]

2.03.09. In Wyoming an immense mass of titaniferous magnetite is known near Chugwater Creek. It is described by Hague as resembling a great dike in granite.[2] Gabbro is in the neighborhood.

2.03.10. Example 12*e*. California Magnetite. Beds of magnetite of lenticular shape in metamorphic slates and limestones on the western slope of the Sierra Nevada. Others of different character are also known. In Sierra and Placer counties lenses of excellent ore are found, accompanying an extended stratum of lime-

---

[1] R. Chauvenet, "Papers on Iron Prospects of Colorado," *Ann. Reps. Colo. State School of Mines*, 1885 and 1887; also *M. E.*, Denver meeting, 1889. Rec. W. B. Devereux, "Notes on Iron Prospects in Pitkin County, Colorado," *M. E.*, XII. 608. B. T. Putnam, *Tenth Census*, Vol. XV., p. 472. Rec. C. M. Rolker, "Notes on Iron Ore Deposits in Colorado," *M. E.*, XIV. 266. Rec.

[2] "Iron Mountain, Wyoming," 40*th Parallel Survey*, Vol. II., p. 14.

stone in chlorite slate. A great ore body of magnetite described as a vein has lately been reported from San Bernardino County. It is said to be from 30 to 150 feet thick, and to lie between dolomitic limestone and syenite.[1] A great bed of a kind not specified is reported from San Diego County.[2]

2.03.11. Example 13. Cornwall, Penn. Immense beds of soft magnetite, associated with green slates, limestones of Cambrian age, and Triassic sandstones. They are pierced by dikes of Triassic diabase. The exact geological relations of these beds have been greatly disputed, the doubtful point being whether they are of Triassic or Siluro-Cambrian age. They occur just at the juncture of schists of the latter with red sandstones of the former. The most probable explanation is that they were originally an immense deposit of Siluro-Cambrian limonite (Example 2a), which has been changed to magnetite by the diabase eruptions. They have been

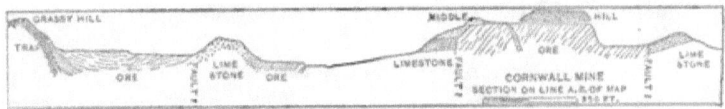

FIG. 28.—*Cross section of the magnetite ore body at Cornwall, Penn. After Bailey Willis, Tenth Census, Vol. XV., p.* 226.

thought eruptive also, an idea far less probable. The ore forms two low hills. It is mined by open cuts. The amount is enormous, 600,000 to 700,000 tons being shipped yearly. Up to 1888 more than 8,500,000 tons had been mined. The ore is rather low in iron but is not high in phosphorus. Considerable amounts of copper ores (chalcopyrite, etc.) occur in fissures crossing the magnetite. Farther southwest, in York County, the Dillsburg group of mines affords specular hematite and magnetite from deposits in Triassic sandstones.[3]

---

[1] *Ann. Rep. State Mineralogist*, 1889, p 235.

[2] *Ibid.*, 1889, p. 154. J. R. Browne, "Mineral Resources West of the Rocky Mountains," 1868. C. King and J. D. Hague, "Mineral Resources West of the Rocky Mountains," 1874, p. 44. H. G. Hanks and W. Irelan, *Ann. Reps. State Mineralogist, California.* (Very little on iron.) F. von Richthofen, private reports quoted in *Tenth Census*, Vol. XV., p. 495. J. D. Whitney, *Geol. Survey Cal.*, Vol. I.

[3] P. A. Fraser, "Study of the Specular and Hematite Ores of Iron of the New Red Sandstone in York County, Pennsylvania," *M. E.*, V. 132.

2.03.12. **Example 14. Iron County, Utah.** Beds of magnetite and hematite bearing evidence of being metamorphosed limonite, in limestones of questionable Silurian age, and associated with eruptive rocks described as trachyte. The limestones have been much upturned, metamorphosed, and pierced by dikes and eruptive masses. The ore forms great projecting ridges and prominent outcrops, locally called "blow-outs." The usual lenticular shape is not lacking. They occur over an area of fifteen by five miles, and are in the southern end of the Wasatch Mountains. The samples show rich ores, which at times exceed the Bessemer limit of phosphorus. In the Star district the ore apparently lies between quartzite and granite. Hematite occurs in large amount, as does quartz, while some streaks have large crystals of apatite. The importance of the deposits lies in the future. They are the largest in the West, and are interesting in their bearing on the general origin of magnetite. Coal, not proved to be good for smelting, is near, but centers of iron consumption are very far away.[1]

2.03.13. **Example 15. Magnetite Sands.** Beds of magnetite sands concentrated on beaches or bars by waves and streams. The magnetite has been derived from the weathering of igneous and metamorphic rocks, through which it is everywhere distributed. When in the sand of a sea beach, it and other heavy minerals tend to be concentrated by the sorting action of the waves. They resist the retreating undertow better than lighter materials. Such deposits are very abundant at Moisie, on the St. Lawrence, below Quebec, and in the United States are known in smaller developments on Lake Champlain; at Quogue, L. I.; on Block Island; in Connecticut; along the Great Lakes, and on the Pacific coast. Grains of garnet, olivine, hornblende, etc., minerals of high

---

Also *Penn. Survey*, Rep. C2. E. V. D'Invilliers, "Cornwall Iron Ore Mines," *M. E.*, XIV. 873. Rec. Lesley and D'Invilliers, *Ann. Rep. Second Penn. Survey*, 1885. J. P. Lesley, Final Report, Vol. I., p. 351, 1892. Rec. T. S. Hunt, "The Cornwall Mines," etc., *M. E.*, IV. 319. H. D. Rogers, *First Penn. Geol. Survey*, II. 718. B. Willis, *Tenth Census*, Vol. XV., p. 223. Rec.

[1] W. P. Blake, "Iron Ore Deposits of Southern Utah," *M. E.*, XIV 809. J. S. Newberry, "Genesis of Our Iron Ores," *School of Mines Quarterly*, March, 1880. Rec. *Engineering and Mining Journal*, April 23, 1881, p. 286. *Proc. National Academy*, 1880. B. T. Putnam, *Tenth Census*, Vol. XV. 486. Rec.

specific gravity, are also in the sands. Many are too high in titanium to be of use, but there is no more difficulty in concentrating them than artificially crushed ore. In Brazil and New Zealand they have attracted attention.[1]

2.03.14. *On the Origin of Magnetite Deposits.*—It is important to note that magnetite deposits are almost always in metamorphic rocks, which owe their character to regional metamorphism, or to the neighborhood of igneous rocks (Pennsylvania and Utah). Gneisses form the commonest walls, but so-called norites, or gabbros, and crystalline limestones also contain them. Where there is lamination, or bedding, the magnetite conforms to it. As the history of the metamorphic rocks is so often uncertain, the magnetites share the same doubt. In igneous rocks magnetite is the most widely occurring of the rock-making minerals. In all explanations the prevailing lenticular shape, the general arrangement in linear order, and the existence of great beds must be considered. The shape is very similar to that of deposits of specular hematite, with which magnetite is often associated. (Examples 9 and 10.) The following hypotheses have been advanced as to their origin : 1. As intruded (eruptive) masses. This supposes an origin for the lenses on the analogy of a trap dike. Though formerly much advocated, it is now generally rejected. 2. As excessively basic portions of igneous rocks. This supposes that large amounts of iron oxide separate in the cooling and crystallizing of basic magmas. There are such occurrences, although seldom, if ever, pure enough or abundant enough for mining. The titaniferous magnetite of the Minnesota gabbros has been alluded to (2.02.25), and also the Brazilian ore and the Cumberland Hill (R. I.) peridotite. (See also Dakyns and Teall, *Q. J. G. S.*, XLVIII., p. 118.) Should such igneous rocks be subjected to regional metamorphism and the stretching action characteristic of it, the ore masses might be drawn out into lenses. 3. As metamorphosed limonite beds. This idea has been most widely accepted in the past. It presupposes limonite beds formed as in Examples 1 and 2, which become buried and subjected to metamorphism, changing the ore to magnetite and the walls to schists and gneisses. Igneous rocks have apparently changed limonites to magnetite at Cornwall, Penn., and in Utah, but such changes by regional metamorphism are less easy

---

[1] T. S. Hunt, *Geol. Survey Canada*, 1866–69, 261, 262; *Canad. Nat.*, VI. 79. A. A. Julien, "The Genesis of the Crystalline Iron Ores," *Acad. Nat. Sci., Phil.*, 1882, 335; *Engineering and Mining Journal*, Feb. 2, 1884.

to demonstrate. The limonite may have resulted from the oxidation of lenses of pyrite. 4. As replaced limestone beds, or as siderite beds subsequently metamorphosed. Such deposits may pass through a limonite stage. The general process is outlined under Example 9*c*, as developed by Irving and Van Hise in the Gogebic district. The lenticular deposits of siderite at the Burden mines (Example 4) are very suggestive, and some such original mass might in instances be metamorphosed to magnetite. 5. As submarine chemical precipitates. This is outlined under Example 9*d*, as applied by the Winchells in Minnesota. 6. As beach sands. The lenses are regarded as having been formed as outlined under Example 15. The same heavy minerals sometimes occur with magnetite lenses as are found on beaches. (See B. J. Harrington, *Can. Geol. Survey*, 1873, 193; A. A. Julien, *Phil. Acad. Sci.*, 1882, 335.) 7. As river bars. This regards the lenses as due to the concentration of magnetite sands in rivers or flowing currents. Hence the overlapping lenses, the arrangement in ranges or on lines of drainage, and the occasional swirling curves found on the feathering edges of lenses, as in the Dickerson mine, Ferromont, N. J. (See H. S. Munroe, *School of Mines Quarterly*, Vol. III., p. 43—an important paper.) It is also reasonable to suppose that lakes or still bodies of water may have occurred along such rivers, and have occasioned the accumulation. (As segregated veins, see Addenda.)

Several other hypotheses with small claims to credibility could be cited. They are outlined at length in *Bull. VI., Minn. Geol. Survey*, p. 224, but in this place there has been no desire to take up any but those deserving serious attention. It may be said that while one or the other of the above seven hypotheses may in instances be applied with reason, yet most candid observers with widened experience have grown less positive in asserting them as axiomatic.

2.03.15. Of importance in connection with iron ore deposits are the recent studies of the distribution of phosphorus along certain lines in the beds, by a knowledge of which it is possible to keep more valuable Bessemer ore distinct from less valuable. Such lines have been found in Michigan, and have been called by Mr. Browne "isochemic lines." Though less marked at the Burden mines (Example 3), the phosphorus was characteristic of certain varieties of the ore. Much work has also been done on the same question at Iron Mountain, Mo.[1]

---

[1] D. H. Browne, "On the Distribution of Phosphorus at the Ludding-

ANALYSES OF MAGNETITES.

(Caution in interpreting analyses is again emphasized as under 2.01.26.)

	Fe.	P.	S.	TiO_2.	SiO_2.	Al_2O_3
Canada (Rideau Canal)..........	50.23			9.80		
Chateaugay mines, N. Y., lump..	49.24	0.029	0.052		18.447	
" " concentrated.	66.00	0.003				
Mineville, N. Y. (Mine 21) ...	62.10	1.198				
Orange County, N. Y. (Forest of Dean)............................	63.00	0.621	0.148			
Putnam County, N. Y............	48.82	0.021	0.08		11.75	3.50
New Jersey (Hibernia)...........	53.75	0.364				
Cornwall, Penn..................	42.7	0.135	0.620			3.411
Cranberry, N. C.................	64.64	0.004	0.115			
Colorado (Calumet).	49.23	0.026			3.85	
" (Iron King)............	58.75	0.044	0.123			
Utah (Iron County).............	62.6		0.12		4.80	
California (Gold Valley).........	60.68				10.87	

PYRITE.

2.03.16. Example 16. Pyrite Beds. Beds (veins) of pyrite, often of lenticular shape and of character frequently analogous to magnetite deposits, in slates and schists of the Cambro-Silurian or Huronian systems, and less often in gneiss of the Archæan. Slates are most common, and gneiss least so. They extend from Canada down the Appalachians to Alabama, being found at Capelton, Quebec; Milan, N. H.; Vershire, Vt.; Charlemont, Mass. (Anthony's Nose, N. Y., and the Gap mine, Pennsylvania, being pyrrhotite, will be mentioned under "Nickel" with other similar occurrences); Louisa County, Virginia; Ducktown, Tenn. (see above, Fig. 6), and at many points less well known in Alabama. Also at Sudbury, north of Lake Superior, recent developments have shown an enormous deposit, specially discussed under "Nickel."

2.03.17. The ore bodies lie interbedded in the slates, and often the different lenses overlap and succeed each other in the footwall and exhibit all the phenomena cited under magnetites. Chalcopyrite is usually present in small amount, and where the copper reaches 3 to 5% they are valuable as copper ores. (See under

---

ton Mines." etc., *M. E.*, XVII., p. 616. I. Olmsted, "The Distribution of Phosphorus in the Hudson River Carbonates," *M. E.*, Colorado meeting, June, 1889. W. B. Potter, Analyses of Missouri ore published in *Mineral Resources*, 1890, p. 47.

"Copper.") At present they are of increasing importance as a source of sulphuric acid fumes for the manufacture of sulphuric acid. Small amounts of lead and zinc sulphide are often present,

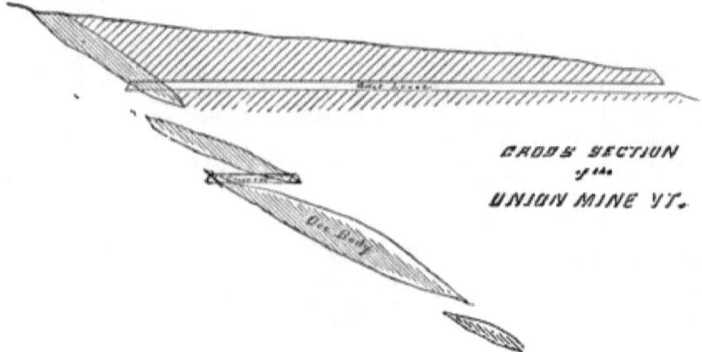

Fig. 29.—*Illustration of overlapping lenses of pyrite. After A. F. Wendt, School of Mines Quarterly, Vol. VII.*, 1886.

rarely a little silver. Nickel and cobalt occur especially in the pyrrhotitic varieties. They are worthless as a source of iron. The auriferous pyrites of the Southern States will be mentioned under "Gold."

2.03.18. They may have accumulated in a way analogous to the bog ore hypothesis, cited under "Magnetite;" but instead of the iron being precipitated as oxide, it has probably come down as sulphide from the influence of decaying organic matter, and has subsequently shared in the metamorphism and solidification of the wall rock. At the same time it must be admitted to be an obscure point. By many they are thought, with more reason, to have originated like a bedded fissure vein whose overlapping, lenticular cavities have been formed by the buckling of folded schists.[1] (Cf. "Gold Quartz," as later described.)

---

[1] W. H. Adams, "The Pyrites Deposits of Louisa County, Virginia." *M. E.*, XII., p. 527. C. R. Boyd, "The Utilization of the Iron and Copper Sulphides of Virginia, North Carolina, and Tennessee," *M. E.*, XIV., p. 81; *Resources of S. W. Virginia*. H. Credner, "At St. Anthony's Nose, Hudson River," *B. und H. Zeit.*, 1866, p. 17; "Pyrite in Virginia, Tennessee, and Georgia," *B. und H. Zeit.*, 1871, p. 370. H. T. Davis, *Mineral Resources of the U. S.*, 1885, p. 501. H. M. Howe, "The Copper Mines of Vermont," *M. E.*, Baltimore meeting, February, 1892. William Martyn, *Mineral Re-*

2.03.19. The relative importance of the different kinds of ore is shown by the following tables for 1880 and 1890. The increase in red hematite is due to the Lake Superior region and to Alabama.

	1880.	Per cent. of Total.	1890.	Per cent. of Total.
Red hematite	2,512,712	31.51	10,527,650	65.65
Magnetite	2,390,389	29.98	2,570,838	16.03
Brown hematite	2,149,417	26.95	2,559,938	15.96
Carbonate	922,288	11.56	377,617	2.36
	7,974,806	100.00	16,036,043	100.00

As indicating the relative importance of the different mining regions, the following figures are of interest. No individual State producing less than 100,000 tons is given.

States.	Total in 1890.	States.	Total in 1890.
Michigan	7,141,656	New Jersey	495,808
Alabama	1,897,815	Tennessee	465,695
Pennsylvania	1,361,622	Georgia	244,088
New York	1,253,393	Missouri	181,690
Wisconsin	948,965	Ohio	169,088
Minnesota	891,910	Colorado	114,275
Virginia	543,583	All the others	326,455
Grand total			16,036,043

sources, 1883-84, p. 877. E. C. Moxham, "The Great Gossan Lead of Virginia" (altered pyrite in Carroll County), *M. E.*, February, 1892. A. F. Wendt, "The Pyrites Deposits of the Alleghanies," *School of Mines Quarterly*, Vol. VII., and separate reprint; also *Engineering and Mining Journal*, June 5, 1886, p. 22 and elsewhere. Rec. H. A. Wheeler, "Copper Deposits of Vermont," *School of Mines Quarterly*, IV 210.

CHAPTER IV.

COPPER.

2.04.01. Copper Ores.

TABLE OF ANALYSES.

	Cu.	S.	Fe.
Native copper (generally with some silver)	100.		
Chalcocite, Cu_2S	79.8	20.2	
Chalcopyrite, $CuFeS_2$	34.6	34.9	30.5
Bornite, Cu_3FeS_3	61.79	25.8	11.7
Tetrahedrite, $4CuS_2Sb_2S_3$ (variable) 26.50 Sb'	36.40	26.7	1.39
Enargite, Cu_3AsS_4 (As.19.1)	48.4	32.5	
Cuprite, Cu_2O	88.8		
Melaconite (tenorite), CuO	79.86		
Malachite, $2CuO+CO_2+H_2O$	57.4		
Azurite, $3CuO+CO_2+H_2O$	55.0		
Chrysocolla, $CuO+SiO_2+2H_2O$	36.1		

2.04.02. **Example 16, Continued.** Pyrite or pyrrhotite beds (veins), with intermingled chalcopyrite. Whether the deposits are true beds or veins parallel with the stratification is as yet a matter of dispute. The resemblance to magnetite argues a bed, and this view is generally taken by German writers. The California mines occur closely associated with the auriferous (pyritous) quartz bodies, which are always esteemed veins. The interbedded lenticular deposits are placed by themselves as the main example. The undoubted veins like Ore Knob, N. C., are then made a sub-example. Pyrites and pyrrhotite (called mundic by the miners) are the principal constituents of such bodies, but often the copper reaches 4 to 5%, and then they are valuable for copper. The ores are often roasted for sulphurous fumes in acid works, and afterward the residues are returned to the copper smelters. They have been or are being worked for this metal at Capelton, Quebec, just north of Vermont. At Milan, N. H., there are several deposits in argillitic schists, and in the same region are numerous other locations. At Vershire, Vt., there is a belt some twenty miles long with three principal mining points. Of these the middle one, containing the Ely mine, is the largest. Two beds of ore occur, separated by from 10 to 20 feet of schists. The lower averages about four feet, but fluctuates; the upper is still more variable, and may reach 25 feet. They are both formed by a succession of thin lenses. The ore is chalcopyrite, mingled with pyrrhotite and quartz. At Ducktown, Tenn., which is in the extreme southeast corner of the State, there are three ranges of ore hills in a width of a mile. The outcrop is marked by great masses of gossan, and below this, and along the contact with the sulphides, was found the rich black ore which gave the early impetus to the mines. When this was exhausted the sulphides alone remained. They consist of chalcopyrite and pyrrhotite, with considerable quartz and country rock, less often calcite. (See Fig. 7, p. 39.)

2.04.03. Example 16a. Ore Knob, N. C. This is described by Kerr as a true fissure vein, extending 2000 feet on the strike, which is parallel to that of the gneiss, but cutting the dip in descent. The width averaged about 10 feet. The gossan extended to a depth of 50 feet, and furnished the usual body of rich ore at the contact with the sulphides. It has not been operated for some years.[1]

---

[1] J. T. Bailey, "The Copper Deposits of Adams County, Pennsylvania," *Engineering and Mining Journal*, Feb. 17, 1883, p. 88. H. Credner, "On Ducktown, Tenn.," *B. und H. Zeit.*, 1867, p. 8. *Engineering and Mining Journal*, Nov. 6, 1886, p. 327 (contains "The Elizabeth Copper Mines, Vermont";) see also April, 1886. P. Fraser, "Some Copper Deposits of Carroll County, Maryland," *M. E.*, IX. 33; "Hypothesis of the Structure of the Copper Belt of the South Mountain, Pennsylvania," *M. E.*, XII. C. H. Henderson, "Copper Deposits of the South Mountain, Pennsylvania," *M. E.*, XII. 85. C. H. Hitchcock, *Geol. of N. H.*, Vol. III., Part III., p. 47. H. M. Howe, "The Copper Mines of Vermont," *M. E.*, February, 1892. T. S. Hunt, "Ore Knob and Some Related Deposits," *M. E.*, II. 125. Kleinschmidt (on Virginia, Tennessee, and North Carolina), *Gangstudien*, Vol. III., p. 256. (A good, short, but old account.) E. E. Olcott, "Ore Knob Copper Mine and Reduction Works," *M. E.*, III. 391. Rec. Richardson, "Copper Ore of Stafford, Vt., *Amer. Jour. Sci.*, I. 21, 383. Tripple and Credner, "Report on the Ducktown Region to the American Bureau of Mines," 1866. M. Tuomey, "A Brief Note of Some Facts Connected with the Ducktown (Tenn.) Copper Mines," *Amer. Jour. Sci.*, II. 19, 181. A. F. Wendt, "The Pyrites Deposits of the Alleghanies," *School of Mines Quarterly*, Vol. VII., 1886; *Engineering and Mining Journal*, July 10 and following, 1886. Rec. H. A. Wheeler, "Copper Deposits of Vermont," *School of Mines Quarterly*, Vol. IV., 219. Rec.

2.04.04. Example 16b. Spenceville, Cal. Beds of pyrites with considerable chalcopyrite, in Jurassic slates, on the western slopes of the Sierra Nevada. The Jurassic slates along the western Sierras contain, in the gold belt, some interbedded deposits of pyrite with considerable chalcopyrite intermingled. The most important are at Newton, Amador County, Copperopolis and Campo Seco, Calaveras County, and Spenceville, Nevada County. At the first named there is a body of sulphides 7 to 8 feet wide and proved about 400 feet. In the adjoining county of Calaveras, the Union mine, at Copperopolis, is on a very large body of sulphides, which impregnate the slates on each side without forming a very sharp foot or hanging wall. The Campo Seco deposits are on the same general line as the Copperopolis. The extension passes also through those at Newton. A long distance north are the mines in Spenceville, Nevada County. The general geology is similar, and the ore body large. It affords from 3.50 to 5.50% copper. All these ores are worked by wet methods.[1]

NOTE.—For Examples 16c and 16d see under "Nickel."

2.04.05. Example 17. Butte, Mont. Veins originally fissures, or shear zones, but greatly enlarged by replacement of the walls with ore, filled with copper sulphides, bornite, chalcopyrite, etc., in a siliceous gangue. Much silver is associated with the copper. At Butte there is a north and south valley six miles wide between high granite ridges on the east and lower rhyolite ridges on the west. The bounding heights north and south are still farther distant. Near the middle of this valley the butte of rhyolite arises which gives the town its name. Silver Bow Creek, which gives the name to the county, flows south along the eastern ridge and then bends westward at a point south of the butte, and, after flowing directly across the valley, leaves it through the western ridge. In the half of the valley east of the meridian of the butte is a very dark basic granite, and also in the extreme west. It consists of quartz, orthoclase, plagioclase, and an unusual amount of mica, augite, and hornblende. In the part west and south of the butte is a highly

---

[1] J. E. Ellis, "On the Spenceville Mines," *Mineral Resources U. S.*, 1884, p. 340. H. G. Hanks, *Rep. California State Mineralogist*, 1884, p. 148. J. B. Hobson, "On Spenceville," *Rep. California State Mineralogist*, 1890, p. 392. William Irelan, Jr., "On Calaveras County Mines," *Rep. California State Mineralogist*, 1888, pp. 150-153. "On the Newton Mines," *ibid.*, p. 106.

acidic, light-colored granite, which consists of quartz and orthoclase feldspar with a very little biotite. Dikes of quartz porphyry penetrate the basic granite, and dikes of rhyolite are also found associated with the ore bodies. Tongues of rhyolite are met, apparently offshoots of the butte.

2.04.06. East of the butte and in the coarse granite are found the older mines along two strongly contrasted east and west zones. A second, later-developed, group is west of the butte in the acidic granite. Of the former, the southern affords argentiferous copper ores, consisting of chalcocite, bornite, chalcopyrite, enargite, and pyrite in a siliceous gangue. The northern zone contains silver ores, chiefly sulphides of silver, lead, zinc, and iron, in a siliceous gangue with much rhodonite. Strangely enough, hardly any copper occurs in the silver zone, and no manganese is met in the copper zone, except in the Gagnon vein, which also contains zinc. The silver zone is mentioned again under "Silver." The zone west of the butte is silver-bearing with manganese. The occurrence of these two parallel systems of fissures in the same country rock and not far from each other, yet filled with such contrasted ores, is a very remarkable phenomenon and points to different and—at least for one of them—deep-seated sources of the ores.

2.04.07. The ore bodies do not, in general, present very sharply defined walls, but the ores fade into the wall rock. From this S. F. Emmons has suggested that they have formed along a series of small fissures marking some line of disturbance, and not from a general faulting, and have enlarged the original channels by replacement of the walls. It would seem probable that the frequent dikes are connected with the butte, or with the same parent body that sent it off. The same eruptive activity probably shattered the rock along the zones. In this event it must have been from an easterly offshoot of the western rhyolite area, and have operated with the same fissuring action which attends earthquakes from the intrusion of subterranean dikes. The Butte district alone rivals the Lake Superior copper mines in output, and has in recent years surpassed them. Reference will again be made to this region under "Silver."[1]

---

[1] "Butte Copper Mines," *Engineering and Mining Journal*, June 19, 1886, p. 445; also April 24, 1886, p. 299. S. F. Emmons, "Notes on the Geology of Butte, Mont.," *M. E.*, XVI. 49. Rec. Richard Pearce, "The Association of Minerals in the Gagnon Vein, Butte City, Mont.," *M. E.*,

2.04.08. Example 17a. Gilpin County, Colorado. Veins of pyrite and chalcopyrite, replacing gneiss (the rock may be granite), and dikes of quartz-porphyry, and felsite along the planes of joints, which cross the gneiss (or granite) perpendicularly to the laminations. The veins are highly auriferous and are worked primarily for gold, the copper being produced as a by-product. The concentrates from the stamps are afterward treated for copper. The veins occupy an area of only about a mile and a half in diameter, centering about Central City. They show little indication of filling a fissure, as usually understood, but follow the cleav-

FIG. 30.—*Cross section of the Bob-tail mine, Central City, Colo. After F. M. Endlich, Hayden's Survey, 1873, p. 286.*

age joints of the gneiss and replace the country rock on each side of them. The joints also cross the porphyry dikes, and the veins are often in the latter rock. They are closely related in structure and origin to the galena veins of the neighboring Clear Creek County, which are referred to under "Silver," but the contrast in mineral contents between the two is very marked. They were the basis of the first extensive deep mining in Colorado, and were located through the placer deposits in the neighboring gulches.[1]

XVI. 62. "On the Occurrence of Goslarite in the Gagnon Mine, Butte City," *Proc. Colo. Sci.*, Vol. II., Part I., p. 12. E. D. Peters, *Mineral Resources of the U. S.*, 1883–84, p. 374. A. Williams and E. D. Peters, "On Butte, Mont.," *Engineering and Mining Journal*, March 23, 1885, p. 208.

[1] S. F. Emmons, *Tenth Census*, Vol. XIII., p. 68. The veins are described as cited above. J. D. Hague, *Fortieth Parallel Survey*, III., p. 493. The veins are called fissure veins by Mr. Hague. A. Lakes, *Ann. Rep.*

2.04.09. **Example 17b. Llano County, Texas.** Impregnations in granite, and veins with quartz gangue in granite, carrying carbonates above, but sulphurets and tetrahedrite with some gold and silver below. Contact deposits between slates and granite are also known. It is not demonstrated as yet whether the ores are to be actually productive.[1]

2.04.10. **Example 18. Keweenaw Point, Mich.** Native copper, with some silver, in both sedimentary and interstratified igneous rocks of the Keweenawan system. The metal occurs as a cement, binding together and replacing the pebbles of a porphyry conglomerate; or filling the amygdules in the upper portions of the interbedded sheets of massive rocks; or as irregular masses, sometimes of enormous size, in veins, with a gangue of calcite, epidote, and various zeolites; or in irregular masses along the contacts between the sedimentary and igneous rocks.

2.04.11. The rocks of the Keweenawan system are most strongly developed on the south shore of Lake Superior, especially in Keweenaw Point, which juts out northwesterly, cutting the lake into two nearly equal portions. They extend some distance east and west and are also known on the north shore. They consist of sandstone and thin beds of conglomerate, interstratified with sheets of diabase, both compact and amygdaloidal, and of melaphyre. They are succeeded on the east by the Eastern Sandstone, which on the south shore in some places abuts unconformably against them, and in others passes under them from an overthrust fault.

On Keweenaw Point they dip northeasterly and pass under Lake Superior to reappear with a southeasterly dip on Isle Royale and the Canadian shore. Western Lake Superior occupies this synclinal trough. In Keweenaw Point the dip is greatest on the southeast, being about 60° at Hancock. To the northwest it gradually flattens to 30° or less on the lake shore. (For the general geology of the neighboring region see under Example 9.)

It is interesting to note that the early investigators of the geology of this country drew a parallel between the sandstones and traps of Lake Superior and the similar Triassic deposits of the

---

Col. State School of Mines, 1887, p. cii. A. W. Rogers, "The Mines and Mills of Gilpin County, Colorado," *M. E.*, II. 29. Further references will be found under "Silver and Gold in Colorado."

[1] T. B. Comstock, *First Ann. Rep. Texas Geol. Survey*, 1889, p. 334. W. H. Streeruwitz, in *Mineral Resources of the U. S.*, 1884, p. 342.

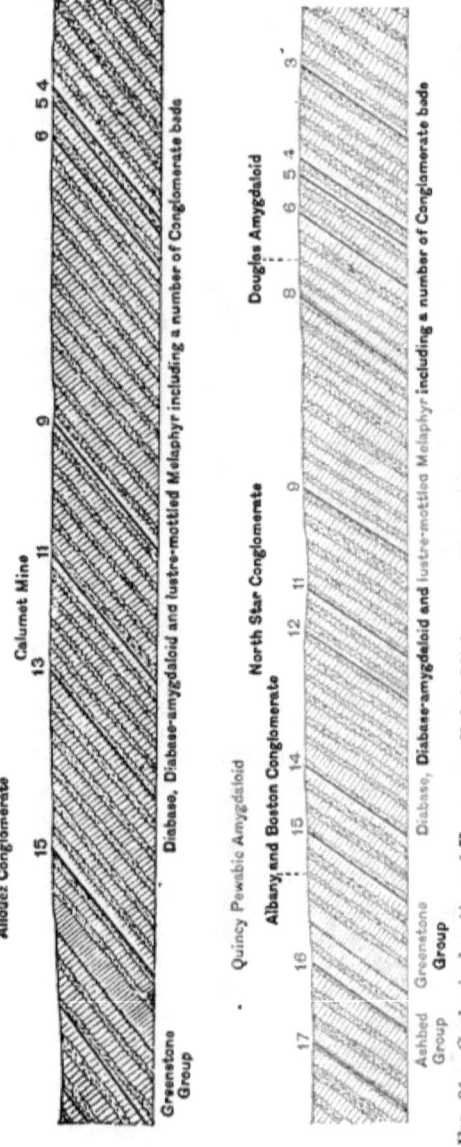

Fig. 31.—*Geological section of Keweenaw Point, Mich., near Portage Lake and through Calumet. After R. D. Irving, Monograph V., U. S. Geol. Survey, Pl.* 18. *The sections run northwest.*

Atlantic coast (see Example 21), even going so far as to regard the former as the western equivalent of the latter.[1]

There are three principal mining districts—the Keweenaw Point, on the end of the Point; the Portage Lake, in the middle; and the Ontonagon, at the western base. Mines have also been worked on Isle Royale, and copper is found in small amounts on the north shore. The Portage Lake district is now the principal and almost the only producer. In the first-named district most of the mines are on original fissures, which have later become much enlarged by the alteration of the walls. They are usually from 1 to 3 feet broad, but may reach 10, 20, and 30 feet, this last in the looser textured rocks. These expansions are also richer in copper. The veins stand nearly vertical and cross the beds at right angles. They were the earliest discovered and the first to be extensively worked. The metallic masses, both large and small, occur distributed through the gangue. The best known mines of the district are the Cliff, the Phœnix, and the Copper Falls.

2.04.12. In the Portage Lake district the mines are either in conglomerate (Calumet and Hecla, Tamarack, Peninsula, etc.) or in amygdaloidal, strongly altered diabase, certain very scoriaceous sheets of which are known as ash-beds (Quincy, Franklin, Atlantic, etc.). In the conglomerates the copper has replaced the finer fragments, so as to appear like a cement, and often the boulders themselves, or particular minerals in them, are permeated with copper. The amygdaloids have copper in their small cavities, but in the open or shattered rock it fills all manner of irregular spaces, often in fragments of great size. It is associated with calcite, the zeolites (often of great beauty), epidote, and chlorite (the last containing Fe_2O_3).

2.04.13. In the Ontonagon district the copper follows planes approximately parallel to the bedding of the sandstones and igneous rocks, and in one case at least (the National mine) along the contact between the two. The copper is quite irregular in its distribution, but has the same associates that are mentioned above.

2.04.14. In their practical bearings the mines are classed as Mass Mines, Amygdaloid Mines, and Conglomerate Mines, according to the size of the masses of copper or to the character of the inclosing rock.

2.04.15. *On the Origin of the Copper.*—The original source of

---

[1] C. T. Jackson, *Amer. Jour. Sci.*, i., XLIX., 1845, pp. 81-93.

the copper was thought by the earlier investigators to be in the eruptive rocks themselves, and that with them it had come in some form to the surface and had been subsequently concentrated in the cavities. Pumpelly has referred it to copper sulphides distributed through the sedimentary, as well as the massive rocks from which the circulating waters have leached it out as carbonate, silicate, and sulphate. Although the traps are said by Irving to be devoid of copper, except as a secondary introduction, it would be interesting to test their basic minerals for the metal in a large way, as has been so successfully done by Sandberger on other rocks. It is probable that these may be its source.

Irving states that the coarse basic gabbros of the system contain chalcopyrite, but they do not occur near the productive mines. The electro-chemical hypothesis of deposition was earliest advocated (Foster and Whitney), and on account of the electrolytic properties of the two metals copper and silver, at first thought it has strong claims to probability. Still the unsatisfactory character of all experiments made in other regions to detect such action militates against it. Pumpelly, however, has worked out an explanation much more likely to be the true one. He found, on studying the mineralogical changes which have taken place in the rocks, that the alteration had been very extensive; that it had proceeded through a series of minerals involving at one stage a change in the iron present from protoxide to sesquioxide (which would occasion a reducing action), and that at this stage the copper was deposited. In two fissure veins near Portage Lake sulphides and arsenides of copper occur, and in a vein near Lac la Belle, on the Mendota property, a little chalcocite has been found. These are the only instances, yet recorded, of sulphides or related compounds of copper in the district. A pocket of melaconite, the black oxide, was opened in the early days at Copper Harbor.

The discovery of copper dates back to the explorations of the French, who in the seventeenth century left the settlements on the lower St. Lawrence and penetrated the Great Lakes. The country was the scene of a great mining excitement in the forties. After many vicissitudes and exploded schemes the district settled down to the largest production of any American region. Within the last few years, however, Butte, Mont., has temporarily exceeded it. Many interesting traces of prehistoric mining were found by the early explorers, for the copper was a much prized commodity among the aborigines.

2.04.16. Some important mining for copper has been done on Isle Royale, along the Canadian shore, and in Minnesota, but although Keweenawan rocks are in great force, no large amount of the metal has been found.[1]

2.04.17. Example 19. St. Genevieve, Mo. Beds of chalcopyrite associated with chert in magnesian limestone of the Lower Silurian system. St. Genevieve is situated on the Mississippi, about forty miles south of St. Louis. The Second Magnesian Limestone of the Lower Silurian outcrops, with the Carboniferous on the north, and more or less Quaternary in the vicinity. There are two nearly horizontal beds of ore, of widths varying between three

---

[1] It would be impossible and undesirable to give in this place complete references to the literature. Such bibliography will be found in Irving's monograph, and in Wadsworth's. The more important papers are given below, with some additions to the lists mentioned above.

Blake, W. P., "Mass Copper of the Lake Superior Mines," etc., *M. E.*, IV. 110.

Credner, H., On the geology, etc., *Neues Jahrbuch*, 1869, p. 1.

Engineering and Mining Journal, "History of Copper Mining in the Lake Superior District," March 18, 1882, p. 141.

Foster and Whitney, Report on the Lake Superior Copper Lands, 1850.

Hall, C. W., "A Brief History of Copper Mining in Minnesota," *Bull. Minn. Acad. Nat. Sci.*, Vol. III., No. 1, p. 105.

Irving, R. D., "The Copper-bearing Rocks of Lake Superior," *Monograph* V., *U. S. Geol. Survey*, especially p. 419. Rec.

Jackson, C. T., "The Great Copper-bearing Belt of Canada," *Boston Soc. Nat. Hist.*, IX. 202.

Lawson, A. C., "Note on the Occurrence of Native Copper in the Animikie Rocks of Thunder Bay," *Amer. Geol.*, V. 174.

Poole, H., "Michipicoten Island and its Copper Mines," *Engineering and Mining Journal*, Aug. 6, 1892, p. 125; Sept. 3, p. 220.

Pumpelly, R., *Geol. Survey of Mich.*, 1873, Vol. I.

"On the Origin of the Copper," *Amer. Jour. Sci.*, i.i., III. 183-195, 243-253, 347-353. Rec. A later and fuller paper is in *Proc. Amer. Acad.*, 1878, Vol. XIII., p. 233.

Wadsworth, M. E., *Notes on the Geology of the Iron and Copper Districts of Lake Superior*. Cambridge, 1880.

Whitney, J. D., "On the Black Oxide of Copper of Lake Superior," *Proc. Boston Soc. Nat. Hist.*, January, 1849, p. 102; *Amer. Jour. Sci.*, ii., VIII. 273.

Metallic Wealth in the United States, p. 245. Rec.

Whittlesey, C., "On Electrical Deposition," *A. A. A. S.*, XXIV. 60.

Wright, C. E., and Lawson, C. D., *Mineral Statistics of Michigan*. Annual.

inches and several feet. They lie between chert seams and are associated with clay and sand. The ore is thought by Nicholson to have been deposited in cavities formed by dolomitization, much

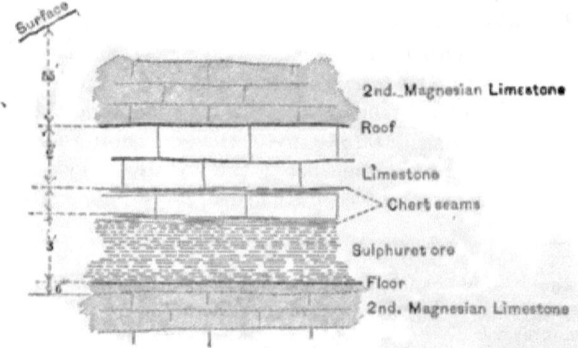

FIG. 32.—*Cross section in the St. Genevieve copper mine, illustrating the relations of the ore. After F. Nicholson, M. E., X., p.* 450.

as is advocated by Schmidt for the lead and zinc deposits of southwest Missouri, and as is described under Example 25. For ten years the mines have not been operated.[1]

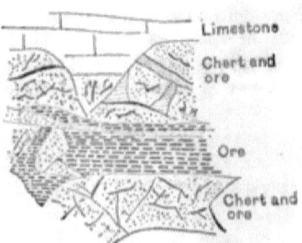

FIG. 33.—*Section at the St. Genevieve mine, illustrating the intimate relations of ore and chert. After F. Nicholson, M. E., X., p.* 451.

2.04.18. **Example 20.** Arizona Copper. Bodies of oxidized copper ores in Carboniferous limestones, associated with eruptive rocks. In addition to these, which are the most important, are

---

[1] F. Nicholson, "Review of the St. Genevieve Copper District," *M. E.*, X. 444. B. F. Shumard, "Observations on the Geology of the County of St. Genevieve, Missouri," *Trans. St. Louis Acad. Sci.*, Vol. I., p. 40; abstract in *Amer. Jour. Sci.*, ii., XXVIII. 126.

veins in eruptive rocks, or in sandstones, or ore bodies of still different character as set forth under the several sub-examples. The copper districts are nearly all in the southeastern part of the territory, but the Black range is near the center.

2.04.19. Example 20a. Morenci. The Morenci district, known also as the Clifton or Copper Mountain, lies in a basin, six to ten miles across, whose high surrounding hills consist of limestone,

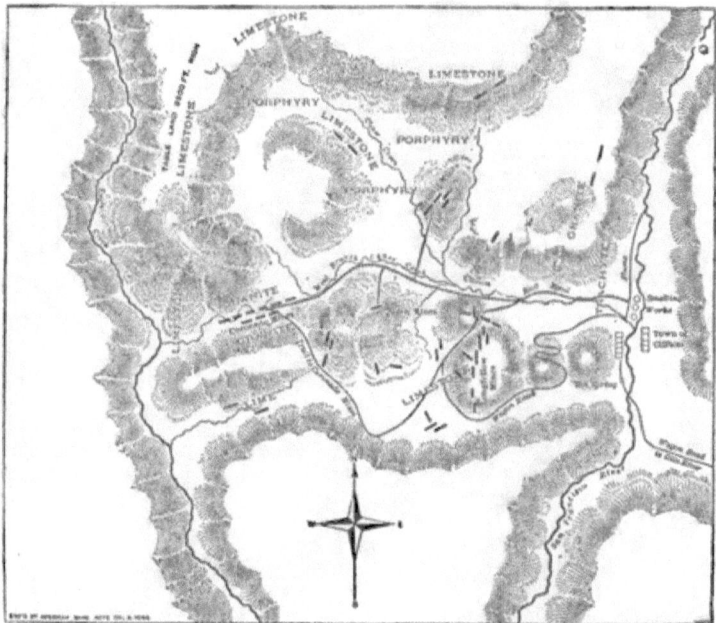

FIG. 34.—*Geological map of the Morenci or Clifton copper district of Arizona. After A. F. Wendt, M. E.*

probably Lower Carboniferous, which rests on sandstone, and this on granite. The principal mines are grouped about the town of Morenci. Clifton is seven miles distant at the point where the smelter of the Arizona Copper Company is located. In the basin is a mass of porphyry, containing frequent great inclusions of limestone. Felsite or porphyry dikes are also abundant in the surrounding sedimentary and granite rocks. Several miles to the east there is an outflow of late trachyte and evidence of recent volcanic action. From this it appears that eruptive phenomena are abundant and widespread.

146 KEMP'S ORE DEPOSITS.

2.04.20. The ores are classified by Henrich as follows:
1. Contact deposits. These occur in a zone of decomposed and

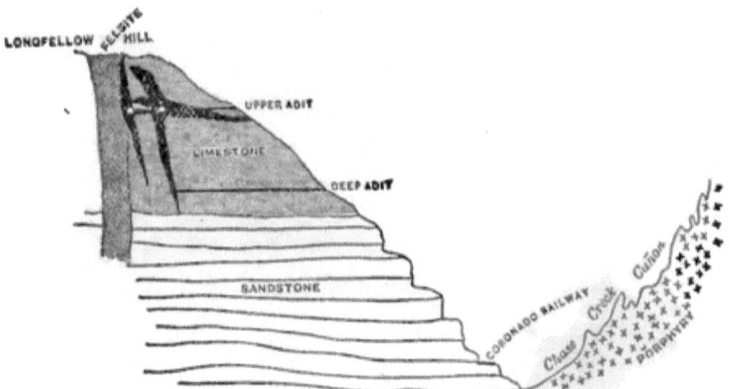

FIG. 35.—*Vertical section of Longfellow Hill, Clifton district, Arizona. After Wendt, M. E., XV. 52.*

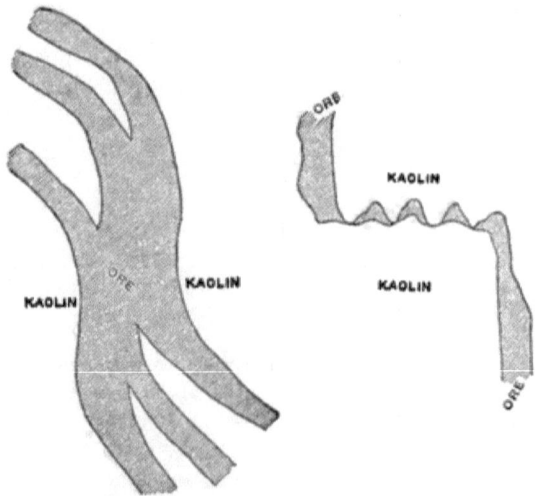

FIG. 36.—*Horizontal sections of Longfellow ore body. After Wendt.*

kaolinized porphyry, between a bluish, fine-grained limestone, and solid porphyry. Many ore bodies, and probably the largest, are directly on the limestone, while others are surrounded by the decomposed porphyry. As included masses of limestone, with associated ore, are found in the decomposed porphyry, it is probable that these ore bodies may have originally replaced such. The ores are malachite, azurite, cuprite with some metallic copper and malachite, in a gangue principally of limonite. Wad is also frequent. Much clay of a residual character occurs with the ores.

2. Deposits in limestone. These are closely associated with

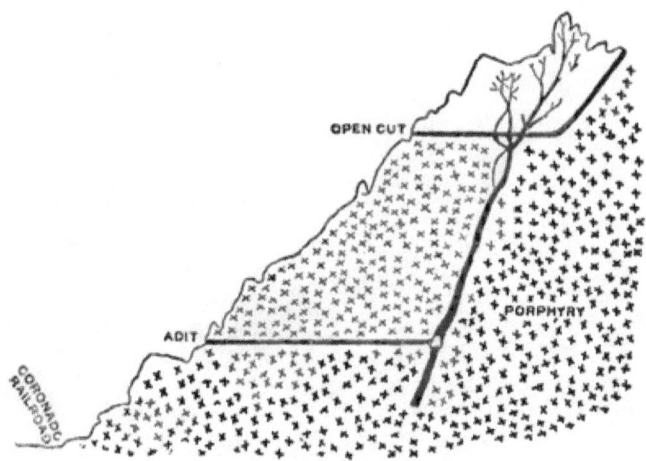

FIG. 37.—*Geological section of the Metcalf mine, Clifton district, Arizona. After Wendt.*

the first class, and have apparently formed as outlying bodies in the limestone, as they are connected by ore channels with the principal lines of circulation along the contact. They appear to contain more wad and lime than the typical contact deposits.

3. Deposits in porphyry. These form sheets and pockets in porphyry, or impregnate the solid rock itself. They are oxidized at the surface, but pass in depth into chalcocite. The principal gangue is kaolinized porphyry.

According to Wendt, the Coronado vein fills a longitudinal fissure in a quartz porphyry dike. It afforded chalcocite above, but passed into chalcopyrite below. Wendt also mentions a group of veins in granite that likewise afforded chalcocite.[1]

2.04.21. Example 20b. The Bisbee district, called also the Warren district, is situated in the Mule Pass Mountains in southern Arizona, near the Mexican line. The range runs east and west and consists of beds of Lower Carboniferous limestone, dipping away from a central mass of porphyritic rock. The ores are found in the cañons on the south side, which have been formed by erosion, along the contact of the limestone and porphyry. They are of the same oxidized character as at Morenci, and in the

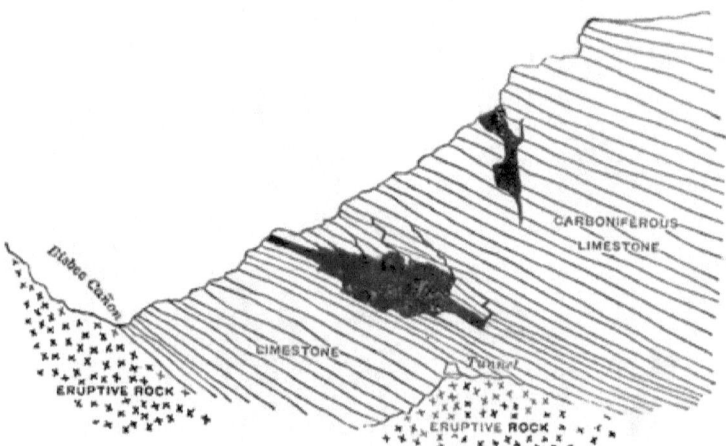

FIG. 38.—*Section of Copper Queen ore body Bisbee district, Arizona. After A. F. Wendt, M. E., XV. 52.*

important mines occur in the limestone. The ore-bearing solutions seem to have spread out chiefly along the bedding planes and to have replaced the limestone at a distance from the contact. The ore bodies leave great chambers when excavated and partake of the nature of bedded veins. Empty caves occur associated with

---

[1] J. Douglass, "Copper Resources of the United States," *M. E.*, September, 1890. Rec. "Arizona Copper and Copper Mines," *Engineering and Mining Journal*, Aug. 13, 1881, p. 103. "Clifton Copper Mines of Arizona," *Ibid.*, Feb. 21, 1880, p. 133. C. Henrich, "The Copper Ore Deposits near Morenci, Ariz.," *Ibid.*, March 26, 1887, pp. 202, 219. Rec. A. Wendt, "Copper Ores of the Southwest," *M. E.*, XV., p. 23. Rec.

FIG. 39.—View of the Copper Queen mine, Bisbee district, Arizona. From a photograph by James Douglass.

them, doubtless formed, as at Eureka, Nev. (Example 36), by surface waters and having no genetic connection with the ore. Evidences of hydrothermal action along the contact are abundant. The Copper Queen and the Arizona Prince are the principal mines. There are minor deposits in the porphyry of oxidized ores above, changing to chalcocite and this to chalcopyrite in depth.[1]

2.04.22. Example 20c. Globe District. As in the other districts, the most productive mines are in limestone near the contact with eruptive rocks.

1. Contact deposits in limestone. At the Globe mines the Carboniferous limestone abuts against a great dike of diorite, while trachyte and granite are near. Along the contact there is abundant evidence of thermal action in the kaolinized rock. The great bodies of oxidized ores are found on this contact and extend out into the limestone. The one on the Globe claim is described by Wendt as resembling a great chimney.

2. A fissure vein in sandstone, containing arsenical and antimonial copper ores and known as the Old Dominion, was formerly worked.

3. Fissure veins in talcose slate and gneiss, and filled by a quartz gangue with bunches of malachite and azurite (New York and Chicago mines), and now no longer worked.

4. Numerous small veinlets forming a stockwork, in gneiss near a dike of diorite, which is crossed by a dike of trachyte. These are known as the Black Copper Group. The ores are too low grade for profitable exploitation. Of greater interest are the bodies of chrysocolla, found in the wash, down the hill from the outcrop of the veins, and evidently due to the superficial drainage of the stockworks. Similar bodies of ore, though not chrysocolla, were found at Rio Tinto, in Spain.[2]

2.04.23. Example 20d. Santa Rita District. Although in New

---

[1] J. Douglass, "Copper Resources of the United States," *M. E.*, September, 1890. Rec. A. Wendt, "Copper Ores of the Southwest," *M. E.*, XV., p. 52. Rec.

[2] J. Douglass, "Copper Resources of the United States," *M. E.*, September, 1890. Rec. "The Globe District," *Engineering and Mining Journal*, April 9, 1881, p. 243. W. E. Newberry, "Notes on the Production of Copper in Arizona," *School of Mines Quarterly*, VI. 370. A. Trippel, "Occurrence of Gold and Silver in Oxidized Copper Ores in Arizona," *Engineering and Mining Journal*, June 16, 1883, p. 435. A. Wendt, "Copper Ores of the Southwest," *M. E.*, XV., p. 60.

Mexico, this district has much in common with those already mentioned. A great dike of felsite cuts limestones, and along the contact, as well as in the felsite itself, copper ores are found.

1. Contact deposits in limestone. These afforded the usual oxidized ores, but were not found to extend to any great depth, and while for a time productive, they were soon exhausted.

2. Deposits in felsite. These consisted of pellets and sheets of native copper in the dike itself, which were oxidized to cuprite near the surface. (Cf. Lake Superior amygdaloids, Example 13.) They were worked by the Mexicans in the early part of the present century.[1]

2.04.24. Example 20e. Black Range District. The mines of this region differ from those described above, and in some respects resemble the pyrite beds of the Alleghanies. (Example 19.) The ore occurs in one or more great fissure veins, along the contact of vertical slates with a great dike of porphyrite. The veins run completely into the porphyrite and into the slate, and afford oxidized ores above, changing into chalcopyrite below. They contain considerable silver and gold, as well as arsenic and antimony. The principal mines are the Hampton and Eureka. They are situated in the valley of the Verde River, twenty miles east of Prescott.[2]

2.04.25. Example 20f. Copper Basin. Beds of closely textured conglomerate and sandstone, resting on granite and gneiss, and having a cement of copper carbonates. Copper Basin lies about twenty miles southwest of Prescott, and is formed by a depression in greatly decomposed granite, which is traversed by numerous small veinlets of copper ores. The granite is pierced by porphyry dikes, and covered by the sedimentary conglomerates and sandstones into which its copper is thought by Blake to have partially leached and precipitated as a cement. Reference, by way of comparison, may be made to the Lake Superior conglomerates, in which, in part, the native copper serves as a cement.[3]

---

[1] A. F. Wendt, "Copper Ores of the Southwest," *M. E.*, XV. 27. Wislizenus, "On the Santa Rita Mines: Memoir of a Tour in Northern Mexico, 1846-47," p. 47; *Amer. Jour. Sci.*, ii., VI. 385, 1848.

[2] J. F. Blandy, "The Mining Region around Prescott, Ariz.," *M. E.*, XI. 286. G. K. Gilbert, "On the General Geology of the Black Mountain District," *Wheeler's Survey*, III., p. 35. A. R. Marvine, "Brief Details of the Verde Valley," *Wheeler's Survey*, III., p. 209. A. F. Wendt, "Copper Ores of the Southwest," *M. E.*, XV. 63. Rec.

[3] W. P. Blake, "The Copper Deposits of Copper Basin, Arizona, and Their Origin," *M. E.*, XVII. 479.

2.04.26. There are numerous other copper districts in Arizona of minor importance or entirely undeveloped, but the examples above cited probably illustrate the occurrences quite fully. Those not referred to above are of sporadic development. Mention should also be made of the mines in Lower California, opposite Guaymas, a brief description of which will be found in Wendt's paper.[1]

2.04.27. Example 20*g*. Crismon-Mammoth, Utah. In the Tintic district, Juab County, Utah, are three great ore belts, in vertically dipping dolomitic limestone, as more fully set forth under "Silver" (Example 35*a*). One of these, the Crismon-Mammoth, contains ores that bear silver, gold, and copper in proportions of about equal value. They have been a very difficult mixture to treat successfully. Of late considerable copper has been produced, placing the ore deposits among those deserving mention. The Crismon-Mammoth vein or belt covers a maximum width of 70 feet, and runs 500 feet on the strike, dipping 75° west. The ores seem to have been deposited along the bedding planes, though often cutting across them. The productive portions are found in richer chutes or chimneys, amid much low-grade material and gangue, and are of all shapes and sizes, from 25 feet in diameter, down. The Copperopolis is thought to be on the same belt, and is a neighboring location of similar geological structure and ores.[2]

2.04.28. Sunrise, Wyoming. Oxidized ores have been exploited to some extent at the Sunrise mines, in the Laramie range, Wyoming, but were never of much importance.

2.04.29. Example 21. Copper ores in Triassic or Permian sandstone. They occur as oxidized ores, native silver, and chalcocite in contact deposits in Triassic and Permian sandstones at their junction with diabase or gneiss, or as disseminated masses replacing organic remains. Copper ores are very common throughout the estuary Triassic rocks of the Atlantic coast, and although formerly much mined, they are now proved valueless, and of scientific interest only.

2.04.30. Example 21*a*. Contact deposits in sandstone at its

---

[1] See also M. E. Saladin, "Note sur les Mines de Cuivre du Boles (Basse Californie)," *Bull. de la Société de l'Industrie Minérale*, 3 Serie, VI. 5, 283.

[2] O. J. Hollister, "Gold and Silver Mining in Utah," *M. E.*, XVI., p. 10. D. B. Huntley, *Tenth Census*, Vol. XIII., p. 456.

junction with diabase. These include the New Jersey ores, vigorously worked before the Revolution. They consist of the carbonates, of cuprite and native copper, disseminated through sandstone near the trap. The Schuyler mines, near Arlington, N. J., and several other openings near New Brunswick, N. J., are best known. These Triassic diabases often show chalcopyrite, and it is probable that the copper came from this or from copper in the augite of the rock, in accordance with Sandberger's investigations. The deposits are unreliable, and except at a very early period have never been an important source of ore.

2.04.31. Example 21b. Contact deposits in sandstones at the

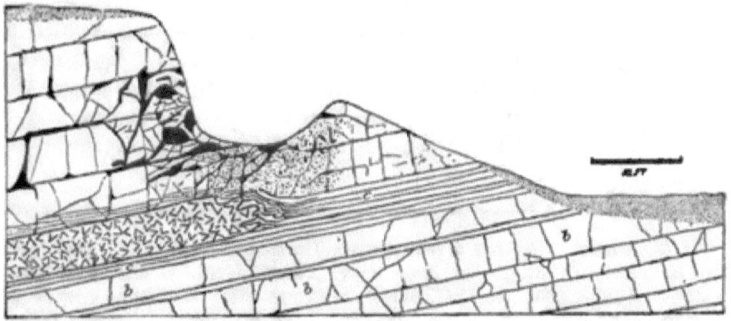

FIG. 40.—*Cross section of the Schuyler Copper mine, New Jersey.* a, *trap;* b, *sandstone;* c, *shales;* the black shading, copper ores. After N. H. Darton, U. S. Geol. Survey, Bull. 67, p. 57.

junction with gneiss. A number of deposits were formerly worked of this character, especially at Bristol, Conn., and at the Perkiomen mine, Pennsylvania. The mine at Bristol, Conn., is a well-marked contact deposit, on the line between the Triassic sandstone and the schistose rocks. The contact runs northeast and southwest, has suffered great decomposition from mineral solutions, and has been largely kaolinized. A broad band of this decomposed material, 30 to 120 feet wide, lies next the sandstone and contains disseminated ore. Then follow micaceous and hornblende slates, often with horses of gneiss. The slates are much broken by movements that have formed cavities for the ores. It is reasonable to connect the stimulation of the ore currents with the neighboring trap outbreaks. Unusually fine crystals of chalcocite and barite have made the mine famous the world over. While at one time a source of copper, for many years it has been unproductive.[1]

2.04.32. Example 21c. Chalcocite and copper carbonates replacing vegetable remains, etc., in the Permian or Triassic sandstones of Texas, New Mexico, and Utah. In the Permian of northern central Texas are three separate copper-bearing zones, forming three lines of outcrop that extend in a general northeasterly direction over a range of about three counties. The ore is largely chalcocite in beds of shale, and often replacing fragments of wood. It may be available in time.[2]

At various places in Utah and New Mexico (Abiquiu, N. M., Silver Reef, Utah) the sandstones, as reported by Newberry and others, have copper ores disseminated through them and deposited on fossils, at times with associated silver (Utah). The copper, whether coming from the waters along the shore line or from subterranean currents, was precipitated by the organic matter. (See also under "Silver," in Utah.) These deposits are not yet sources of copper.[3]

---

[1] L. C. Beck, "Notice of the Native Copper Ores, Copper, etc., near New Brunswick, N. J.," *Amer. Jour. Sci.*, i., XXXVI. 107. G. H. Cook, *Geol. of N. J.*, 1868, p. 675; also L. C. Beck, *Ibid.*, 218-224. J. G. Percival, *Rep. on Geol. of Conn.*, p. 77. C. A. Shaeffer, "Native Silver in New Jersey Copper Ore," *Engineering and Mining Journal*, February, 1882, p. 90. C. Shepherd, *Geol. of Conn.*, 1837, p. 47. B. Silliman and J. D. Whitney, "Notice of the Geological Position and Character of the Copper Mine at Bristol, Conn.," *Amer. Jour. Sci.*, ii., XX. 361. J. D. Whitney, *Metallic Wealth.* Rec.

[2] W. F. Cummins, "Report on the Permian of Texas and its Overlying Beds," *First Ann. Rep. Texas Geol. Survey*, p. 196. J. F. Furman, "Geology of the Copper Region of Northern Texas and Indian Territory," *Trans. N. Y. Acad. Sci.*, 1881-83, p. 15.

[3] F. M. F. Cazin, "The Origin of the Copper and Silver Ores in Triassic Sand Rock," *Engineering and Mining Journal*, April 30, 1880; Dec. 11, 1880, p. 331. "The Nacemiento Copper Deposits," *Ibid.*, Aug. 22, 1885, p. 124. A. W. Jackson, *Rep. Director of the Mint*, 1880, p. 334. J. S. Newberry, "Copper in Utah, Triassic Sandstones," *Engineering and Mining Journal*, Vol. XXXI., p. 5. Also Oct. 23, 1880, p. 269; Jan. 1, 1881, p. 4. See also *Tenth Census*, Vol. XIII., Precious Metals, pp. 40, 478. C. M. Rolker, "The Silver Sandstone District of Utah," *M. E.*, IX. 21. R. P. Rothwell, Quoted in *Tenth Census*, Vol. XIII., p. 478. B. Silliman, "The Mineral Regions of Southern New Mexico," *M. E.*, XVI. 427.

2.04.33. Copper production in 1882 and 1890, in tons of 2000 pounds each.

	1882.	1890.
Lake Superior	28,578	50,372
Montana	4,529	56,490
Arizona	8,992	17,398
Colorado	747	441
New Mexico	434	425
California	413	11
Utah	303	503
Elsewhere	1,412	3,906
	45,408	129,546

The figures indicate in general a vast increase in production, and, above all, the advance of Montana. For detailed statistics the volume on "Mineral Industry" issued by the Scientific Publishing Company (New York, 1893) is most available.

CHAPTER V.

LEAD ALONE.

2.05.01. The deposits of lead are treated in three different classes, according as they produce or have produced lead alone, lead and zinc, or lead and silver. Of late years the lead-silver ores have been the great source of the metal. Only the southeast Missouri region is of much importance among the others, although considerable lead is also obtained in association with zinc.

LEAD SERIES.

	$Pb.$	$S.$
Galena, PbS	86.6	13.4
Cerussite, $PbCO_3$	77.5	
Anglesite, $PbSO_4$	67.7	
Pyromorphite, $Pb_3P_2O_8 + 1/3 Pb.Cl_2$	75.36	

Earthy mixtures of these last three and limonite.

2.05.02. Example 22. Atlantic border. Veins of galena in the Archæan rocks of the States along the Atlantic border; also others into Paleozoic strata, as described in the sub-examples.

2.05.03. Example 22a. Veins in gneiss and crystalline limestone, sometimes with a barite or calcite gangue. These deposits were vigorously exploited forty years ago or more, but have since been of small importance other than scientific. They may be described best by districts, as they hardly deserve a greater prominence.

2.05.04. (1) St. Lawrence County, New York. Veins with galena in a gangue of calcite in Archæan gneiss. Those near Rossie are perhaps best known, especially for their unusually interesting calcite crystals. There are numbers of veins in the district which are notable in that the galena is without zinc or iron associates. The lead carries a very small amount of silver, not enough to separate. Hornblende and mica schists occur in the same region and the Potsdam sandstone is not far removed. A few minor veins cut the Trenton limestone near Lowville, Lewis County, sometimes with fluorite for a gangue.[1]

2.05.05. (2) Massachusetts, Connecticut, and eastern New York. Veins of galena with more or less chalcopyrite and pyrite in a quartz gangue in gneiss, slates, limestones, or mica schists. The mines near Northampton, Mass., were formerly well known, although never productive of a great deal of metal; but as there is a large, prominent vein, it attracted attention. There are numerous others in the same region. Veins also occur at Middletown, Conn., where much silver is said to be found in the galena. More recently (*circa* 1873) at Newburyport, Mass., argentiferous galena attracted attention, but was not of any importance. Other veins are known in Lubeck, Me., and in various parts of New Hampshire and Vermont. For a time small lodes in the slates of Columbia County, New York, were unsuccessfully exploited, of which the Ancram mine is of historic interest. Although these galena veins are numerous, they are not to be taken too seriously.[2]

2.05.06. (3) Southeastern Pennsylvania. Veins on the contact of Archæan gneiss and Triassic sandstone and diabase. These were referred to under Example 21*b*. As noted by Whitney, the copper is especially strong in the sandstone, and the lead in the gneiss. Trap dikes are abundant, and the eruptive phenomena in connection with them doubtless occasioned the activity of circulation which filled the veins. The Wheatley mine is best known. It has afforded a great variety of lead minerals, especially pyromorphite. They have not been worked in years.[3]

2.05.07. (4) Davison County, North Carolina. Veins in talcose slate were formerly exploited, but are now little known,

---

[1] L. C. Beck, *Mineralogy of New York*, p. 45. E. Emmons, "Geology of the Second District," *N. Y. Geol. Survey*, 1842. G. Hadley, "Crystallized Carbonate of Lead at Rossie," *Amer. Jour. Sci.*, ii., II. 117. F. L. Nason, "Calcite from Rossie," *Bull.* 4, *N. Y. State Museum*, 1888. J. D. Whitney, *Metallic Wealth*. Rec.

[2] C. A. Lee, "Notice of the Ancram Lead Mine," *Amer. Jour. Sci.*, i., VIII. 247. A. Nash, "Notice of the Lead Mines and Veins in Hampshire County, Massachusetts," *Amer. Jour. Sci.*, i., XII. 238. R. H. Richards, "The Newburyport Silver Mines," *M. E.*, III. 442. J. D. Whitney, *Metallic Wealth*.

[3] H. D. Rogers, *Geol. of Penn.*, II. 701; also *Amer. Jour. Sci.*, ii., XVI. 422. J. D. Whitney, *Metallic Wealth*, p. 396.

except as having furnished beautiful crystals of oxidized lead minerals.[1]

2.05.08. *Example 22b.* Sullivan and Ulster Counties, New York. Veins along a line of displacement on the contact between the Hudson River slates and the sandstones of the Medina stage (Shawangunk grit), carrying galena and chalcopyrite in a quartz gangue; or else gash veins filled with the same in the grit. These mines formerly produced considerable lead and copper, but are now best known for the excellent quartz crystals which they have furnished to all the mineralogical collections of this and other lands.[2]

2.05.09. *Example 23.* Southeast Missouri. Galena accompanied by nickeliferous pyrite, disseminated through beds of the Third or Lower Magnesian limestone of the Missouri geologists, which is doubtless Cambrian in age. The mines are at Bonne Terre, Mine La Motte, and Doe Run, twenty-five miles west of the Mississippi River and forty to one hundred miles south of St. Louis. The strata lie almost horizontal, and are known to carry lead through over 200 feet in thickness. The productive places fade out into barren rock and appear to be local enrichments of the limestone, of which the galena forms an integral part. At Bonne Terre they are of enormous size, one working running 3000 feet, and being 100 to 200 feet broad and 25 to 60 feet high. No zinc, however, occurs with the lead, and the silver contents are very small, being about four ounces to the ton of lead. At Mine La Motte some copper is found, and considerable nickel and cobalt. The rare mineral siegenite, a variety of linnacite, impregnates a sandstone supposed to be the equivalent of the Potsdam. Pyrite accompanies the galena both at Mine La Motte and at the other mines, and carries the nickel and cobalt, which is obtained as a by-product in the lead smelting. All the ore bodies are crossed by small faults, adjoining which the rock is invariably barren. Knobs of Archæan granite, containing diabase dikes, crop out near the mines both at Mine La Motte and at Doe Run. But the dikes never penetrate the limestone, and were evidently intruded before it was deposited.

---

[1] J. C. Booth, "Analyses of Various Ores of Lead, etc., from King's Mine, Davison County, North Carolina," *Amer. Jour. Sci.*, i., XLI. 348. W. C. Kerr, *Geol. of North Carolina*, p. 289.

[2] J. D. Whitney, *Metallic Wealth.* W. W. Mather, *New York State Survey, Report on First District,* 358.

The ore must have been deposited with the limestone or it must have been introduced since the latter was formed, and by the percolation of ore-bearing solutions through the rock, with no marked fissure vein development. The first view has been advanced by J. F. Kemp (1887), it being thought that decaying marine vegetation had precipitated the ores from solution in sea-water, as is outlined for another region under Example 24, but this explanation has been practically disproved. W. P. Jenney has considered the ore to have come in ascending solutions through the small fault fissures referred to above, and from these to have spread outward, replacing the limestone (privately communicated). Places where several fissures cross are said to be specially favorable. It is a curious fact, however, that as the ore bodies are followed up to the faults they invariably become lean or run out. Their place of formation has apparently some connection, as recent explorations seem to indicate, with low folds at right angles to the faults. The ore bodies favor the anticlinal bends.

This whole region of Cambrian and Lower Silurian rocks, over nearly 3000 square miles, contains lead, and within a year or so past some new mines, not yet under way (1892), have been started. These disseminated deposits are in no way to be confused with the mines of the Upper Mississippi in Wisconsin and Iowa. They are now large producers of lead and the only mines worked in the United States for lead alone. The ore affords an average of about eight per cent. galena. Except at Mine La Motte, lead was also obtained at this region, previously to 1865, from small gash veins like those of Example 24, but the workings were never in any degree commensurate with the present mines of disseminated ore. The history of Mine La Motte dates back to the early part of the eighteenth century, and it is said to have furnished lead for bullets used in the Revolution.[1]

2.05.10. The great increase in lead production in the United States came about 1880, with the opening of the Leadville ore

---

[1] G. C. Broadhead, "The Southeastern Missouri Lead District," *M. E.*, V. 100. Rec. J. R. Gage, "On the Occurrence of Lead Ores in Missouri," *M. E.*, III. 116. Rec. *Geol. Survey of Missouri*, 1873–74, pp. 30, 603. J. F. Kemp, "Notes on the Ore Deposits, etc., in Southeastern Missouri," *School of Mines Quarterly*, October, 1887. Rec.

Several other papers have been published on the metallurgical treatment and methods of ore dressing in the *Trans. Inst. of Mining Engineers* and the *School of Mines Quarterly*.

bodies. From 1877 until 1881 Eureka, Nev., was an important source, but since then it has greatly declined. Utah has preserved a fairly uniform production since the early seventies. Lead from all sources is here mentioned, although lead-silver ores are subsequently treated. The amounts are in tons of 2000 pounds. For detailed statistics see the volume on "Mineral Industry." (New York: Scientific Publishing Company, 1893.)

	1880.	1890.
Missouri, Kansas, Wisconsin, Illinois	27,690	55,000
Colorado	35,674	60,000
Nevada	16,659	2,500
Utah	15,000	24,000
Idaho, Montana	—	24,000
Elsewhere	2,802	15,994
	97,825	181,494

From 80 to 85% of the total product is from lead-silver ores.

CHAPTER VI.

LEAD AND ZINC.

2.06.01. **Example 24. The Upper Mississippi Valley.** Gash veins and horizontal cavities (flats), limited to the Galena and Trenton limestones of the Upper Mississippi Valley, and containing galena, zincblende, and pyrite (or marcasite), with calcite, barite, and residual clay. The deposits are found in southwest Wisconsin, eastern Iowa, and northwestern Illinois. The greater portion of the productive territory lies in Wisconsin, and covers an area which would be included in a circle of sixty miles' radius, whose limits would pass a few miles into Illinois and Iowa. A low north and south geanticline runs through central Wisconsin dating back to Archæan times and called by Chamberlain "Wisconsin Island." On its western slope the Cambrian and Lower Silurian rocks are laid down, and these in the western limit of the lead district pass in the adjoining States under the Upper Silurian. They are folded also in low east and west folds, but in the aggregate the whole series dips very gradually westward. The chief east and west fold forms the south bank of the Wisconsin River, and may have been the cause that deflected it from a southerly course. The easterly part of the lead region is 350 feet higher than the western, and the northern is 500 feet above the southern. The general slope is thus southwesterly.

2.06.02. The Galena limestone is a dolomite reaching 250 feet in thickness. On the hilltops left by erosion Maquekota (Hudson River) shales are seen. The Galena has shaly streaks, which have largely furnished the residual clay of the cavities. There are also cherty layers and sandy spots. Under the Galena lies the Trenton, from 40 to 100 feet thick, and made up of an upper blue portion, which is a pure carbonate of lime, and a lower buff portion that is magnesian. The upper portion of the blue has a band of shale locally called the "Upper Pipe Clay," and the pure, cryptocrystalline limestone under this is called "Glass Rock." The blue contains much bituminous matter. The buff is locally called "Quarry Rock" and is prolific in fossils. Under the Trenton lies the St. Peter's sandstone, 150 feet below which is the Lower Magnesian, 100 to 250 feet, and still lower the Potsdam, averaging 700 to 800 feet. The Potsdam rests on the quartzites and schists of the Archæan. The ore beds especially favor the shallow, synclinal depressions of the east and west folds. They occur in crevices, the great majority of which run east and west. The productive ground comes in spots which are separated by stretches

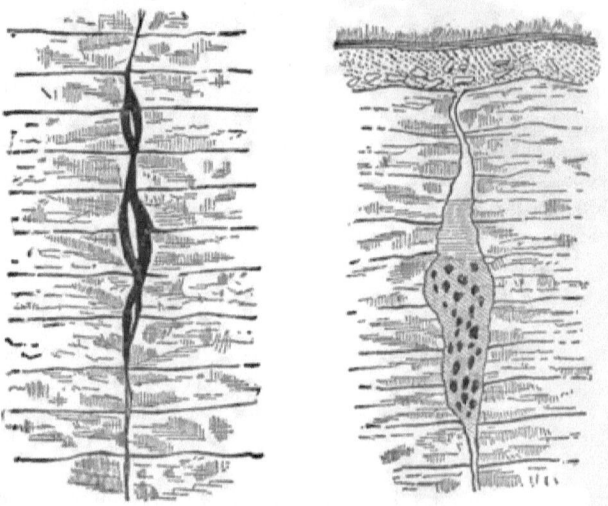

FIG. 41.—*Gash veins, fresh and disintegrated. The heavy black shading indicates galena. After T. C. Chamberlain, Geol. Wis., Vol. IV., p. 454.*

of barren ground. The lead ores are chiefly produced by the crevices in the Upper Galena. In the Lower Galena the zinc ores become relatively more abundant, and they are also in the Trenton. The ores do not extend in any appreciable amounts either above or below these horizons. The upper deposits favor the vertical gash vein form; the lower tend rather to horizontal openings, called flats, which at the ends dip down (pitches) and often connect with a second sheet (flat) lying lower. There are several minor varieties of those two main types of cavity, which mainly depend for their differences on the grade of decomposition, which the walls have undergone, and whether there was an original opening, or only a brecciated and crushed strip. Chamberlain cites twelve varieties in all, some of which are based on rather fine distinctions.

2.06.03. The cavities were referred by J. D. Whitney to joints, formed either by the drying and consolidating of the rock or by gentle oscillations of the inclosing beds. The later work has largely corroborated this, and they are generally thought to be chiefly caused by the cracks and partings formed by the gentle synclinal foldings. Such cavities have usually been enlarged by subsequent alteration of the walls. Whitney also essentially outlined the explanation of origin, which has been more fully elabo-

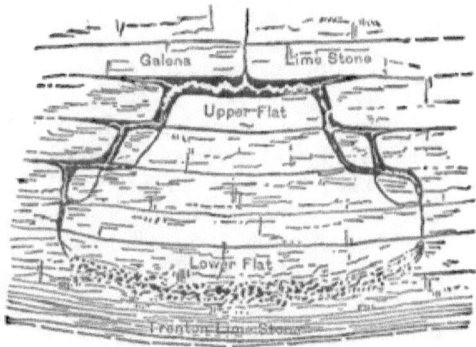

FIG. 42.—*Idealized section of "flats and pitches," forms of ore bodies in Wisconsin.* After T. C. Chamberlain, Geol. Wis., Vol. IV., p. 458.

rated by Chamberlain. Both these writers have urged that the ores could not have come from below, for the lower rocks are substantially barren of them. The conclusion therefore follows that they were deposited in the limestones at the time of their formation. The source of the ores is placed in the early Silurian sea, from which it is thought they were precipitated by sulphuretted hydrogen, exhaled by decaying seaweeds, or similar dead organisms on the bottom. In carrying the idea further, Chamberlain has endeavored to reproduce the topography of the region in the Lower Silurian times and to indicate the probable oceanic currents. These are conceived to have made an eddy in the lead district and to have collected there masses of seaweed, etc., resembling the Sargasso Sea. While interesting, this must be considered very hypothetical. When the sulphides became precipitated they were doubtless finely disseminated in the rock and were gradually segregated in the crevices. The sulphurous exhalations from the bituminous limestones may have aided in their second precipitation. The paragenesis of the minerals shows the following succession: (1) Pyrite, (2) Galena, (3) Pyrite; or (1) Pyrite, (2) Blende, (3) Galena, (4) Pyrite; or (4) Calcite. The ores, especially of zinc, are often oxidized, and afford considerable calamine and smithsonite. Some oxidized copper ores are produced at Mineral Point, formed by the alteration of chalcopyrite. In the early mines lead alone was sought, but of late years the zinc has been produced in greater quantities and is more valuable than the lead.

Dr. W. P. Jenney, whose work in the region of southwest Missouri is later referred to (2.06.07), has also written of these mines, and his views are quite different from those of any of the writers mentioned above. The courteous permission to read his manuscript has made possible the following abstract. The east and west fissures, already mentioned as crevices, are regarded as faulting planes. They are usually not far from the vertical, but in a few instances dip 35° to 45°. The smaller north and south series are considered to be due to the same cause, but to an earlier period of disturbance, as they are faulted by the east and west set. The latter exhibit but little vertical displacement, although some considerable horizontal. The ore is principally in and along the east and west fissures, but these seem to be locally enriched at their intersections with the north and south series. The deposits are described as runs; that is, lateral enrichments along a fissure. The ores are thought to have come up from below through the chief fissures, and in this respect Dr. Jenney's views radically differ from those of the earlier writers. The solutions are said to have favored particular beds for the following reasons. The beds were cellular from long exposure to atmospheric agents, or they were chemically (being dolomitic) and physically of a nature to occasion it, or they were soft and permeable shaly beds.[1] (See also under 2.06.07.)

---

[1] WISCONSIN.

J. A. Allen, "Description of Fossil Bones of Wolf and Deer from Lead Veins," *Amer. Jour. Sci.*, iii., II. 47. T. C. Chamberlain, *Wis. Geol. Survey*, Vol. IV., 1882, p. 367. Rec. E. Daniels, "Geology of the Lead Mines of Wisconsin," *A. A. A. S.*, VII. 290; *Engineering and Mining Journal*, July 6, 13, 20, 27, Aug. 3, 10, 24, Oct. 5, 1878; *Wis. Geol. Survey*, 1854.

2.06.04. Example 24*a*. Washington County, Missouri. Gash veins in the Lower Magnesian limestones of eastern Missouri in the same region as the disseminated ores of Example 23, and containing galena, barite (locally called "tiff"), calcite, and residual clay. The cavities are described by Whitney as resembling in all respects the gash veins farther north, which, however, lie in rocks higher in the geological series. These mines were the earliest worked, but have been given up since the price of lead has been at present figures (1875 and subsequently). The ore was obtained from pockets, caves, irregular cavities, and from the overlying residual clays. This whole region has been exposed and above water since the close of Carboniferous times and has suffered enormous surface decay (see R. Pumpelly, *Tenth Census*, Vol. XV., p. 12, and *Geol. Soc. Amer.*, Vol. II., p. 20), which has left a mantle of residual clay spread widely over its extent. In this, more or less float mineral occurred. The mines were located in Washington, Franklin, Jefferson, and St. François counties.[1]

2.06.05. Example 24*b*. Livingston County, Kentucky. Veins in limestone of the St. Louis stage of the Lower Carboniferous, containing galena in a gangue of fluorite, calcite, and clay. The

---

James Hall, "Notes on the Geology of the Western States," *Amer. Jour. Sci.*, i., XLII. 51. J. T. Hodge, "On the Wisconsin and Missouri Lead Region," *Amer. Jour. Sci.*, i., XLIII. 35. R. D. Irving, "Mineral Resources of Wisconsin," *M. E.*, VIII. 478. E. James, "Remarks on the Limestones of the Mississippi Valley Lead Mines," *Phil. Acad. Sci.*, V., Part I., p. 51. J. Murrish, Report on the lead regions, 1871, as commissioner for their survey. D. D. Owen, "Report on the Lead Region," *U. S. Senate Documents*, 1844. J. G. Percival, *Wis. Geol. Survey*, 1856. Squier and Davis, Historical account, *Smithsonian Contributions*, Vol. I., p. 208. M. Strong, *Wis. Geol. Survey*, 1877, I. 637; II. 645, 689. J. D. Whitney, *Wis. Geol. Survey*, 1861-62, I. 221. Rec. *Metallic Wealth*, p. 403, 1856. "On the Occurrence of Bones and Teeth in the Lead-bearing Crevices," *A. A. A. S.*, 1859.

ILLINOIS.

J. Shaw, *Geol. Survey of Illinois*, 1873, Vol. II., p. 340. J. D. Whitney, *Geol. Survey of Illinois*, 1866, Vol. I. 153.

IOWA.

C. A. White, *Iowa Geol. Survey*, 1870, Vol. II., p. 339. J. D. Whitney, *Iowa Geol. Survey*, 1858, Vol. I., p. 422.

[1] Compare the older references under Example 23, and the following: A. Litton, *Second Ann. Rep. Missouri Geol. Survey*, 1854. J. D. Whitney, *Metallic Wealth*, p. 419.

ore bodies have never been well described and no very accurate diagnosis can be given. They are found in Livingston, Crittenden, and Caldwell counties, Kentucky, in that portion of the State lying south of the Ohio River and east of the Cumberland. While limestone always forms one wall, a sandstone of geological relations not well determined forms the other. The veins run from two to seven feet wide and in instances are richer in their upper portions than in the lower. As yet they are of greater scientific than practical importance. Some galena occurs also in irregular cracks in the limestone. As a possible indication of a stimulating cause for the formation of the veins, the interesting dike of mica-peridotite may be cited, which has been described by J. S. Diller.[1] The dike occurs in the same fissure with a vein of fluorspar.[2]

2.06.06. **Example 25. Southwest Missouri.** Zincblende and very subordinate galena with their oxidized products, associated with chert, residual clay, calcite, a little pyrite and bitumen, in cavities of irregular shape and in shattered portions of Subcarboniferous limestone. Across Missouri, from a point south of St. Louis, and including the country as far to the northwest as Sedalia and Glasgow, a broad belt, called the Ozark uplift, extends southwesterly into Arkansas. It has formed a great plateau in central and southern Missouri and consists largely of Silurian rocks. These have a fringe of Devonian on the edges and dip under the Lower Carboniferous. The plateau reaches 1500 feet above the sea in Wright County, but on the limit is succeeded by lower country. To the southwest it drops somewhat, with Lower Carboniferous strata outcropping, which in Kansas are overlain by the coal measures. The surface then rises again in the prairies. At the edge of the plateau is a trough, in whose bottom the Lower Carboniferous strata are cut by the Spring River, which flows southwesterly from Missouri across the western State line into Kansas and has a general direction parallel to the western limits of the uplift. It receives tributary streams on each bank, which cut the strata in strongly marked valleys and afford good exposures. Those on the east bank, from south to north, are Shoal

---

[1] "Mica-Peridotite from Kentucky," *Amer. Jour. Sci.*, October, 1892.

[2] S. F. Emmons, "Fluorspar **Deposits of Southern Illinois**," *M. E.*, February, 1892. C. J. Norwood, "Report on the Lead Region of Livingston, Crittenden, and Caldwell Counties," *Kentucky Geol. Survey*, 1875, New Series, Vol. I., p. 449.

Creek, Short Creek, Turkey Creek, and Center Creek, while from the west come the Brush, Shawnee, and Cow creeks, all in Kansas. Along the first mentioned creeks the principal mining towns are situated, but others are found on the minor streams. They extend through an area fifteen miles broad from east to west and twenty-five miles from north to south. Newton and Jasper are the most productive counties in Missouri, while Cherokee County, in Kansas, also contains notable mines. Undeveloped districts are recorded in Arkansas, but apparently at a lower geological horizon. The ore occurs in the Keokuk or Archimedes limestone of the Lower Carboniferous. A generalized section of the rocks, according to F. L. Clerc, is as follows. On the higher prairie, some 15 feet of clay or gravel; 10 feet of flint or chert beds; 40 feet of limestone with thin beds of chert; 60 feet of alternating layers of limestone and chert; 100 feet and more of chert, sometimes chalky with occasional beds of limestone; 225 feet in total. In basins and extensive pockets in these rocks, deposits of slates with small coal seams are found, of undetermined geological relations. The large bed of limestone of the section affords a datum of reference in relation to which the ores may be described. A few minor, shallow deposits occur in the flints over it. In the limestone the ores are associated with a gangue of dolomitic clay and residual flint. They occupy irregular cavities or openings, locally known as circles, spar openings, and runs. (Clerc.) Below the limestone the ore is found in "sheets, bands, seams, and pockets," and filling in the interstices of a breccia of chert, which has been formed by the breaking down of the chert layers on the solution and removal of the interbedded limestones. There are districts where the overlying bed of limestone has also disappeared, and they then lack it for a capping. The deposits extend to considerable depths below the position of the limestone. The present mines have not demonstrated as yet their limit of depth. At times the ore is associated with a later formed quartz rock that has coated and filled the cavities of the breccia.

2.06.07. The removal of the interbedded layers of limestone and the caving in of the associated cherts have been the principal causes of the formation of cavities. Adolph Schmidt referred the shrinkage to the dolomitization of pure lime carbonate, an idea that has had extended adoption, and has also had an important part in causing the general porosity. Schmidt traced five periods in the geological history of the ore bodies: 1. Period of deposition of the rocks. 2. Period of dolomitization of certain strata and of principal ore deposition. 3. Period of dissolution of part of the limestone, of breaking down of chert, and of continued but diminishing ore deposition. 4. Period of regeneration, secondary deposition of carbonate of lime and quartz, and continued ore deposition. 5. Period of oxidation.

Schmidt's work was done in 1871-72. Since then the increased

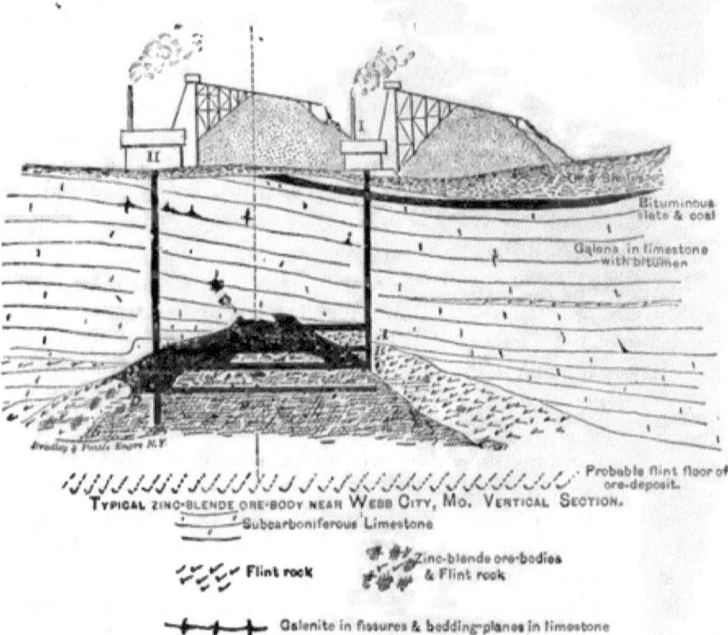

Fig. 43.—*Vertical section of a typical zincblende ore body, near Webb City, Mo. After C. Henrich, M. E., June,* 1892.

development of the mines has afforded greater opportunities for observation. Haworth, in 1884, referred, with much reason, the shattering of the chert in certain areas to oscillations of the strata, and Clerc, in 1887, emphasized particularly the dissolving action of water. A forthcoming description by W. P. Jenney, of the United States Geological Survey, is awaited with great interest. While the formation of the cavities and the method of introduction of the ore are not so difficult to understand, it is a hard problem to discover the original source of the metals. No published account

FIG. 44.—View of the Motley mine, Webb City, Mo. After a photograph of W. P. Jenney.

brings them up from below. Haworth discusses a possible precipitation from the ocean, as is outlined under Example 24, and Clerc mentions the pockets of slate and coal as a probable source. Further and more extended study of the mines has been much needed.

A brief outline of Dr. Jenney's views is here given, which has been abstracted from manuscript that he has kindly allowed the writer to see in advance of its publication. In the forthcoming report all the lead or lead and zinc regions of the Mississippi Valley are considered together. They are described as occurring along three lines of upheaval. The region of Wisconsin and Iowa is on the flanks of the Archæan "Wisconsin Island" of Chamberlain, referred to above under 2.06.01. The southeast and southwest Missouri regions are on the Ozark uplift, while a minor argentiferous galena district is on the line of the Ouachita uplift of Arkansas and Indian Territory. The formation of the ore bodies in the first three of these is regarded as having been in general the same. They are thought to have originated from uprising solutions, which came through certain principal fissures, and spread laterally into strata favorable to precipitation. In southwest Missouri this was the Cherokee limestone of the Lower Carboniferous. In its unaltered state it is an extremely pure carbonate of lime. It has a maximum thickness, where not eroded, of 165 to 200 feet, and contains many interbedded layers of chert. Much organic matter, and more or less bitumen, are also at times present. The limestone seems to have been raised above the ocean level at the close of the Lower Carboniferous and to have remained for a long period exposed to the atmospheric agents. Much caving in of unsupported layers of chert and much attendant brecciation resulted. The general stratum became quite open and cellular in certain portions. At a later period, supposed from several indications to be at the close of the Cretaceous, dynamic disturbance occurred, which along certain lines produced fissures, sometimes parallel, sometimes intersecting. Solutions arose through these which dolomitized much of the remaining limestone and caused additional porosity. Zinc and lead ores were afforded, and where the conditions were favorable they spread laterally from the fissures and deposited the sulphides in the cellular rock or replaced the limestone itself. The intersection of crossing fissures is a frequent point of deposition, and at times parallel master fissures have given a wide area of impregnation. This form of ore deposit is called a run. The runs are from 5 to 50 feet in height, 100 to 300 feet long, and 10 to 50 feet across. At Webb City they are even larger. As a general thing the ore is in the interstices of the brecciated chert, but it is also in limestone and dolomite, and associated with a silicified form of the insoluble residue left by the solution of the limestone, which Dr. Jenney calls "cherokite." All the ores require concentration. Galena usually occurs near the surface, while blende is more abundant in depth. Cadmium is at times present in the blende in notable amount.

2.06.08. Some interesting alterations of the minerals have occurred, which have changed the blende to smithsonite and calamine. In one case a secondary precipitation of zinc sulphide has occurred as a white amorphous powder which is of very recent date. With the original precipitation of the blende the asphaltic material may have had something to do. In the matter of production Dr. Jenney fixes the ratio of the blende, galena, and pyrite at about 1000 : 80 : 0.5.[1]

2.06.09. Other zinc and lead deposits are known in central Missouri generally resembling the above quite strongly, but of less economic importance. Some, however, are described by Schmidt as conical stockworks. They sometimes are found in Lower Silurian strata.

---

[1] G. C. Broadhead, "Geological History of the Ozark Uplift," *Amer. Geol.*, III. 6. H. M. Chance, "The Rush Creek (Arkansas) Zinc District," *Trans. Amer. Inst. Min. Eng.*, Washington meeting, 1890. F. L. Clerc, Geological description of the mines in a statistical pamphlet on the *Lead and Zinc Ores of Southwest Missouri Mines*, p. 4, published by J. M. Wilson, Carthage, Mo., 1887. Rec. See also *Engineering and Mining Journal*, June 4, 1887, p. 397. "Zinc in the United States," *Mineral Resources*, 1882, p. 368. *Engineering and Mining Journal*, Nov. 3, 1888, p. 389 ; March 8, 1890, p. 286. E. Haworth, *A Contribution to the Geology of the Lead and Zinc Mining District of Cherokee County, Kansas*, Oskaloosa, Iowa, 1884. C. Henrich, "Zincblende Mines and Mining near Webb City, Mo.," *M. E.*, February, 1892 ; *Engineering and Mining Journal*, June 4, 1892. R. W. Raymond, "Note on the Zinc Deposits of Southern Missouri," *M. E.*, VIII. 165 ; *Engineering and Mining Journal*, Oct. 4, 1879. J. D. Robertson, "A New Variety of Zinc Sulphide from Cherokee County, Kansas," *Amer. Jour. Sci.*, iii., XL., p. 160. A. Schmidt and A. Leonhard, *Missouri Geol. Survey*, 1874. A. Schmidt, "Forms and Origin of the Lead and Zinc Deposits of Southwest Missouri," *Trans. St. Louis Acad. Sci.*, III. 246 ; *Amer. Jour. Sci.*, iii., X., p. 300. *Die Blei und Zink Erzlagerstätten von Südwest Missouri*, Heidelberg, Germany, 1876. W. H. Seamon, "Zinciferous Clays of Southwest Missouri," *Amer. Jour. Sci.*, iii., XXXIX., p. 38.

2.06.10. Both the mines of Example 25 and those of Example 24 were originally worked for lead, and the zinc minerals were regarded as a nuisance ; of late years the zinc has been much more of an object than the lead. The deposits in southwest Virginia (Example 26) also produce lead, but are best known for zinc.

2.06.11. Example 26. Wythe County, Virginia. Veins or beds of oxidized ores, probably changing to blende below in crystalline limestone or dolomite, just above the Calciferous but as yet not sharply determined in their stratigraphy. The ore-bearing terrane is known over a considerable extent of country, running from near Roanoke one hundred miles westward. The largest mines are in Wythe County, and of these the Bertha is best known. According to Boyd, there are in one section 486 feet of strata impregnated with lead and zinc in varying amounts. Farther east, other openings of considerable promise have lately been made at Bonsacks. The zinc ore bodies are at times of great size (40 feet wide), and are associated with more or less of lead minerals and iron pyrites. It would appear as if the region must be an important producer of zinc in the future.[1]

Dr. Jenney's paper may be expected in the *Trans. Amer. Inst. Min. Eng.*, 1893.

[1] C. R. Boyd, *Resources of Southwest Virginia*, p. 71 ; "Mineral Wealth of Southwest Virginia," *M. E.*, V. 81 ; *Ibid.*, VIII. 340. Rec. H. Credner, *Zeitschr. für die gosammten Naturwissenschaften*, 1870, Vol. XXXIV., p. 24. F. P. Dewey, "Note on the Falling Cliff Zinc Mine," *M. E.*, X. 111. A. v. Groddeck, *Typus Austin, Lehre von den Lagerstätten der Erze.*, p. 103.

FIG. 44a.—*View of the Bertha Zinc Mines, Wythe County, Virginia. From a photograph by A E. W. Miller.*

CHAPTER VII.

ZINC ALONE, OR WITH METALS OTHER THAN LEAD.

2.07.01. Zinc ores commonly occur in association with lead, but there are one or two exceptional deposits in this country which are without lead and which have no parallel in other parts of the world. The minerals containing zinc at Franklin Furnace and Ogdensburg, N. J., are known elsewhere only as rarities, although they are found in vast amounts in New Jersey.[1]

ZINC SERIES.

	Zn.	S.	Fe.	SiO_2.	Mn.
Sphalerite (commonly called blende) ZnS	67	33			
Zincite, ZnO	80.3				
Franklinite, $(Fe.Zn.Mn)O,(Fe.Mn)_2O_3$ (variable)	5.54		51.8		7.5
Willemite, $2ZnO.S.O_2$	58.5			27.1	
Calamine, $2ZnO.SiO_2,H_2O$	54.2			25.0	
Smithsonite, $ZnOCO_2$	51.9				

2.07.02. Example 27. Saucon Valley, Pennsylvania. Zincblende and its oxidation products, calamine and smithsonite, filling innumerable cracks and fissures in a disturbed, magnesian limestone, thought to belong to the Chazy stage. The ore bodies occur in the Saucon Valley near the town of Friedensville, about four miles south of Bethlehem. The limestone is inclosed between two northerly spurs of the South Mountain, and has apparently been tilted and shattered by the upheaval of the latter. The shattering and disturbances decrease as the South Mountain is left and the dip decreases. There are three principal mines, the Ueberroth, the Hartman, and the Saucon, the first named being in the portion which is tilted nearly to a vertical dip and is much disturbed, while the next is where the dip has gradually decreased to 35°. The mines are on a belt some three quarters of a mile long. At the Ueberroth an enormous quantity of calamine was found on

---

[1] F. L. Clerc, "Zinc in the United States," *Mineral Resources*, 1882, p. 358.

the surface, but it passed in depth into blende and was clearly an oxidation product. In the others the blende came nearer the surface. The ore follows the bedding planes and the joints normal to these throughout a zone varying from 10 to 40 feet across and fills the cracks. At their intersection the largest masses are found. Six larger parallel fissures were especially marked at the Ueberroth. This mine proved in development to be very wet, and a famous pumping engine, the largest of its day, was built to keep it dry. The Hartman and Saucon are less wet. A little pyrite occurs with the blende, and thin, powdery coatings of greenockite sometimes appear on its surface, but it is entirely free from lead and a very high grade spelter is made from it. The mines were strong producers from 1853 to 1876, but little has been done since. It is reported (1891) that the great pumping engine has been started, and they may once more furnish considerable quantities of ore.

2.07.03. The mines were evidently filled by circulations from below that brought the zinc ore to its present resting place in the shattered and broken belt. Drinker considers it to have been derived from a disseminated condition in the limestone.[1]

2.07.04. Example 28. Franklin Furnace and Sterling, N. J. A bed consisting of franklinite, willemite, zincite, etc., in crystalline limestone, in many respects analogous to the magnetite of Example 13. The franklinite and zincite beds are in a belt of white, crystalline limestone which runs southwesterly from Orange County, New York, across northwestern New Jersey. It was considered metamorphosed Lower Silurian by H. D. Rogers, but its association with Archæan gneiss is so close that it has with some reason been regarded as of the same age with the gneiss. Beyond the gneiss to the west a blue limestone supposed to be Lower Silurian outcrops, and the same rock appears again to the southeast. F. L. Nason, of the New Jersey Survey, has recently argued, after careful and praiseworthy field work, and after discovering in unmetamorphosed portions some fossils which belong to the Olenellus fauna of the Cambrian, that the blue and white limestones are of the same age, and that the latter owes its character to a great dike of granite,

---

[1] F. L. Clerc, *Mineral Resources*, 1882, 361. Rec. H. S. Drinker. "On the Mines and Works of the Lehigh Zinc Company," *M. E.*, I. 67. C. E. Hall, in Rep. D3, *Second Geol. Survey Penn.*, p. 239. *Die Gruben und Werke der Lehigh Zink Gesellschaft in Pennsylvanien*, B. und H. Zeit., 1872, p. 51.

which appears at various points. The granite is not always continuous, and is often in isolated masses or horses, as in the Trotter mine. There is also in some portions of the belt a curious scapolite rock, regarded as igneous. It is not unlikely that the great dikes may have been a factor in the formation of the ore, although this is not demonstrated. At Franklin Furnace the crystalline limestone forms a low hill (Mine Hill) east of the upper waters of the Wallkill, and again at Ogdensburg, two miles south, another (Sterling Hill), on the west bank. There is a valley and unexposed strip between, so that the unbroken continuity without a possible intervening fault cannot be established. The bed at Franklin outcrops on the west side of the hill. It begins on the north just across the Hamburg road and runs south 30° west as a

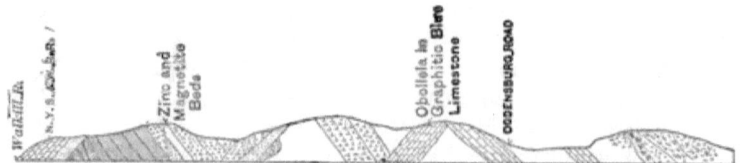

Fig. 45.—*Section at Franklin Furnace, N. J., showing the geological relations of the franklinite ore body. After F. L. Nason, Geol. of N. J., 1890, XIV., p. 50. The ore body is in white limestone with underlying gneiss.*

continuous bed for about 2500 feet. This portion is called the Front vein. It contains on the north the old Hamburg mine, then the Trotter mine, and in the southern portion belongs to the New Jersey Zinc and Iron Company. It runs from 8 to 30 feet broad at the outcrop, but swells below. It dips southeast 40 to 60° into the hill, and is interbedded in the limestone. In the Trotter mine a wedge or horse of hornblende, augite, plagioclase, and various other silicates enters the bed a short distance. In this horse some of the most interesting minerals have been found, such as fluorite, rhodonite, blende (var. cleiophane), smaltite (var. chloanthite), axinite, etc. At the end of the Front vein—or, more properly, bed—a branch or bend strikes off at an angle of 30 to 40° to the west. This more easterly branch, which is called the Buckwheat mine, outcrops on the surface some 500 feet, and then, after being cut by a trap dike 22 feet wide, pitches down at an angle of 27° and passes under the limestone. The portion of the mine northeast of the dike furnishes the most and best ore. The surface outcrop of the Buckwheat was 25 to 30 feet across, but it swelled below to 52 feet, and in the second level, about 200 feet from the surface, it was penetrated by a cross-cut 125 feet without finding the wall. The character of the ore varies; for while it is excellent at the point of the cross-cut, at 125 feet nearer the intersection with the front bed it becomes lean, while preserving its width lower. Beyond the dike the bed is likewise broad, and is mined out for 40 to 50 feet across. The workings are now some distance down on the pitch. The impression made by the arch of the roof and by the curving beds is that this is the crest of an anticline whose axis pitches north 27°, and whose central portion is formed by the franklinite bed being doubled up together on itself before the two parts diverge in depth. Its western portion probably is continuous in a synclinal trough with the front bed, and its eastern portion dips east at some unknown angle. If this is true, it would doubtless be struck by drilling in the surface to the eastward. Mining has generally been followed along the entire outcrop except at the junction of the two branches. At present the most active work is being done on the front bed at the Trotter mine and on the rear bed of the Buckwheat, the latter being much the larger.

2.07.05. The ore consists of franklinite in black crystals, usually rounded and irregular, but at times affording quite a perfect octahedron combined with the rhombic dodecahedron and set in a matrix of zincite, willemite, and calcite. The richest ore lacks the calcite and consists of the other three in varying proportions. This best ore is in largest amount in the Buckwheat mine, beyond the trap dike which cuts it. The limestone containing the ore has a notable percentage of manganese replacing the calcium, and where it is exposed to the atmosphere it weathers a characteristic brown. An analysis of a sample occurring with the ore at Sterling Hill afforded F. C. Van Dyck:

$CaCO_3$	82.23
$MnCO_3$	16.57
Fe_2O_3	0.50
SiO_2	0.20
H_2O	1.0
	100.50

The percentage of manganese is very high for a limestone.

2.07.06. The Sterling Hill outcrop is less extensive. It begins on the north with the New Jersey Zinc and Iron Company's property and runs south 30° west for 1100 feet. It then branches or bends around to the west and runs north 60° west for 300 feet, bending again to north 30° east, and pitches beneath the surface. Thus the general relations between the front and back beds are somewhat the same as at Mine Hill, and the dip and pitch are similar. The principal workings are on the Front vein, where there are two veins (beds), according to the older descriptions, one rich in franklinite and the other in zincite. It is doubtful if there really are two distinct beds, but probably one portion is richer in zincite than the other. The part mined is from two to ten feet. The footwall is corrugated and causes many pinches and swells, whose troughs pitch north. The limestone between the front and back outcrop is charged with franklinite and various silicates (jeffersonite, augite, garnets, etc.), and has been mined out in large open cuts now abandoned. A deposit of calamine was found in the interval about 1876, and has furnished many fine museum specimens.

2.07.07. It is not clear that the Sterling Hill and Mine Hill deposits were once continuous. The bed at Mine Hill runs in the front portion close to the contact of the white limestone and the gneiss. The Sterling Hill bed is much farther away from the gneiss, and this would indicate that it is at a higher horizon. The evidence, too, of a pitching syncline is strong, but a pitching S-fold is not as clear. A faulting of the Archæan rocks in an east and west line across their strike, and a subsequent tilting so as to give them a northerly pitch, is a very widespread phenomenon in the Highlands, and lends weight in this instance to the idea that a fault intervenes between the two hills. Such faulting is shown even in New York City.

2.07.08. The origin of these beds is very obscure. They are so unique in their mineralogical composition that very little direct aid is furnished by deposits elsewhere. At Mine Hill, below the fork of the franklinite bed, there was formerly a large lense of magnetite that has now been mined out. It was in the white limestone. There are many points of analogy between the franklinite beds and extended magnetite deposits. They are both minerals of the spinel group, and the spinels are a common result of metamorphic action. The presence of zincite and willemite complicates matters, however, and while an original ferruginous deposit might be conceived with a large percentage of manganese, yet such abundance of zinc is beyond previous experience. It is, however, suggestive that no inconsiderable amount of zinc is found in the Low Moor (Va.) limonites, as shown by the flue dust (see E. C. Means, "The Dust of the Furnaces at Low Moor, Va.," Buffalo meeting Amer. Inst. Min. Eng., October, 1888), and this in the course of a protracted blast may amount to many tons, but it does not approach the Franklin Furnace ores. None the less, in the absence of a better explanation, the franklinite bed may be thought of as perhaps an original manganese, zinc, iron deposit in limestone, much as many Siluro-Cambrian limonite beds are seen to-day, and that in the general metamorphism of the region it became changed to its present condition. Minerals of the spinel group occur all through this limestone belt, and in Orange County, New York, to the north, there is an old and prolific source of them.

2.07.09. If it is possible to demonstrate the connection between the white limestone and a granite dike, as Mr. Nason argues, this may have been an important factor in the ore formation. It is very reasonable that the igneous intrusion should start ore-bearing currents along a certain stratum in the limestone, which would replace it with ore. Subsequent folding and metamorphism must then have changed these ores, whatever they were, to the present unusual minerals.[1]

2.07.10. Blende is known in numerous places in the Rocky Mountains and is often argentiferous, but it is not as yet profitably smelted for zinc, and is a drawback to the lead-silver process.

2.07.11. A large amount of zinc ore is turned directly into zinc white and employed as a pigment. For this reason later statistics of the metal do not indicate all the ore mined. The accompanying figures are short tons. For detailed statistics see the volume on "Mineral Industry" of the *Engineering and Mining Journal*, 1893.

---

[1] F. Alger, "On the Zinc Mines of Franklin, Sussex County, N. J.," *Amer. Jour. Sci.*, i., XLVIII. 252. Bemis and Woolson, An unpublished thesis in the School of Mines Records, 1885. H. Credner, "On the Franklinite Beds," *B. und H. Zeit.*, 1866, 29, and 1871, 369. *Geol. of New Jersey*, 1868 (with a map), and subsequent Annual Reports. F. L. Nason, *Ann. Rep. State Geol. N. J.*, 1890, p. 25. Rec. *Amer. Geol.*, VIII. 166. Vanuxem and Keating, "On the Geology and Mineralogy of Franklin, Sussex County, N. J.," *Phil. Acad. Sci.*, Vol. II., p. 277.

	1882.	1890.
Illinois	18,201	26,243
Kansas	7,366	15,199
Missouri	2,500	13,127
Eastern and Southern States	5,698	9,114
	33,765	63,683

These amounts are from the *Mineral Resources of the United States*, 1889-90, p. 89.

CHAPTER VIII.

LEAD AND SILVER.

2.08.01. There are two general methods of extracting silver from its ores, the one indirectly, by smelting with and for lead; the other by amalgamation, chlorination, or some such process. Hence under silver there are two classes of mines—lead-silver and high-grade silver ores. Both have almost always varying amounts of gold. The lead-silver mines furnish also, as noted above, by far the greater portion of the lead produced in the United States. Ores adapted to lead-silver metallurgical treatment form, in general, the oxidized alteration products of the upper parts (above permanent water level) of deposits of galena and pyrites. They may be well-marked fissure veins, chimneys, chambers, or contact deposits. Ores which of themselves are adapted to other processes are often worked in with the lead ores, and unchanged sulphides are artificially oxidized by roasting preparatory to smelting. The localities are taken up geographically from east to west.

2.08.02. LEAD-SILVER DEPOSITS IN THE ROCKY MOUNTAIN REGION AND THE BLACK HILLS.—The mines are described in order from south to north, beginning with New Mexico.

NEW MEXICO.

2.08.03. Example 29. The Kelley Lode. Oxidized lead ores, with some blende, calamine, etc., forming a contact deposit between slates and porphyry. The ore body is in the Magdalena Mountains, thirty miles west of Socorro, and has supplied the Billings smelter at that point. Numerous other ore bodies along the contact between sedimentary and eruptive rocks occur in the same region.

2.08.04. Example 29a. Lake Valley. Farther south, in Doña Aña County, the mines of Lake Valley afford lead ores on the bedding planes of limestone and along the contact between it and various igneous rocks. Considerable carbonate of iron is associated with them, and a variety of rare minerals, including vanadinite, descloizite, etc., occur. The limestones are probably Subcarboniferous. There are other districts in the territory of minor importance.[1]

COLORADO.

2.08.05. Example 30. Leadville. Bodies of oxidized lead-silver ores, passing in depth into sulphides, deposited in much faulted Carboniferous limestone, in connection with dikes and sheets of porphyry. Leadville is situated in a valley which is formed by the head waters of the Arkansas River. The valley runs north and south, being confined below by the closing in of the hills at the town of Granite. It is about twenty miles long and sixteen broad, and even to superficial observation is seen to be the dried bottom of a former lake. The mountains on the east form the Mosquito range, a part of the great Park range, while those on the west are the Sawatch, and constitute the Continental Divide at this point. Leadville itself is on the easterly side, upon some foothills of the Mosquito range. The eastern slope of the Mosquito range rises quite gradually from the South Park to a general height of 13,000 feet. The range then forms a very abrupt crest, with steep slopes looking westward, which are due to a series of north and south faults whose easterly sides have been heaved upward as much as 7500 feet. The faults pass into anticlines along their strike. The Mosquito range consists of crystalline Archæan rocks, foliated granites, gneisses, and amphibolites, and of over 5000 feet of Paleozoic sediments and igneous rocks. The former include Cambrian quartzite, 150 to 200 feet; Silurian white limestone, 160 feet, and quartzite, 40 feet; Carboniferous blue limestone, 200 feet (the chief ore-bearing stratum); Weber shales and sandstones, 2000 feet; and Upper Carboniferous limestones, 1000 to 1500 feet. The igneous rocks are generally porphyries. The sedimentary rocks were laid down in Paleozoic time on the shores of the Archæan Sawatch island, and were penetrated by the igneous rocks, probably at the close of the Cretaceous. They were all upheaved, folded, and faulted in the general elevation of the Rocky Mountains, about the beginning of the Tertiary period. The intrusion of the igneous rocks was the prime mover

---

[1] *Rep. Director of the Mint*, 1882, pp. 341, 376. B. Silliman, "Mineral Regions of New Mexico," *M. E.*, X. 224.

in starting ore deposition, and the solutions favored the under side of the sheets, along their contacts with the blue Carboniferous limestone.

2.08.06. The early history of Leadville will be subsequently referred to in speaking of auriferous gravels. The lead-silver ores first became prominent in 1877, although discovered in 1874, and by 1880 the development was enormous. The region grew at once to be the largest single producer of these ores, and has remained such ever since. The mines are situated east of the city on the three low hills, Fryer, Carbonate, and Iron, but recently a deep shaft in the city itself has found the extension of the ore chutes and opened up great future supplies. The ores have chiefly come in the past from the upper oxidized portions of the deposits. Of late years, however, the older and deeper workings have been showing the unchanged sulphurets. The ores are chiefly earthy carbonate of lead, with chloride of silver, in a clayey or siliceous mass of hydrated oxides of iron and manganese. In the Robert E. Lee mine silver chloride occurred without lead. Some zinc is also found, and a long list of rare minerals. Where the ore is in a hard, siliceous, limonite gangue it is called hard carbonate, but where it is sandy and incoherent it forms a soft carbonate, or sand carbonate. All the mines produce small amounts of gold, which in one case (the Printer Boy) has been of more importance than the silver. A few ore bodies are found at other horizons than the Carboniferous. They also run in instances as much as 100 feet from the contact, and may likewise be found in the porphyry, doubtless replacing included limestone. They were all deposited as sulphides, and, according to Emmons, when the rocks were at least 10,000 feet below the surface.

2.08.07. In the valuable monograph on the region, which is now a classic on the subject and which is cited below, Emmons endeavors to prove the following points :

I. That the ores were deposited from aqueous solution.

II. That they were originally deposited mainly in the form of sulphides.

III. That the process of deposition involved a metasomatic interchange with the material of the rock in which they were deposited.

IV. That the mineral solutions or ore currents were concentrated along natural water channels, and followed, by preference, the bedding planes at a certain geological horizon, but that they

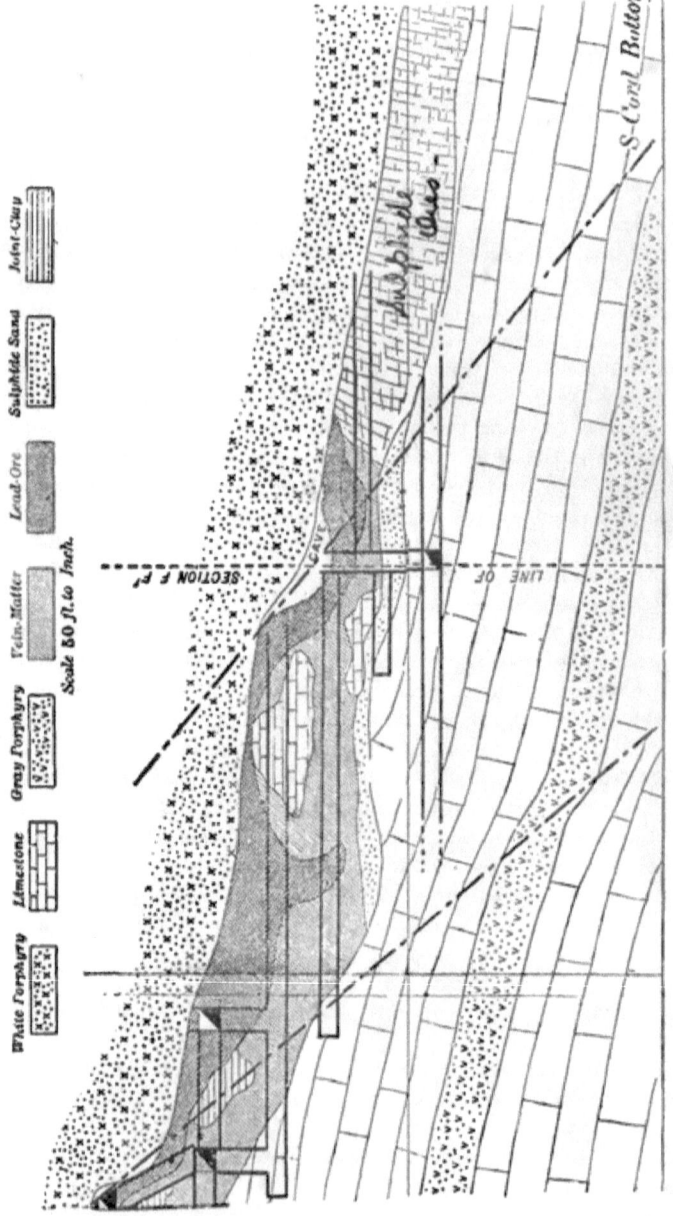

FIG. 46.—Section of the White Cap chute, Leadville, showing the geological relations of the ore, and its passage into unchanged sulphides in depth. After A. A. Blow, M. E., Vol. XVIII, Plate V.

also penetrated the adjoining rocks through cross joints and cleavage cracks.

These additional points are also advanced:

I. That the solutions came from above.

II. That they were derived mainly from the neighboring eruptive rocks.

2.08.08. The first four points are doubtless correct, and No. III. is an important application of the theory of replacement, frequently referred to in the introduction. The last two propositions merit less confidence. Seven additional years of mining have brought many new facts to light and have led others (A. A. Blow in particular, whose valuable paper is cited below) to refer the ores to upward rising currents. Emmons foresaw this possibility and mentioned it on p. 584 of his monograph. The amount of the adjacent, igneous rocks is quite insufficient to afford the ore. In alteration the galena has passed through an intermediate stage of sulphate before changing to carbonate. These mines have been important not alone in their own metallic products, but in furnishing the smelters with oxidized lead ores they have supplied a means of reduction for many other more refractory ones, which could be conveniently beneficiated through the medium of lead.[1]

2.08.09. Example 30*a*. Ten Mile, Summit County. Bodies

---

[1] F. M. Amelung, "The Geology of the Leadville Ore District," *Engineering and Mining Journal*, April 16, 1880, p. 25. "On the Origin of the Ore," *Ibid.*, Dec. 20, 1879. A. A. Blow, "The Geology and Ore Deposits of Iron Hill, Leadville, Colo.," *M. E.*, June, 1889. Rec. *Ann. Rep. Colo. School of Mines*, 1887, p. 62. S. F. Emmons, "Geology and Mining Industry of Leadville," *Monograph* 12, *U. S. Geol. Survey*. Rec. *Second Ann. Rep. Director of U. S. Geol. Survey*. Rec. *Tenth Census*, Vol. XIII., p. 76. F. T. Freeland, "The Sulphide Deposits of South Iron Hill, Leadville," *M. E.*, XIV. 181. C. Henrich, "The Character of the Leadville Ore Deposits," *Engineering and Mining Journal*, Dec. 27, 1879, p. 470. "Origin of the Leadville Deposits," *Engineering and Mining Journal*, May 12, 1888, p. 33. "On the Evening Star Mine," *Ibid.*, May 7, 1881, p. 361. "Leadville Geology," *Ibid.*, June 3 and 10, 1882; Historical, May 30, April 6, 13, 20, 27, 1878; also many other allusions, 1879-81. R. W. Raymond, Rep. on the Little Pittsburg Mine, *Engineering and Mining Journal*, June 28, 1879. L. D. Ricketts, *The Ores of Leadville*, Princeton, 1883. C. M. Rolker, "Notes on Leadville Ore Deposits," *M. E.*, XIV. 273, 949. F. L. Vinton, "Leadville and the Iron Mine," *Engineering and Mining Journal*, Feb. 15, 1879, p. 110; also June 28, p. 465.

of argentiferous galena, pyrite, and blende, in beds of Upper Carboniferous limestone, on their contact with overlying, micaceous sandstones, or with sheets and dikes of porphyry. The Carboniferous limestones that contain the ores at Leadville extend both north and south, and their equivalents occur also on the west flank of the Sawatch range. Ten Mile is another productive portion, north of Leadville and at a higher altitude. The strata are enormously disturbed, and pierced even more than at Leadville by sheets and dikes of porphyry. The ores are less oxidized and more rebellious. The Robinson is the principal mine.[1]

2.08.10. Example 30b. Monarch District, Chaffee County. Oxidized lead-silver ores in limestone. The belt of limestones south from Leadville contains some notable ore bodies in Chaffee County. The Monarch district is the most important. It is situated at the head waters of a branch of the South Arkansas River. The ore lies in limestones whose age is not yet accurately determined. The Madonna mine is the best known and has shipped much ore to Pueblo.[2]

2.08.11. Example 30c. Eagle River, Eagle County. Galena and its alteration product, anglesite, in Carboniferous limestone, on the contact between it and quartzite or porphyry. The mines lie in the valley of Eagle River, on the western slope of the Continental Divide. The galena has changed to the sulphate, instead of carbonate, probably having been less completely oxidized than at Leadville, and marking the intermediate stage in the process. The wall rocks lie quite undisturbed, having a low dip of 15° north, and not being faulted. Lying lower than the lead-silver deposits, and in Cambrian quartzite, on the contact with an overlying sandstone are found chutes carrying gold in talcose clay.[3]

2.08.12. Example 30d. Aspen, Pitkin County. Bodies of

---

[1] S. F. Emmons, *Tenth Census*, Vol. XIII., p. 73; also a forthcoming monograph of the U. S. Geol. Survey.

[2] S. F. Emmons, *Tenth Census*, Vol. XIII., p. 79. *Rep. Director of the Mint*, 1884, p. 191.

[3] S. F. Emmons, "Notes on Some Colorado Ore Deposits," *Colo. Sci. Soc.*, Vol. II., Part II., p. 100. E. E. Olcott, "Battle Mountain Mining District, Eagle County, Colorado," *Engineering and Mining Journal*, June 11, 1887, pp. 417, 436; *Ibid.*, May 21, 1892, p. 545. G. C. Tilden, "Mining Notes from Eagle County," *Ann. Rep. Colo. State School of Mines*, 1886, p. 129.

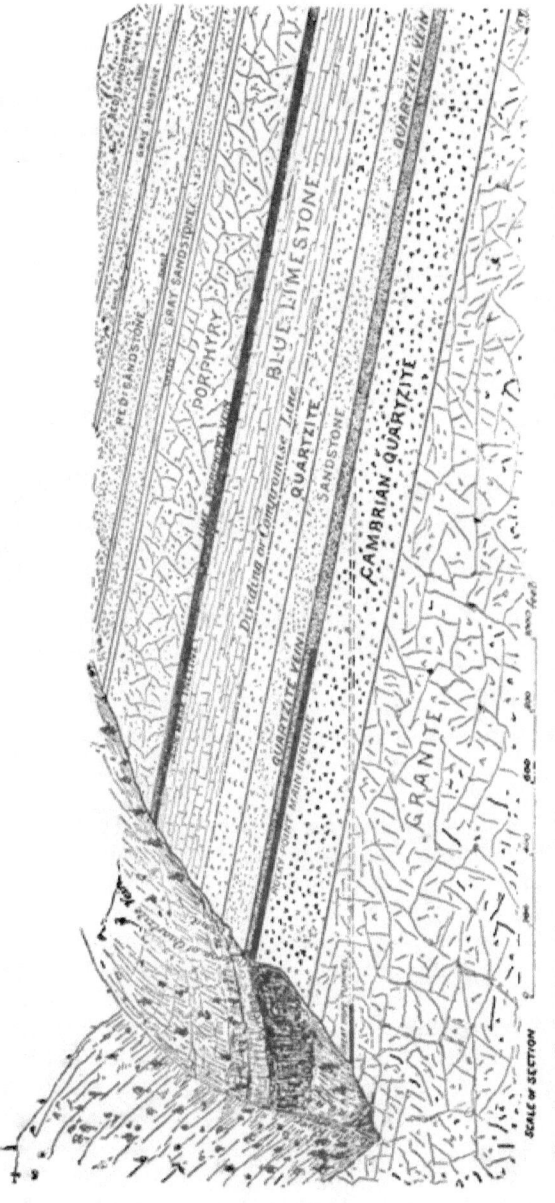

Fig. 47.—Geological section at the Eagle River mines, Colo. After E. E. Olcott, Engineering and Mining Journal, May 21, 1892, p. 544.

lead-silver ores, largely oxidized, occurring with much barite, chiefly along the contact between extensively faulted, Lower Carboniferous, blue limestone and a brown, dolomitized, underlying portion of the same; but also in fissures and less regular deposits in these and older limestones and quartzite. Aspen is on the western slope of the Continental Divide, in the valley of the Roaring Fork, just at the point where it crosses the contact of crystalline Archæan gneisses and Paleozoic sediments. The stream cuts them at right angles to the strike. Aspen Mountain lies on the south side and Smuggler Mountain on the north. The limestone belt continues north and south, and is prospected over a stretch of nearly forty miles. At Aspen there is evidence of a faulted, syn-

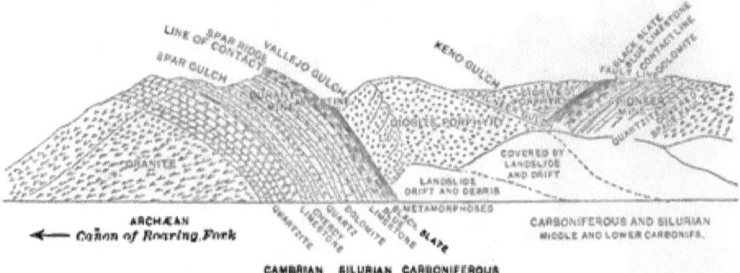

FIG. 48.—*Geological section at Aspen, Colo. After A. Lakes, Ann. Rep. Colo. School of Mines, 1886.*

clinal fold, with many minor disturbances. The westerly dipping rocks by the faulting are repeated to the west and are pierced by a great granite intrusion and much porphyry. Still farther west the Red Jura-Trias sandstones are in great force. The faulted repetitions of the Paleozoic rocks are eroded into a narrow ridge, between Castle Creek and the Roaring Fork, just below the town. Over beyond Aspen Mountain, and to the south, lies Tourtelotte Park, in a small synclinal basin of the limestones, and eight or ten miles farther is Ashcroft. The dips in Tourtelotte Park are low, but they increase going down the mountain toward Aspen, and are steepest of all at its foot, where the strata at 60° run under the stream gravels and glacial deposits. The geology when closely viewed is very complicated, and involves the following sections according to D. W. Brunton. (See papers of W. E. Newberry and S. F. Emmons, cited on p. 191.)

Brunton.

Archæan.

1. Cambrian quartzite, 400 feet.
2. Silurian quartzite and limestone, 460 feet.
3. Lower Carboniferous dolomite, 225 feet.
4. Lower Carboniferous blue limestone, 110 feet.
5. Middle Carboniferous (Weber) shales, 50 to 450 feet.
6. Intruded diorite, maximum 400 feet.
7. Middle Carboniferous limestone, 10 to 160 feet.
8. Jura-Trias sandstone.

Emmons.

Archæan.

1. White quartzite of Upper Cambrian, 200 feet.
2. Silurian limestone and sandstone, 340 feet.
3. Lower Carboniferous brown and blue limestone, 240 feet.
4. Middle Carboniferous clays (Weber shales), 425 feet.
5. Middle Carboniferous green and red sandstone, of the Weber grits, with thin limestone.
6. Jura-Trias sandstone.

2.08.13. Two other sections by different writers (Lakes and Henrich) have been published; but as fossils are almost unknown, the strata can be divided more or less at will. The blue limestone is certainly Lower Carboniferous, for fossils gathered by J. F. Kemp from the same horizon on Lime Creek, twenty-five miles north, where they are plentiful, were pronounced by authorities in the East to be such.

2.08.14. On Smuggler Mountain the same section is shown, but it is not broken by igneous rocks; and although there is a faulting along planes striking parallel with the beds and cutting the dip at a sharp angle, the geology is less complicated.

2.08.15. On Aspen Mountain the ore bodies favor the contact between the blue limestone and the brown dolomite. The former is a very pure limestone, while the latter contains from 20 to 28% magnesium carbonate. The ore replaces and impregnates the blue limestone, often with very little change in its appearance, but it fills the numerous cracks in the more broken dolomite, coating larger and smaller blocks. The ore occurs also in minor fissures. On Smuggler Mountain the ore especially follows the fissure veins.

To the north the mines at Woody are on a great fissure, according to W. E. Newberry (private communication), which carries ore where not filled by a porphyry dike. They are of recent development, but promise to be rich. Although further systematic study is needed, it is quite clear that the ore bodies of Aspen Mountain have originated by replacement of the blue limestone and by coating the fragments of brown dolomite. The solutions doubtless came up along the fault fissures and selected the contact for the chief point of deposition. The United States Geological Survey has had a party in the region.[1]

2.08.16. Example 30e. Rico, Dolores County. Contact deposits of lead-silver ores, in Carboniferous limestones, along intrusive porphyries. Considerable base bullion has been shipped. There are coals in the vicinity, but the operation of the smelter has been somewhat intermittent. The Newman Hill mines are mentioned under "Silver."[2]

NOTE.—Example 30f will be found after Example 31, which has been inserted for geographical reasons.

2.08.17. Example 31. Red Mountain, Ouray County. Oxidized lead-silver ores passing in depth into sulphides, in large and small cavities, in knobs of silicified andesite. The cavities have a close resemblance to caves, but differ from ordinary caves in not being in limestone. They permeate the mountain in an irregular way, and mark the courses of old hot spring conduits. The andesite is generally altered to a mass of quartz, but the process is thought by Mr. Emmons to have taken place at a considerable depth, and that the quartz is a residual deposit left by the removal of more soluble elements of the andesite. T. B. Comstock regards them as hot spring deposits.[3]

---

[1] D. W. Brunton, "Aspen Mountain: Its Ores and Mode of Occurrence," *Engineering and Mining Journal*, July 14 and 21, 1888, pp. 22, 42. S. F. Emmons, "Preliminary Notes on Aspen," *Proc. Colo. Sci. Soc.*, Vol. II., Part III., p. 251. Rec. C. Henrich, "Notes on the Geology and on Some of the Mines of Aspen Mountain," *M. E.*, XVII. 156. A. Lakes, "Geology of the Aspen Mining Region," *Ann. Rep. Colo. School of Mines*, 1886. W. E. Newberry, "Notes on the Geology of the Aspen Mining District," *M. E.*, June, 1889. Rec. L. D. Silver, "Geology of the Aspen (Colo.) Ore Deposits," *Engineering and Mining Journal*, March 17 and 24, 1888.

[2] M. C. Ihlseng, "Review of the Mining Regions of the San Juan," *Ann. Rep. Colo. School of Mines*, 1885, p. 43.

[3] T. B. Comstock, "Hot Spring Deposits in Red Mountain, Colorado."

SOUTH DAKOTA.

2.08.18. Example 30*f*. Galena (town), in the Black Hills. Contact deposits of galena, in part altered to carbonate, in Carboniferous limestone along intruded porphyries. The ore occurs in the Carboniferous limestone, which overlies the Potsdam sandstone near Deadwood, in the northerly flank of the Black Hills. The principal localities are the towns of Galena and Carbonate. Sheets and dikes of igneous rocks penetrate the limestones and have occasioned the ore deposits. Several smelters have had somewhat desultory campaigns.[1]

MONTANA—IDAHO.

2.08.19. Example 32. Glendale, Beaver Head County. Ore bodies of argentiferous galena, zincblende, copper and iron pyrites, and their oxidation products, occurring parallel with the stratification planes of a blue-gray limestone, of age not yet determined. These deposits constitute the Hecla mines, and are in the southwestern part of the State. They offer some parallel features with those of southeastern Missouri. (Example 23.) They differ from Example 30 in not being associated, so far as known, with igneous rocks.[2]

2.08.20. Example 32*a*. Wood River, Idaho. Bodies of argentiferous galena and alteration products, irregularly distributed in limestone, of age as yet undetermined. Southwestern Idaho is largely formed of granite, southeastern is covered by the immense fissure outpourings of basalt along the Snake River. North of these, and on the flanks of the granite, are slates and limestones, especially on the Wood River. The latter contain the lead-silver ores. They are not in immediate association with igneous rocks, and from published descriptions appear to be somewhat irregu-

---

M. E., XVIII. 261. S. F. Emmons, "Notes on Some Colorado Ore Deposits," *Proc. Colo. Sci. Soc.*, Vol. II., Part II., p. 97. M. C. Ihlseng, "Review of the Mining Interests of the San Juan Region," *Ann. Rep. Colo. School of Mines*, 1885, p. 46. T. E. Schwartz, "The Ore Deposits of Red Mountain, Colorado," *M. E.*, June, 1889; *Proc. Colo. Sci. Soc.*, Vol. III., Part I., p. 77.

[1] F. R. Carpenter, "Ore Deposits of the Black Hills of Dakota," *M. E.*, 1879, New York meeting. See also report by Dr. Carpenter on the geology, etc., of the Black Hills, to the trustees of the Dakota School of Mines, 1888, p. 124. S. F. Emmons, *Tenth Census*, Vol. XIII., p. 91.

[2] S. F. Emmons, *Tenth Census*, Vol. XIII., p. 97.

larly distributed, although possibly connected with fissures. The structural relations with Example 23 may again be referred to. The neighboring slates and granite contain gold and silver veins, which are taken up later on. Several small smelters have been erected in the region, and have been intermittently operated. The country is really in the northern end of the Great Basin.[1]

2.08.21. Example 33. Wickes, Jefferson County, Mont. Fissure veins near the contact of granite and liparite, but cutting both rocks and carrying in a gangue of quartz the ores, galena, zincblende, copper and iron pyrites, and mispickel. The liparite is said by Lindgren to be Cretaceous or Tertiary. Wickes is just south of Helena, and was one of the first places in the West to establish successful concentration. There are two companies, the Helena and the Gregory, both large producers.

2.08.22. Example 34. Cœur d'Alene, Idaho. Galena and very subordinate alteration products, in a mineralized zone having a well-marked quartzite footwall and an impregnated, brecciated hanging of the same rock. The ore is in large chutes, which fill innumerable small fractures in the rock. The mines are in Wardner Cañon, in the Bitter Root Mountains, northern Idaho. The rocks are quartzite and thin beds of schists, much folded along east and west axes. In this way they became faulted and shattered, and in the principal mineral belt afforded an opportunity for the ore to deposit. The gangue is siderite. The mines are extremely productive and are the chief sources of ore supply for lead smelters in Montana and on the Pacific coast.[2]

THE REGION OF THE GREAT BASIN.

UTAH.

2.08.23. Example 35. Bingham and Big and Little Cottonwood Cañons, Utah. Bed veins, often of great size, containing oxidized lead-silver ores above and galena and pyrite below the water level, in Carboniferous limestones, or underlying quartzite, or on the contact between the two. The mines are situated in the Oquirrh and Wasatch Mountains, southwest and southeast of Salt Lake City, in cañons well up toward the summits. The region is

---

[1] G. F. Becker, *Tenth Census*, Vol. XIII., p. 55. *Engineering and Mining Journal*, July 2, 1887, p. 2. *Rep. Director of the Mint*, 1882, p. 198.

[2] J. E. Clayton, "The Cœur d'Alêne Silver-lead Mines," *Engineering and Mining Journal*, Feb. 11, 1888, p. 108.

FIG. 49.—View of the Bunker Hill and Sullivan Mines, Wardner, Idaho. The vein dips toward the observer, and outcrops on each hill, right and left. From a photograph furnished by E. E. Olcott.

much disturbed, and there are great faults and porphyry dikes and knobs of granite associated with the sedimentary rocks. The ores occur in belts, extending considerable distances, and these in places have the rich chutes or chimneys of oxidized products. In Bingham Cañon an immense bed of auriferous quartz is found, overlying the lead zone and next the hanging. Some peculiarity about the gold prevents its easy treatment, but much of the rock is very low grade. Other fissure veins in the massive rock of the region are known, but are of less importance. The general geological relations suggest the deposits mentioned under Example 30 and subtypes. The mines were the occasion of the first development of the lead-silver smelters in the West, and have made Salt Lake City an important center of the industry. The Telegraph group, the Emma, Flagstaff, and others were famous mines in their day. As will appear, nearly all the Utah mines are productive of lead-silver ores.

2.08.24. Example 35*a*. Tooele County. Bedded veins in limestone, or between it and quartzite, and containing lead-silver ores with ochers, in rich chutes. The deposits occur in the west side of the Oquirrh range, in Ophir and Dry cañons, over the divide from Bingham. The principal mine is the Honorine. Fissure veins also occur in the region, but are of less importance. The Deep Creek district, near the Nevada line, is mentioned under 2.11.04.[1]

2.08.25. Example 35*b*. Tintic District. Ore beds or belts, three in number, and one to three miles long, generally parallel with the stratification of vertical blue limestones, but sometimes running across them. The ore-bearing zone is from 300 to 600 feet wide in at least one belt, and bears in places rich chutes of carbonate ore. The Crismon-Mammoth has been referred to under

---

[1] W. P. Blake, "Brief Description of the Emma Mine," *Amer. Jour. Sci.*, ii., II. 216. C. E. Fenner, "The Telegraph Mine." *School of Mines Quarterly*, July, 1893. O. J. Hollister, "Gold and Silver Mining in Utah," *M. E.*, XVI. 3. Rec. D. B. Huntley, *Tenth Census*, Vol. XIII., p. 407. G. Lavagnino, "The Old Telegraph Mine," *M. E.*, XVI. 25. "Little Cottonwood and Bingham, Utah," *Engineering and Mining Journal*, Aug. 14, 1880, p. 106; also July 19, 1889. J. S. Newberry, *School of Mines Quarterly*, 1884, p. 329. R. W. Raymond, *Mineral Resources West of the Rocky Mountains*, 1868-76, and J. R. Brown, *Ibid.*, 1867-68. *Ann. Reps. of Director of the Mint.* B. Silliman, "Geological and Mineralogical Notes on Some Mining Districts of Utah," *Amer. Jour. Sci.*, iii., III. 195.

"Copper" (Example 20h), as it contains much copper. The ore is thought by Hollister to have replaced the limestone.[1]

Passing mention should also be made that lead-silver ores occur in Summit County, at the Crescent and other mines.

2.08.26. Example 30g. Horn Silver Mine, Beaver County. A great contact fissure between a rhyolite hanging wall and a limestone footwall, and carrying, at the Horn Silver mine, oxidized lead-silver ores, chiefly anglesite, with considerable barite, and with many other rarer minerals. The town of Frisco, containing the mine, is at the southern end of the Grampian Mountains. The great fissure is known for two miles, but is proved valuable only between the lines of the Horn Silver mine. It strikes north and south and dips 70° east. In the neighborhood of the vein the rhyolite is largely altered to residual clay. The mine is very dry and the entire region lacks good water. The vein in general varies from 20 to 60 feet, but has pinched twice in going down, and of late years has largely ceased producing, although there may yet be much ore below. The ores are smelted near Salt Lake, and the base bullion is refined at Chicago. Some free milling-ore has been afforded.[2]

2.08.27. Example 33a. Carbonate Mine, Beaver County. A fissure vein in hornblende andesite, filled with rounded fragments of wall rock, which are cemented by residual clay and galena. Some oxidized products occur near the surface. The mines are two and a half miles northeast of Frisco, but are in a different eruptive rock from that forming the walls of the Horn Silver. The literature is the same as for Example 30g, especially Hooker, l. c. p. 470.

2.08.28. Example 32b. Cave Mine, Beaver County. Chambers irregularly distributed in the limestone, and more or less filled with limonite and oxidized lead-silver ores. Small leaders of ores, which mark old conduits, connect the chambers. Up to 1880 five large and fifteen small chambers had been found. They are of very irregular shape, and have a vacant space of from one to ten feet between the ore and the roof. This deposit was the typical one cited by Newberry as illustrating the chamber or cave

---

[1] D. B. Huntley, as above (footnote, p. 134); also O. J. Hollister, as above, under Example 35a. J. S. Newberry, *Engineering and Mining Journal*, Sept. 13 and 20, 1879.

[2] O. J. Hollister, "Gold and Silver Mining in Utah," *M. E.*, XVI. 3. Rec. W. A. Hooker, Report quoted in the *Tenth Census*, Vol. XIII., p. 464.

form of deposit. According to this view, the chambers were formed before the ore was brought in. It is also possible that the ore bodies have been deposited by replacement of the limestone with sulphides, as is known abundantly elsewhere, and that the alteration of these to oxides has occasioned the apparent caves. The products of the mine afford but 5 to 7¾ lead, but are valuable as an iron flux to the neighboring smelters. The mines are in the Granite range, seven miles southeast of Milford.[1]

NOTE.—Although the larger part of the Utah mines are for lead and silver, several others of great importance will be taken up under "Silver" itself.

NEVADA.

2.08.29. Example 36. Eureka. Bodies of oxidized lead-silver ores in much faulted and fractured Cambrian limestone, with great outbreaks of eruptive rocks near. The Eureka geological section is one of the most interesting in the entire country, and involves some 30,000 feet of Paleozoic strata, divided as follows: Cambrian quartzite, limestone, and shale, 7700 feet; Silurian limestone and quartzite, 5000 feet; Devonian limestone and shale, 8000 feet; Carboniferous quartzite, limestone, and conglomerate, 9300 feet. These have afforded some extremely valuable materials for comparative studies with homotaxial strata in the East. The ore occurs especially in what is called the Prospect Mountain limestone of the Cambrian, one smaller deposit being also known in Silurian quartzite. The limestone has been crushed and shattered along a great fault, and through its substance ore solutions have circulated, replacing it in part with large bodies of sulphides which have afterward become oxidized to a depth of 1000 feet. The ore bodies were puzzling as regards their classification, and a famous mining suit, with many interpretations from various experts, resulted. The alteration of the ore has caused shrinkage and the formation of apparent caves over it. But there are many empty caves, formed by surface waters long after the ore was deposited, and Mr. Curtis very clearly shows that the ore bodies originated by replacement. All are connected

---

[1] O. J. Hollister, "Gold and Silver Mining in Utah," *M. E.*, 1887. D. B. Huntley, *Tenth Census*, Vol. XIII., p. 474. J. S. Newberry, *School of Mines Quarterly*, March, 1880. Reprint, p. 9. Cf. also J. P. Kimball, "The Silver Mines of Santa Eulalia, Chihuahua," *Amer. Jour. Sci.*, ii., XLIX. 161.

with more or less strongly marked fissures which formed the conduits. Mr. Curtis made a careful series of assays of the neighboring igneous rocks to find some indication of the source of the ore. A quartz porphyry gave significant results and to this the metals are referred, but the portions of the mass at a great

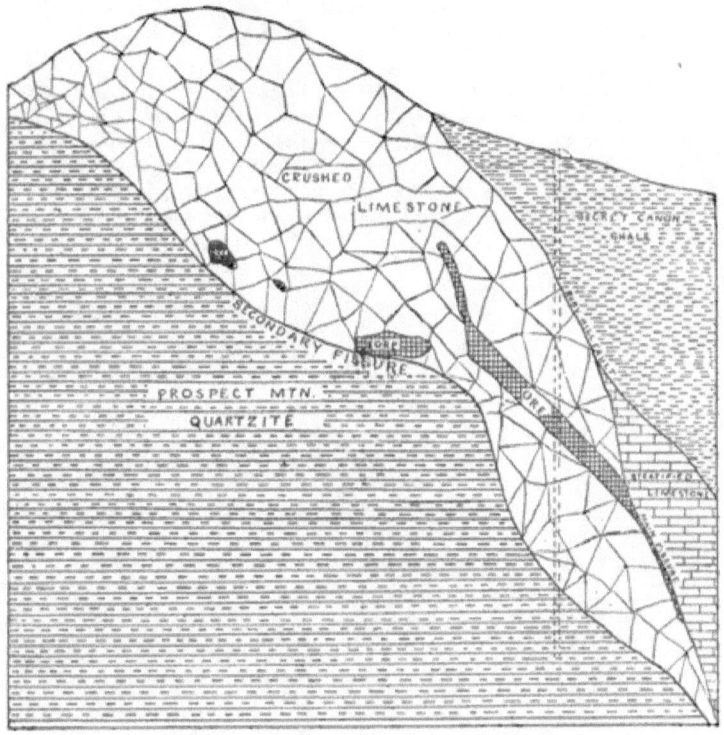

FIG. 50.—*Section at Eureka, Nev. Reproduced in line work after colored plate by J. S. Curtis, Monograph VI., U. S. Geol. Survey.*

depth are considered to have furnished them. Eureka was one of the first places in this country where the hypothesis of replacement was applied to ores in limestone. The district is now far less productive than it was ten or fifteen years ago.[1]

---

[1] G. F. Becker, *Tenth Census*, Vol. XIII., p. 32. Rec. W. P. Blake, "The Ore Deposits of the Eureka District, Nevada," *M. E.*, VI. 554. J. S. Curtis, "Silver-lead Ore Deposits of Eureka, Nev.," *Monograph VII*,

ARIZONA—CALIFORNIA.

2.08.30. Some $98,000 worth of lead ores were shipped from Arizona in 1889, chiefly from Cochise County (Tombstone region) and Pima County (Tucson region). They will be mentioned under "Silver." Insignificant amounts are also afforded by California (about $2000 in 1889), mostly from Inyo County. (See *Eleventh Census, Bull. No.* 80, June 18, 1891.)

U. S. Geol. Survey. A. Hague, "Geology of the Eureka District," *Monograph XX., U. S. Geol. Survey.* Abstract in *Third Ann. Rep. Director U. S. Geol. Survey.* W. S. Keyes, "Eureka Lode of Eureka, Nev.," *M. E.*, VI. 344. J. S. Newberry, *School of Mines Quarterly*, March, 1880. R. W. Raymond, "The Eureka-Richmond Case," *M. E.*, VI. 371. C. D. Walcott, "Paleontology of the Eureka District," *Monograph VIII., U. S. Geol. Survey.*

CHAPTER IX.

SILVER AND GOLD.—INTRODUCTORY: EASTERN SILVER MINES AND THE ROCKY MOUNTAIN REGION OF NEW MEXICO AND COLORADO.

2.09.01. The two "precious" metals are so generally associated that they cannot be separately treated. While endeavoring to preserve the distinctive impression given by examples, it is practically impossible to set forth all the widely varying phenomena of the silver-gold veins of the West in any other than an approximate way. Hence geographical considerations are placed first, and where markedly similar ore bodies in different States are to be grouped together cross references are given. The following general examples have been made because their individual features are based on those geological relations which are most vitally concerned with questions of origin.

2.09.02. Example 37. Veins containing the precious metals usually with pyrite, galena, chalcopyrite, and less common sulphides, sulpharsenides, sulphantimonides, etc., in igneous rocks. No special subdivision is made on the character of the gangue, which may be quartz, calcite, barite, fluorite, etc., one or all. The first named is commonest. A great and well-defined original fissure is not necessarily assumed, but some crack, or joint, or crushed strip must have directed the ore-bearing solutions, which may have then replaced the walls in large measure. For other structural features see the discussion of veins (1.05.01); compare also Example 17, Butte, Mont.

Example 37a. Replacements more or less complete of igneous dikes, which have usually been described as porphyry. Compare Example 17a under "Copper" (Gilpin County, Colorado), and Example 20d (Santa Rita, N. M.). Ore and gangue (where the matrix is not the dike rock) as in Example 37.

Example 38. Contact deposits between two kinds of igneous rock or between two different flows. Ore and gangue as in Example 37.

Example 39. Agglomerates of rounded, eruptive boulders, bombs, etc., in abandoned volcanic necks or conduits, and coated with ores. The mines of Custer County, Colorado, are the only examples of ore deposits of this kind yet identified.

Example 40. Contact deposits between igneous and sedimentary rocks. No subdivisions are made on the kind of rocks. Ore and gangue as in Example 37. Compare also Example 20, "Arizona-Copper;" Example 21a, "Triassic Copper;" Example 30, "Leadville;" and Example 30g, "Horn Silver Mine."

Example 41. Veins in sedimentary rocks, generally cutting the bedding, but at times parallel with it. Lateral enlargements are frequent. The ore body may be largely due to replacement. Ore and gangue as in Example 37.

Example 42. Veins cutting both sedimentary and igneous rocks, and therefore due to disturbances after the intrusion of the latter. Ore and gangue as in Example 37.

No special examples are made for metamorphic rocks.

2.09.03. Silver minerals.

	$Ag.$	$S.$	$As.$	$Sb.$	$Cl.$
Native silver............................	100.				
Argenite (silver glance), Ag_2S............	87.1	12.9			
Prousite (light ruby silver), $3Ag_2S.As_2S_3$..	65.5	19.4	15.1		
Pyrargerite (dark ruby silver), $3Ag_2S.Sb_2S_3$.	59.8	17.7		22.5	
Stephanite (brittle silver), $5Ag_2S.Sb_2S_3$....	68.5	16.2		15.3	
Cerargerite (horn silver), $AgCl$............	75.3				24.7

Silver also occurs with galena (Cf. "Lead") and with tetrahedrite (Cf. "Copper"). Gold occurs combined with tellurium in a few rare tellurides, mechanically mingled with pyrites, and as the uncombined native metal. From a metallurgical point of view the ores of the precious metals are divided into two classes. 1. Those whose amount of precious metal amalgamates readily with mercury, and is thus obtained with comparative ease—the free milling ores. 2. Those which require roasting or some previous treatment before amalgamation, chlorination, or similar process, or which must be smelted primarily for lead or copper, from which the precious metals are afterward extracted—the rebellious ores. In the subsequent description the endeavor has been made to work from the distinctively silver mines to those of gold where geographically possible.[1]

---

[1] *Ann. Reps. Directors of the Mint.* Rec. W. P. Blake, "The Vari-

2.09.04. Example 22a. Atlantic Border. Already mentioned (2.05.02), the region is only of historical interest as affording silver, although lately some attention has been directed to Sullivan, Me., where the veins have pyrite and probably stephanite, in a quartz gangue, in slates, associated with granite knobs and trap dikes which are of later age than the veins. Some silver is generally found in the galena of the Eastern States, but the ores have never yet proved abundant enough to be important.[1]

Mention may also be made at this point of the argentiferous galena veins along the Ouachita uplift of Arkansas. A few are known, usually with Trenton shales or slates for walls. They are low grade, and though once the basis of a small excitement, their production has never been serious. Additional reference to the region will be found under "Antimony." Some mines of the latter metal are stated by W. P. Jenney to show low-grade, argentiferous ores in depth.[2]

2.09.05. Example 42. Silver Islet, Lake Superior. A fissure vein carrying native silver, argentite, tetrahedrite, galena, blende, and some nickel and cobalt compounds in a gangue of calcite, in flags and shales of the Animikie (Cambrian) system, and cutting a large trap dike, within which alone the vein is productive. Silver Islet is or was originally little more than a bare rock some

---

ous Forms in which Gold Occurs in Nature," *Rep. Director of the Mint*, 1884, p. 573. Rec. Brown, Raymond, and others, 1868 to 1876, "Mineral Resources West of the Rocky Mountains." Annual. T. C. Chamberlain, "On the Geological Distribution of Argentiferous Galena," *Geol. of Wis.*, Vol. IV. Clarence King, "Production of the Precious Metals in the United States," *Second Ann. Rep. Director U. S. Geol. Survey*, p. 333. A. G. Lock, *Gold*, 1882. *Mineral Resources of the U. S.;* annual publication of the Geological Survey. R. I. Murchison, "General View of the Conditions under which Gold is Distributed," *Quar. Jour. Geol. Soc.*, VII. 134. Also in *Siluria* and *Amer. Jour. Sci.*, ii., XVIII. 301. J. S. Newberry, "On the Genesis and Distribution of Gold," *School of Mines Quarterly*, III., No. 1, and *Engineering and Mining Journal*, Dec. 24 and 31, 1881, pp. 416, 437. R. Pearce, "On the Ores of Gold," etc., *Colo. Sci. Soc.*, III., p. 237. J. A. Phillips, *Ore Deposits*, 1884. *The Mining and Metallurgy of Gold and Silver*, 1867. *Tenth Census Report on the Precious Metals*.

[1] C. W. Kempton, "Sketches of the New Mining District at Sullivan, Me.," *M. E.*, VII. 349. M. E. Wadsworth, "Theories of Ore Deposits," *Proc. Boston Soc. Nat. Hist.*, 1884, p. 205. *Engineering and Mining Journal*, May 17, 1884. *Bull. Mus. Comp. Zoöl.*, 3, Vol. VII., 181.

[2] T. B. Comstock, *Ann. Rep. Geol. Survey of Arkansas*, 1888, Vol. I., "Gold and Silver."

90 feet square, lying off the north shore of Lake Superior just outside of Thunder Bay, and within the Canadian boundaries. The native silver was detected outcropping beneath the water. The vein was productive to a depth of 800 or 1000 feet, but below this it yielded little. The trap dike has usually been called diorite, but is determined to be norite by Wadsworth (*Bull.* 2, *Minn. Geol. Survey*, p. 92), and gabbro by Irving (*Monograph* V., *U. S. Geol. Survey*, p. 378). Some $3,000,000 was obtained from the mine, yet the expenses were so great in keeping up the surface works against winter gales and ice that but little profit was realized. The vein has been traced 9000 feet but is nowhere else productive. Considerable graphite has been found in the workings, and some curious pockets of gas.[1]

2.09.06. Example 42. Thunder Bay, Canada. The mainland near Silver Islet contains many similar veins. They have furnished considerable silver as argentite in a gangue of quartz, barite, calcite, and fluorite, and associated with zincblende, galena, and pyrite.[2]

THE REGION OF THE ROCKY MOUNTAINS AND BLACK HILLS.

NEW MEXICO.

2.09.07. *Geology.*—The general topography and geology of New Mexico were outlined in the introduction. Much remains to be done in developing its geology. The eastern part belongs to the prairie region and is very dry. A few rivers, notably the Pecos and the Rio Grande, afford water for irrigation, the former of which is now being utilized on a grand scale, and for the latter plans have been prepared. In the central portion many subordi-

---

[1] R. Bell, *Engineering and Mining Journal*, Jan. 8 and 15, 1887. See also May 14, 1887. W. M. Courtis, "On Silver Islet," *Engineering and Mining Journal*, Dec. 21, 1873, and *M. E.*, V. 474. E. D. Ingall, *Ann. Rep. Can. Geol. Survey*, 1887-88, Part II., p. 14. F. A. Lowe, "The Silver Islet Mine and its Present Development," *Engineering and Mining Journal*, Dec. 16, 1882, p. 321. T. MacFarlane, "Silver Islet," *M. E.*, VIII. 226. *Geol. of Canada*, 1863, 717. *Canadian Naturalist*, Vol. IV., p. 37. McDermott, *Engineering and Mining Journal*, Vol. XXIII., Nos. 4 and 5.

[2] R. Bell, "Silver Mines of Thunder Bay," *Engineering and Mining Journal*, Jan. 8 and 15, 1887. E. D. Ingall, *Ann. Rep. Can. Survey*, 1887-88, Part II., p. 1H. Rec. See also *Engineering and Mining Journal*, May 14, 1887; Feb. 18, 1888, p. 123; May 26, 1888, p. 383.

nate north and south ranges of mountains are found, which are less elevated than those of Colorado. The Colorado ranges virtually die out at the northern boundary. The northwestern portion comes in the great Colorado Plateau, and has been quite fully described by Captain Dutton (*Eighth Ann. Rep. Director U. S. Geol. Survey*). In numerous localities throughout the Territory volcanic action has been rife and in places is but recently extinct. The eastern part is largely Cretaceous, and also the northeastern plateau, which contains much valuable coal. The mountain ranges often have nuclei of Archæan crystalline rocks, with successive strata of Carboniferous, Permian, Triassic, Jurassic, and Cretaceous on the flanks. The mining regions are in these ranges of mountains.[1]

2.09.08. The southwestern county is Grant, whose lead-silver deposits have been briefly referred to. North of Silver City are quartz veins of gold and silver ores, in diabase and quartz porphyry (Example 37), and again, west of Silver City, are ferruginous deposits with chlorides and sulphides of silver in limestone. In the Burro Mountains are silver ores in limestones, apparently Lower Silurian. The Santa Rita Mountains contain, in addition to the copper (Example 20*d*), silver and gold in quartz veins in

---

[1] W. P. Blake, *Proc. Bost. Soc. Nat. Hist.*, 1859, Vol. VII., p. 64 "Geology of the Rocky Mountains in the Vicinity of Santa Fé," *A. A. A. S.*, 1859. A. R. Conkling, "Report on Certain Foothills in Northern New Mexico," *Wheeler's Survey, Rep. of Chief of U. S. Engineers*, 1877, II. 1298. E. D. Cope, "Report on the Geology of a Part of New Mexico," *Wheeler's Survey*, 1875; Appendix G1. C. E. Dutton, "Mount Taylor and the Zuni Plateau," *Sixth Ann. Rep. U. S. Geol. Survey*, pp. 111-205. S. F. Emmons, *Tenth Census*, Vol. XIII. 100. O. Loew, "Report on the Geology and Mineralogy of Colorado and New Mexico," *Wheeler's Survey*, 1875; Appendix G2, p. 27. J. Marcou, "The Mesozoic Series of New Mexico," *Amer. Geol.*, IV. 155, 216. R. E. Owen and E. J. Cox, "Report on the Mines of New Mexico," Washington, 1865, 60 pp., *Amer. Jour. Sci.*, ii., XL. 391. G. F. Runton, "On the Volcanic Rocks of New Mexico," *Quar. Jour. Geol. Soc.*, Vol. VI., p. 251, 1850. B. Silliman, Jr., "The Mineral Regions of Southern New Mexico," *M. E.*, X. 424. F. Springer, "Occurrence of the Lower Burlington Limestone in New Mexico," *Amer. Jour. Sci.*, iii., XXVII. 97. J. J. Stevenson, "Geological Examinations in Southern Colorado and Northern New Mexico," *Wheeler's Survey*, 1881. "Geology of Galisteo Creek," *Amer. Jour. Sci.*, iii., XVIII. 471. "On the Laramie Group of Southern New Mexico," *Amer. Jour. Sci.*, iii., XXII. 370.

eruptive rocks (Example 37). Lake Valley, in Doña Aña County, has been mentioned (2.08.04). In Lincoln County gold ores are reported from the White Oak district. The principal mines of Socorro County have been mentioned (Example 29), and the copper in Permian sandstone under Example 21c. There are other silver-bearing lodes in the Socorro Mountains near the town of Socorro. Henrich has described (l. c.) a curious deposit of quartz carrying gold and silver (the Slayback Lode) on the contact between the older bedded eruptions and a later siliceous dike in the Mogollon range (Example 38). In Santa Fé County are important placer mines (Example 44) and thin veins of galena in rhyolite. In Bernalillo County are placers on the slopes of the Sandia Mountains. In Colfax County, in the Rocky Mountains, are other placers, and reported gold and silver mines.¹

COLORADO.

2.09.09. *Geology.*—The eastern portion contains prairies and is a region lacking water. It consist of Quaternary and Cretaceous rocks. The plains rise in the foothills, which are chiefly upturned Jura-Triassic and Cretaceous strata. The Paleozoic is relatively limited, although known. It rests on the crystalline rocks of the Archæan. There are some minor uplifts, running out at right angles to the Front range, that divide the foothill country into basins, and are especially important in connection with coal. Next come the easterly ranges of the Rocky Mountains, in linear north and south succession. They consist largely of dome-shaped peaks of granite, with great local developments of volcanic rocks. To the west follow the several parks, largely consisting of Mesozoic strata. They are bounded by ranges again on the west, some of which, like the Mosquito range (see under Example 30), mark great lines of post-Cretaceous upheaval, and are accompanied by immense igneous intrusions. On the east and west flanks of the Sawatch range (the granitic Continental Divide) are Paleozoic

---

[1] W. P. Blake, "Gold in New Mexico," *Proc. Bost. Soc. Nat. Hist.*, VII., p. 16, July, 1859. "Observations on the Geology, etc., near Santa Fé," *A. A. A. S.*, X. 1859. S. F Emmons, *Tenth Census*, XIII., p. 101. C. Henrich, "The Slayback Lode, New Mexico," *Engineering and Mining Journal*, July 13, 1889, p. 27. R. E. Owen and E. T. Cox, *Rep. on the Mines of New Mexico*, Washington, 1865. *Rep. Director of the Mint*, 1882, p. 339. B. Silliman, "Mineral Resources of Southern New Mexico," *M. E.*, X. 424. *Engineering and Mining Journal*, Oct. 14 and 21, 1882, pp. 199, 212.

strata in considerable thickness, but to the west they dip under the vastly greater development of Mesozoic terranes which shade out into the Colorado Plateau. In northern, central, and southwestern Colorado are vast developments of igneous rocks that have attended the geological disturbances.[1]

2.09.10. The San Juan region includes several counties in southwestern Colorado, in whole or in part, viz.: Ouray, Hinsdale, San Juan, Dolores, and La Plata. The chain of the San Juan Mountains consists of great successive outflows of eruptive rocks, porphyry, diabase, diorite, basalt, etc., which cover up the Archæan and later sedimentary terranes, except in a few scattered ex-

---

[1] G. L. Cannon, "Quaternary of the Denver Basin," *Proc. Colo. Sci. Soc.*, III. 48. See also III. 200. W. Cross, "The Denver Tertiary Formation," *Amer. Jour. Sci.*, iii., XXXVII. 261. G. H. Eldredge, "On the Country about Denver, Colo.," *Proc. Colo. Sci. Soc.*, III. 86. See also 140. S. F. Emmons, "Orographic Movements in the Rocky Mountains," *Geol. Soc. of America*, I. 245-286. F. M. Endlich, "On the Eruptive Rocks of Colorado," *Tenth Ann. Rep., Hayden's Survey*. H. Gannett, "Report on the Arable and Pasture Lands of Colorado," *Hayden's Survey*, 1876, p. 313. H. C. Freeman, "The La Platte Mountains," *M. E.*, XIII. 681. G. K. Gilbert, "Colorado Plateau Province as a Field for Geological Study," *Amer. Jour. Sci.*, iii., XII. 16, 85. J. D. Hague, *Fortieth Parallel Survey*, Vol. III., p. 475. F. V. Hayden, *Reports of Hayden's Survey*, 1873, 1874, p. 40; 1875, p. 33; 1876, pp. 5, 70. R. C. Hills, "Preliminary Notes on the Eruptions of the Spanish Peaks," *Proc. Colo. Sci. Soc.*, III. 24, p. 224. "The Recently Discovered Tertiary Beds of the Huerfano River Basin," *Proc. Colo. Sci. Soc.*, III., pp. 148, 217. "Jura-Trias of Southeastern Colorado," *Amer. Jour. Sci.*, iii., XXIII., p. 243. A. Lakes, "Extinct Volcanoes in Colorado," *Amer. Geol.*, January, 1890, p. 38. Oscar Loew, "Report on the Minerals of Colorado and New Mexico," *Wheeler's Survey*, 1875, p. 97. "Eruptive Rocks of Colorado," *Wheeler's Survey*, 1873. C. A. H. McCauley, "On the San Juan Region," *Rep. Chief of U. S. Engineers*, 1878, III., p. 1753. C. S. Palmer, "On the Eruptive Rocks of Boulder County," etc., *Proc. Colo. Sci. Soc.*, III., p. 230. A. C. Peale, "On the Age of the Rocky Mountains in Colorado," *Amer. Jour. Sci.*, iii., XIII., p. 172; Reply to the above by J. J. Stevenson, *Amer. Jour. Sci.*, iii, XIII. 297. S. H. Scudder, "The Tertiary Lake Basin at Florissant," *Hayden's Survey*, 1878, p. 271; see also 1877. J. A. Smith, *Catalogue of the Principal Minerals of Colorado*, Central City, 1870. J. J. Stevenson, "Notes on the Laramie Group of Southern Colorado," *Amer. Jour. Sci.*, iii., XVIII. 129. "The Mesozoic Rocks of Southern Colorado," *Amer. Geol.*, III., p. 391. P. H. Van Diest, "Colorado Volcanic Cones," *Proc. Colo. Sci. Soc.*, III., p. 19. C. A. White, "On Northwestern Colorado," *Ninth Ann. Rep. Director U. S. Geol. Survey*, 683-710.

posures. Considerable masses of rocks formed of fragmental ejectamenta are also known. All these are crossed by immense vertical veins, largely with quartz gangue, and containing argentiferous minerals of the usual species, galena, tetrahedrite, pyrargerite, and native silver, as well as bismuth compounds. Gold has been quite subordinate, although late developments near Ouray have shown some peculiar and interesting deposits. R. C. Hills, in the *Proc. Colo. Sci. Soc.*, 1883, traced three systems of veins. (1) Silver-bearing, narrow (six inches to three feet), nearly vertical veins, with base metal ores and no selvage. (2) Large, strong, gold-bearing veins dipping 60° with selvages and intersecting (1). (3) Like (1), but larger and more persistent, and carrying occasionally bismuth and antimonial ores with gold and little or no silver. T. B. Comstock (*M. E*, XV. 218) has classified the veins in three radiating systems. (1) The northwest, with tetrahedrite (freibergite). (2) The east and west, with bismuth and less often nickel and molybdenum. (3) The northeast, with tellurides and antimony and sulphur compounds of the precious metals. Quite recently a series of small caves near Ouray, in quartzite overlaid by bituminous shales, have been found to contain native gold, and have excited great interest. It is thought by Endlich that they represent inclusions of shale, now dissolved away, and that the gold was precipitated on the walls. If this view is correct, they mark one of the very few illustrations of chamber deposits which are known. More extended mining work has proved them to be in all cases connected with a supply fissure from which small leaders guide the miners to the chambers.

Placer gold mines (Example 44) are quite extensively worked in San Miguel County. J. B. Farish has recently described the veins at Newman Hill, near Rico, in a valuable paper cited below. The lowest formation exposed is magnesian limestone, supposed to be Carboniferous. It contains large ore bodies of low grade, and is also, strangely enough, heavily charged with carbonic acid gas. Above this for 500 feet are alternating sandstones and shales, and then a narrow stratum of limestone 18 to 30 inches thick. This is followed by about 500 feet additional of shales and sandstone, regarded as Carboniferous. Fifty feet above the lowest limestone a laccolite of porphyrite has been intruded. Two sets of fissures are present—one nearly vertical and striking northeast, the second dipping 30 to 45° northeast and striking northwest. The former are the richest, are banded (see Fig. 5) and persistent, being worked in one case for 4000 feet. The flatter fissures are less rich. The principal ore bodies, however, occur as horizontal enlargements of both these sets of veins. Just over the thin bed of limestone mentioned above the ores have spread out into sheets from 20 to 40 feet wide and from a few inches to three feet thick. They consist of solid masses of the common sulphides, galena, pyrite, gray copper, etc., and are very rich. Above them the fissures apparently cease, or at least are tight. Two hundred feet down from them the vein filling becomes nearly barren, glassy quartz. These are most remarkable ore bodies, and would appear to have been formed by uprising solutions, which met the tight place and spread sidewise, depositing their minerals; but as Mr. Farish advances no explanation, it is hardly justifiable for others, less familiar than himself with the phenomena, to do so.

The lead-silver ores of Red Mountain and Rico have already been mentioned (2.08.17). Silverton and Ouray are the principal towns of the San Juan.[1]

2.09.11. The new mining region of Creede, now decided to be in Saguache County, should be mentioned in this connection. It

---

[1] T. B. Comstock, "The Geology and Vein Structure of Southwestern Colorado," *M. E.*, Vol. XV., 218; also XI. 165, and *Engineering and Mining Journal*, numerous papers in 1885. "Hot Spring Formation in the Red Mountain District, Colorado," *M. E.*, XVII. 261. Rec. S. F. Emmons, "On the San Juan District," *Engineering and Mining Journal*, June 9, 1883, p. 332. "Structural Relations of Ore Deposits," *M. E.*, XVI. 804. Rec. *Tenth Census*, Vol. XIII., p. 60. F. M. Endlich, "Origin of the Gold Deposits near Ouray," *Engineering and Mining Journal*, Oct. 19, 1889. "San Juan District," *Hayden's Survey*, 1874, p. 229. *Ibid.*, 1875, Bull. III.; *Amer. Jour. Sci.*, iii., X. 58. J. B. Farish, "On the Ore Deposits of Newman Hill, near Rico, Colo.," *Colo. Sci. Soc.*, April 4, 1892. Rec. R. C. Hills, *Proc. Colo. Sci. Soc.*, 1883. Rec. W. H. Holmes, "La Plata District," *Hayden's Survey*, 1875; *Amer. Jour. Sci.*, iii., XIV. 420. M. C. Ihlseng, "Review of the Mining Interests of the San Juan Region," *Rep. Colo. State School of Mines*, 1885, p. 27. G. E. Kedzie, "The Bedded Ore Deposits of Red Mountain Mining District, Ouray County, Colorado," *M. E.*, XV. 570. Rec. G. A. Koenig and M. Stocker, "Lustrous Coal and Native Silver in a Vein in Porphyry, Ouray County, Colorado," *M. E.*, IX. 650. T. E. Schwartz, "The Ore Deposits of Red Mountain, Ouray County, Colorado," *M. E.*, 1889. J. J. Stevenson, "On the San Juan," *Wheeler's Survey*, III., p. 376. "The San Juan Region," *Engineering and Mining Journal*, Aug. 27, 1881, p. 136; Sept. 24, 1881, p. 201; July 17, 1880; Dec. 20, 1879; and many other references in 1879 and 1880. P. H. Van Diest, "On the San Juan District," *Proc. Colo. Sci.*, January, 1886.

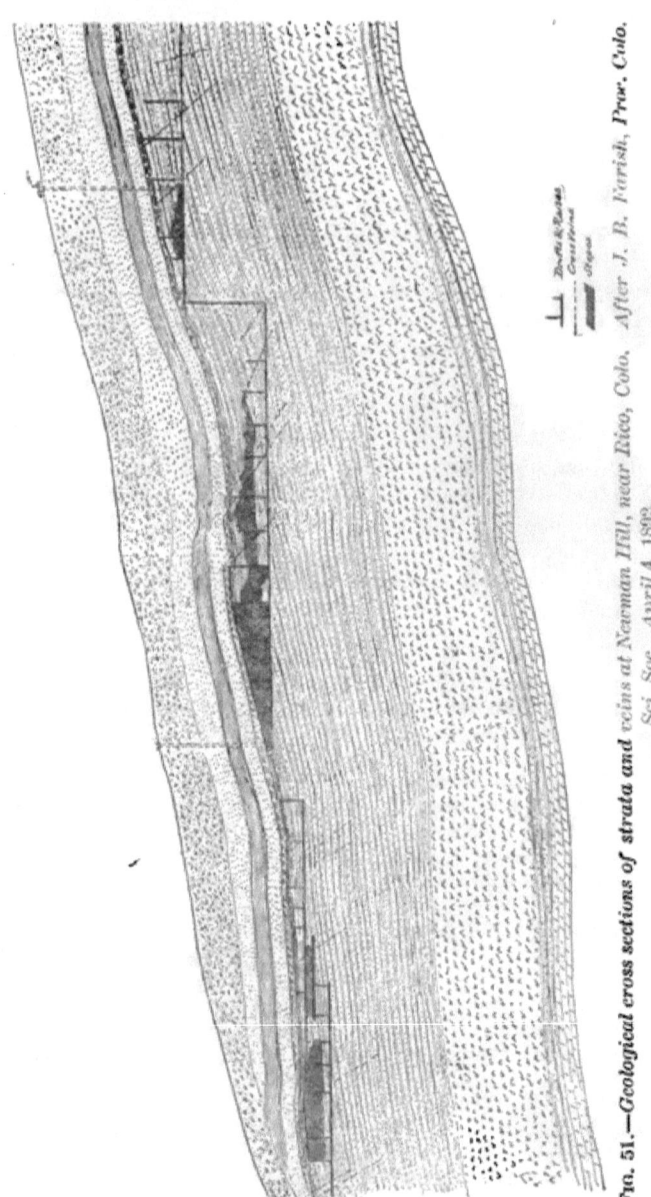

FIG. 51.—*Geological cross sections of strata and veins at Newman Hill, near Rico, Colo.* After J. B. Farrish, Proc. Colo. Sci. Soc., April 4, 1892.

is situated near the junction of Saguache, Ouray, and Hinsdale counties, and some ten or twelve miles from Wagon Wheel Gap. There is a great development of igneous rocks as well as of Carboniferous limestone, but the veins as yet developed are in the former. They appear to be fissure veins and have quartz, in large

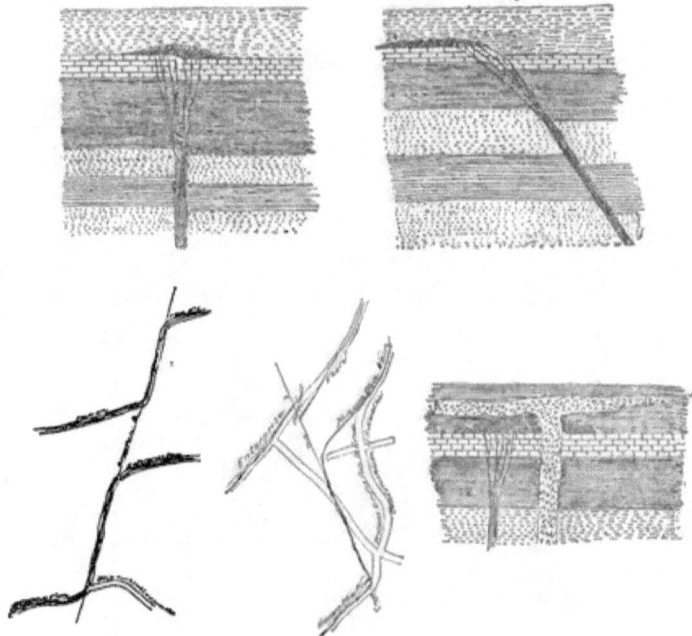

FIG. 52.—*Geological cross sections of strata and veins at Newman Hill, near Rico, Colo. After J. B. Farish, Proc. Colo. Sci. Soc., April 4, 1892. See also Figures 5 and 6.*

part amethyst, with some manganese minerals as a gangue, and, with these, oxidized silver ores. The mines are on two mountains, Bachelor and Campbell, which are on opposite sides of Willow Creek Cañon.[1]

---

[1] E. B. Kirby, "The Ore Deposits of Creede and Their Possibilities," *Engineering and Mining Journal*, March 19, 1892, p. 325. Rec. T. R. MacMechen, "The Ore Deposits of Creede," *Engineering and Mining Journal*, March 12, 1892, p. 301. Rec.

FIG. 63.—*View of Lower Creede, Colo. From the Engineering and Mining Journal, March 19, 1892.*

2.09.12. The Gunnison region lies on the western slope of the Continental Divide and embraces both mountains and plateaus. West of the main and older range are the later Elk Mountains, in which several mining districts are located. Aspen has already been mentioned, and the long series of ore bodies in the Carboniferous limestones. The other principal districts are Independence, Ruby, Gothic, Pitkin, and Tin Cup. The ores at Independence are sulphides with silver, in the Archæan granite rocks. In the Tin Cup district the Gold Cup mine is in a black limestone and contains argentiferous cerussite and copper oxide. In the Ruby district the ores are in the Cretaceous rocks, and in the Forest Queen they are ruby silver and arsenopyrite, partly replacing a porphyry dike. On Copper Creek, near Gothic, a series of nearly vertical fissures traverse eruptive diorite. They contain sulphide of silver and native silver. The Sylvanite is one of the principal mines.[1]

2.09.13. Eagle County. The lead-silver mines of Red Cliff have already been mentioned (Example 30c), and also the underlying gold deposits. The Homestake mine, northwest of Leadville, over toward Red Cliff, is on a vein of galena in granite, and was one of the first openings made in the region.[2]

2.09.14. Summit County. The Ten Mile district, which is the principal one, has been mentioned under Example 30a. The Pride of the West mine, on Jacque Mountain, is peculiar, being on a quartz porphyry dike which is partly replaced by ferruginous quartz and barite. Lake County, containing Leadville, has been treated under Example 30. Mention should also be made of the placer deposits in California Gulch, which first attracted prospectors to the region in 1860. In its eastern part Summit County borders on Clear Creek County, and at Argentine are some veins related to those of the latter. They are high up on Mount McClellan, and are remarkable for the veins of ice that are found in them.[3]

---

[1] F. Amelung, "Sheep Mountain Mines, Gunnison County," *Engineering and Mining Journal*, Aug. 28, 1886, p. 149. F. M. Chadwick, "The Tin Cup Mines, Gunnison County, Colorado," *Engineering and Mining Journal*, Jan. 1, 1881, p. 4. See also Example 12d for iron mine-.

[2] Guiterman, "On the Gold Deposits of Red Cliff," *Proc. Colo. Sci. Soc.*, 1890. "On the Battle Mountain Quartzite Mines," *Mining Industry*, Denver, Jan. 10, 1890, p. 28. E. E. Olcott, "Battle Mountain Mining District, Eagle County," *Engineering and Mining Journal*, June 11 and 18, 1887, pp. 417, 436; May 21, 1892. G. C. Tilden, "Mining Notes from Eagle County," *Ann. Rep. Colo. State School of Mines*, 1886, p. 129.

[3] E. L. Berthoud, "On Rifts of Ice in the Rocks near the Summit of Mount McClellan," etc., *Amer. Jour. Sci.*, iii., II. 108.

2.09.15. Park County, which lies east of Lake County and embraces the South Park, has some mines on the eastern slope of the Mosquito range, and in the Colorado range, to the northwest. The latter are similar in their contents to the Georgetown silver ores, mentioned under Clear Creek County, but the former are bodies of argentiferous galena and its alteration products in limestone and quartzite. Pyrite is also abundant, and at times a gangue of barite appears. The mines are in the sedimentary series, resting on the granite of the Mosquito range, and are pierced by porphyry instrusions, as at Leadville. The placer deposits at Fairplay deserve mention, as it was from these that the prospectors spread over the divide to the site of Leadville in 1860.[1]

2.09.16. Chaffee County, on the south, contains the iron mines referred to under Example 12d. There are some other gold-bearing veins near Granite and Buena Vista. The lead-silver deposits of the Monarch district are mentioned under Example 30b. In Huerfano County, in the Spanish Peaks, veins of galena, gray copper, etc., are worked to some extent.[2]

2.09.17. Rio Grande County. In the Summit district are a number of rich gold mines, of which the Little Annie is the best known. The gold occurs in the native state, in quartz on the contact between a rhyolite and trachyte breccia and andesite. The deposits are thought by R. C. Hills to be due to a silicification of the rhyolite along those lines, probably by the sulphuric acid, which brought the gold. Then the rocks were folded. Oxidation and impoverishment of the upper parts followed, forming bonanzas below. The paper has a very important bearing on the formation of many replacements.[3]

2.09.18. Conejos County. Some deposits of ruby silver ores have recently been developed in this county, near the town of Platoro. The county lies near the middle of the southern tier.

2.09.19. Custer County affords some of the most interesting deposits in the West. Rosita and Silver Cliff are the principal towns and are situated in the Wet Mountain Valley, between the Colorado range on the north and the Sangre de Cristo on the

---

[1] J. L. Jernegan, "Whale Lode of Park County," *M. E.*, III. 352.

[2] R. C. Hills, "On the Eruption of the Spanish Peaks," *Proc. Colo. Sci. Soc.*, III., pp. 24, 224.

[3] R. C. Hills, *Proc. Colo. Sci. Soc.*, March, 1883. Abstract by S. F. Emmons in the *Engineering and Mining Journal*, June 9, 1883, p. 332.

south. At Silver Cliff an outbreak of pinkish rhyolite occurs, impregnated with silver chloride. It affords a free-milling although low-grade ore. This forms a unique deposit. There is a great thickness of tuffs beneath it, as shown in the Geyser mine, and some remarkable forms of spherulitic crystallizations.

2.09.20. Example 39. Bull Domingo and Bassick. The first named is two miles north of Silver Cliff, and the latter seven miles east, near Rosita. The Bull Domingo is in Archæan, hornblende gneiss, and consists of what appear to be pebbles or boulders of the wall rock, which are coated with argentiferous galena and an outer shell of quartz. The ore body is 40 to 60 feet across. The Bassick is in andesite, and likewise consists of what appear to be boulders and pebbles of the country rock, coated by concentric shells of rich ores, and in an elliptical chimney 20 to 100 feet across. The first coat is a mixture of lead, antimony, and zinc sulphides, and is always present. A second, somewhat similar, but of lighter color and richer in lead and the precious metals, is sometimes seen. A third is chiefly zincblende, rich in silver and gold, and is the largest of all. A fourth, of chalcopyrite, sometimes occurs, and lastly a fifth, of pyrite. Various other minerals are found, and, curiously enough, carbonized wood on the outer limits. Both these deposits have been thought to be the tubes of geysers, in which boulders have been tossed about, rounded, and finally cemented together. Mr. Emmons has argued against this view, and in a forthcoming monograph will present the results of extended study. A brief account of these results has, however, been published by Dr. Whitman Cross. The region is shown to be one containing numerous, although not always large, volcanic outbreaks. One of them furnished the sheet of rhyolite at Silver Cliff, while others had for their conduits the chimneys of the Bull Domingo and the Bassick. The ore bodies thus occur in volcanic necks, and make a new form for the science.

2.09.21. Humboldt-Pocahontas. These mines are near Rosita on fissure veins in andesite, but of a different flow and kind from the walls of the Bassick. They are filled with gray copper and chalcopyrite, in a barite gangue. Other mines of less importance occur in the district, but the three above cited are given prominence because of their own intrinsic interest and because they have often been referred to in discussions about the origin of ores.[1]

---

[1] R. N. Clark, "Humboldt-Pocahontas Vein," *M. E.*, VII. 21. "Sil-

2.09.22. Gilpin County has already been mentioned under "Copper" (2.04.08). The general geology of the veins is much like that of Clear Creek, although the ores are quite different. R. Pearce has shown the existence of bismuth in the ore, and gives reasons for believing that the gold is in combination with it. Clear Creek County contains veins on a great series of jointing planes in gneiss (granite), and in large part replacements of the wall. Others are replacements of porphyry dikes or of pegmatite segregations. The ores are chiefly galena, tetrahedrite, zincblende, and pyrite, and the gangue is the wall rock. The curious decrease of value in depth of a series of parallel veins in Mount Marshall was referred to (1.05.05). Georgetown is the principal town and mining center. Others of importance are Idaho Springs and Silver Plume.[1]

2.08.23. Boulder County contains veins along joints or faulting planes in gneiss, or granite, or associated with porphyry dikes, or pegmatite segregations, and carrying tellurides of the precious metals more or less as impregnations of the country rock. The prevalent country rock is called by Emmons a granite-gneiss. Van Diest distinguishes four successive terranes of massive and schistose rocks along three principal axes and two side ones, and states that the mines are on the sides of the folds. The country is very generally pierced by porphyry dikes, with which the ore bodies are often associated. A large number of species of telluride minerals have been determined from the region, especially by the late Dr. Genth of Philadelphia. The mines afford very rich ores, somewhat irregularly distributed.[2]

---

ver Cliff, Colorado," *Engineering and Mining Journal*, Nov. 2, 1878, p. 314. W. Cross, "Geology of the Rosita Hills," *Proc. Colo. Sci. Soc.*, 1890, p. 269. Rec. S. F. Emmons, "The Genesis of Certain Ore Deposits," *M. E.*, XV. 146. *Tenth Census*, Vol. XIII., p. 80. L. C. Graybill, "On the Peculiar Features of the Bassick Mine," *M. E.*, XI., p. 110; *Engineering and Mining Journal*, Oct. 28, 1882, p. 226. Rec. O. Loew and A. R. Conkling, "Rosita and Vicinity," *Wheeler's Survey*, 1876, p. 48. See also Stevenson in the Report for 1873.

[1] S. F. Emmons, *Tenth Census*, Vol. XIII., p. 70. Rec. F. M. Endlich, *Hayden's Survey*, 1873, p. 293; 1876, p. 117. P. Fraser, *Hayden's Survey*, 1869, p. 201. J. D. Hague, *Fortieth Parallel Survey*, Vol. III., p. 589. Rec. R. Pearce, *Proc. Colo. Sci. Soc.*, Vol. III., pp. 71, 210. "The Association of Gold with Other Metals," *M. E.*, 1890. J. J. Stevenson, *Wheeler's Survey*, Vol. III., p. 351. F L. Vinton, "The Georgetown (Colo.) Mines," *Engineering and Mining Journal*, Sept. 13, 1879, p. 184.

[2] A. A. Eilers, "A New Occurrence of the Telluride of Gold and Sil-

2.09.24. The resources of the remaining counties of Colorado are chiefly in coal.

---

ver," *M. E.*, Vol. I., p. 16. S. F. Emmons, *Tenth Census*, Vol. XIII., p. 64. J. B. Farish, "Interesting Veins Phenomena in Boulder County, Colorado," *M. E.*, September, 1890. F. A. Genth, "On Tellurides," *Amer. Jour. Sci.*, ii., XLV., p. 305, and other papers in the same journal. C. S. Palmer, "Eruptive Rocks of Boulder and Adjoining Counties," *Proc. Colo. Sci. Soc.*, Vol. III., p. 230. P. H. Van Diest, "The Mineral Resources of Boulder County," *Ann. Rep. Colo. State School of Mines*, 1886, p. 25.

CHAPTER X.

SILVER AND GOLD, CONTINUED.—ROCKY MOUNTAIN REGION, WYOMING, THE BLACK HILLS, MONTANA, AND IDAHO.

WYOMING.

2.10.01. *Geology.*—The southeastern part of Wyoming is in the Prairie region, the southwestern in the Plateau. The Rocky Mountains shade out more or less on leaving Colorado, but are again strongly developed in northern Wyoming. The northwestern portion contains the great volcanic district of the National Park, and the northeastern, a part of the Black Hills. The Cretaceous and Tertiary strata chiefly form the plains and plateaus. Granite and gneiss constitute the central portion of some of the greater ranges. Paleozoic rocks are very subordinate. The resources in precious metals are small, consisting chiefly of gold in quartz veins in the gneisses, schists, and granites of Stillwater County. The great mineral wealth of the State is in coal. The iron mines have already been mentioned (2.03.09), and the copper (2.04.27).[1]

THE BLACK HILLS.

2.10.02. *Geology.*—The Black Hills lie mostly in South Dakota. They consist of a somewhat elliptical core of granite and metamorphic rocks, with a north and south axis, and on these are laid down successive strata of Cambrian, Carbonifer-

---

[1] H. M. Chance, "Resources of the Black Hills and Big Horn Country, Wyoming," *M. E.*, XIX., p. 49. T. B. Comstock, "On the Geology of Western Wyoming," *Amer. Jour. Sci.*, iii., VI. 426. S. F. Emmons, *Tenth Census*, Vol. XIII., p. 86. F. M. Endlich, "The Sweetwater District," *Hayden's Survey*, 1877, p. 5; "Wind River Range Gold Washings," p. 64. A. Hague, "Geological History of the Yellowstone National Park," *M. E.*, XVI. 783. See also F. V. Hayden, *Amer. Jour. Sci.*, iii., III. 105, 161. F. V. Hayden, *Rep. for* 1870–72, p. 13; also *Amer. Jour. Sci.*, ii., XXXI. 229. A. C. Peale, "Report on the Geology of the Green River District," *Hayden's Survey*, 1877, p. 511. Raymond's *Statistics West of the Rocky Mountains.*

ous, Jura-Trias, and Cretaceous rocks. There are some igneous intrusions. The principal product of the Black Hills is gold. The lead-silver deposits have already been described (2.08.18), and the tin, mica, etc., will be mentioned later.[1]

2.10.03. The gold occurs in placers of Quaternary and recent age, as well as in Potsdam sandstones, which are old shore beaches now hardened to rock; in pyritous beds in schistose rocks, and in segregated quartz veins. The Quaternary and recent placers are the usual gravels, which are more fully described under "California." The Potsdam sandstone is an extremely interesting deposit. It has resulted from the wearing action of the waves of the Potsdam ocean on the Archæan schists. The Potsdam also carries other deposits in the vicinity of porphyry sheets and dikes,

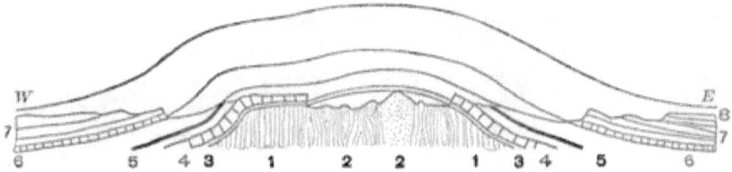

FIG. 54.—*Geological section of the Black Hills. After Henry Newton. Report on the Black Hills, p.* 206.

1. Schists. 2. Granite. 3. Potsdam sandstone. 4. Carboniferous. 5, 6. Jura-Trias. 7. Cretaceous.

which consist of auriferous pyrite, sometimes oxidized. This has replaced the original calcareous cement of the quartzite. The pyritous beds are in a great impregnation zone 2000 feet broad (Carpenter), of slates and schists, with portions especially rich in auriferous pyrites. They occur near the town of Lead City, not far from Deadwood, in the northern hills. The deposits present many analogies with Example 16, and also are like fahlbands (1.06.10). The ore is not high grade, running $3 to $4 per ton, but it is treated at great profit by mining it in enormous quantities. There are also many so-called segregated quartz veins in the

---

[1] F. R. Carpenter, "Ore Deposits in the Black Hills," *M. E.*, XVII. 370. *Prelim. Rep. on the Geol. of the Black Hills*, Rapid City, So. Dak., 1888. W. O. Crosby, "Geology of the Black Hills," *Bost. Soc. Nat. Hist.*, XXIII., p. 89. Newton and Jenney, *Report on the Black Hills*, Washington, 1880. C. R. Van Hise, "The Pre-Cambrian Rocks of the Black Hills," *Bull. Geol. Soc. Amer.*, I. 203–244. N. H. Winchell, "Report on the Black Hills," *Rep. Chief of U. S. Engineers*, 1874, Part II., p. 630.

schists and slates. These are lenticular masses of limited extent, horizontally and below, somewhat like a magnetite lens (Example 12) in shape, and carrying a small amount of gold with little or no pyrites.[1]

MONTANA.

2.10.04. *Geology.*—The eastern part of the State belongs to the Prairie region, which is, however, in portions greatly scarred by erosion, forming the so-called Bad Lands. The approaches to the Rocky Mountains are not abrupt and sudden as in Colorado, but are marked by numbers of outlying ranges of both eruptive and sedimentary rocks. The chain of the Rockies takes a northwesterly trend in Wyoming, and so continues across Montana. It is rather the prolongation of the Wasatch than of the Colorado Mountains, whose strike is for the Black Hills. The character of the ranges is also very different. They are less elevated and have broad and well-watered valleys between, that admit of considerable agriculture. Geologically the country is in marked contrast with Colorado. While in the latter the Paleozoic is feebly developed, in the former it reaches great thickness. In the eastern ranges W. M. Davis gives Lower Cambrian 10,000 to 15,000 feet; Silurian and Devonian, not yet recognized; Carboniferous limestones, 3500 feet; Trias, not definitely recognized; Jurassic and Cretaceous sandstones, shales, and thin limestones, 15,000 feet. This more closely resembles the Wasatch and Great Basin sections (see 2.08.29, and 2.11.01). Much granite of a basic or dioritic character is present (Example 17), and great developments of eruptive rocks of extremely interesting character. No more interesting field for geological work awaits the investigator.[2]

---

[1] A. J. Bowie, "Notes on Gold Mill Construction," *M. E.*, X. 1881. W. B. Devereux, "The Occurrence of Gold in the Potsdam Formation," *M. E.*, 465; *Engineering and Mining Journal*, Dec. 23, 1882, p. 334. H. O. Hofman, "Gold Mining in the Black Hills," *M. E.*, XVII. 498; also in preliminary report cited under Carpenter, under Geology.

[2] S. Calvin, "Iron Butte: Some Preliminary Notes," *Amer. Geol.*, IV. 95. G. E. Culver, "A Little Known Region of Northwestern Montana," *Wis. Acad.*, Dec. 30, 1891. W. M. Davis, "The Relation of the Coal of Montana to the Older Rocks," *Tenth Census*, Vol. XV., p. 697. Rec. J. Eccles, "On the Mode of Occurrence of Some of the Volcanic Rocks of Montana," *Quar. Jour. Geol. Sci.*, XXXVII. 399. G. H. Eldridge, "Montana Coal Fields," *Tenth Census*, Vol. XV., p. 739. S. F. Emmons, *Tenth Census*, Vol. XIII., 97. Rec. *Hayden's Survey*, Ann.

2.10.05. Montana took the lead of all the States in 1887 in the production of silver, was second in gold, and first in the total production of the two. It is now second. In its mineral wealth it yields to no other State in the Union. The mining districts are mostly in the western central and western portions. Developments have progressed so rapidly that all the desirable data are not available.

2.10.06. Madison County. Veins in gneiss containing galena and pyrite in a quartz gangue. Virginia City is the principal town, and the veins are north of it in the northern part of the county.[1]

2.10.07. Beaverhead County. Near Bannack City quartz veins with auriferous pyrite on the contact between the limestone and so-called granite. At Glendale, in the northern part of the county, are the Hecla mines, referred to under "Lead-silver" (Example 32). Auriferous quartz veins are reported farther north.[2]

2.10.08. Jefferson County. This county contains ore bodies chiefly auriferous quartz, in gneiss, porphyry, or limestone. The

---

Rep., 1871-72. W. S. Keyes, in Brown's first report on mineral resources, etc., last part, *Amer. Jour. Sci.*, II. 46, 431. Rec. W. Lindgren, "Eruptive Rocks," *Tenth Census*, Vol. XV., p. 719, forming Appendix B of Davis's first paper. See also *Proc. Cal. Acad. Sci.*, Second Series, Vol. III., p. 39. J. S. Newberry, "Notes on the Surface Geology of the Country bordering on the Northern Pacific Railroad," *Annals N. Y. Acad. Sci.*, Vol. III. 242; *Amer. Jour. Sci.*, iii., XXX. 337. "The Great Falls Coal Fields," in Geol. Notes, *School of Mines Quarterly*, VIII. 327. F. Rutley, "Microscopic Character of the Vitreous Rocks of Montana," *Quar. Jour. Geol. Sci.*, XXXVII. 391. See Eccles, above. W. H. Weed, "The Cinnabar and Bozeman Coal Fields of Montana," *Bull. Geol. Soc. Amer.*, II. 349-364. *Engineering and Mining Journal*, May 14 and 21, 1892. "Montana Coal Fields," *Bull. Geol. Soc. Amer.*, III. 301-330. C. A. White, "Existence of a Deposit in Northwestern Montana and Northeastern Dakota that is Possibly Equivalent with the Green River Group," *Amer. Jour. Sci.*, iii., XXV. 411. R. P. Whitfield, "List of Fossils from Central Montana," *Tenth Census*, Vol. XV., p. 712; Appendix A to Davis's paper. J. E. Wolff, "Notes on the Petrography of the Crazy Mountains," etc., *Northern Trans. Survey*. "Geology of the Crazy Mountains," *Bull. Geol. Soc. Amer.*, III. 445. H. Wood, "Flathead Coal Basin," *Engineering and Mining Journal*, July 16, 1892, p. 57. H. R. Wood, "Mineral Zones in Montana," *Engineering and Mining Journal*, Sept. 24, 1892, p. 292.

[1] S. F. Emmons, *Tenth Census*, Vol. XIII., p. 97.
[2] *Ibid.*

lead-silver mines near Wickes have been referred to. (Example 33.) Red Mountain lies at the head of a valley like Wickes and contains many narrow argentiferous veins. A concentrator was at work on them in 1892.[1]

2.10.09. **Silver Bow County.** The copper mines and the general geology of the Butte City region were referred to under "Copper" (Example 17). In the basic granite, and north of the copper zone, is a belt carrying sulphides of silver, lead, zinc, and iron in a siliceous gangue, but abundantly associated with manganese compounds of various sorts, especially rhodochrosite.

No manganese is known in the copper belt, nor **any** copper in the silver belt—most striking phenomena in veins in the same wall

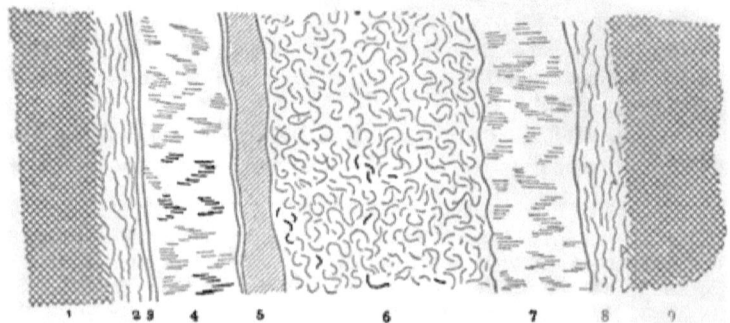

FIG. 55.—*Cross section of vein at the Alice mine, Butte, Mont. The width of vein is 40 feet. After W. P. Blake, M. E., XVI., p. 72.*

1. Granite country. 2. Softened granite with small veins. 3. Clay wall with decomposed granite. 4. Quartz, broken and seamed. 5. Clay and decomposed granite. 6. Quartz and manganese spar—"curly ore." 7. Quartz and ore—"hard vein." 8. Soft granite with veinlets. 9. Dark-colored, hard granite of the hanging-wall country.

rock. The line of outcrop has a crescentic sweep, and it was therefore called by J. E. Clayton the Rainbow Lode. It includes from west to northeast six claims, all but two of which are controlled by the Alice Company. There are as many as four distinct veins present in the Magna Charta. All the mines show that the ore and gangue have replaced the granite along a shattered strip, for cross sections exhibit alternations of quartz with ore, rhodochrosite, crushed wall rock, residual clay, occasional horses of granite, etc. In the more siliceous granite west of the butte is another silver belt with the same ores as in the Rainbow

---

[1] S. F. Emmons, *Tenth Census*, Vol. XIII., p. 97. J. S. Newberry, "On Red Mountain," *Annals N. Y. Acad. Sci.*, III., p. 251.

Lode and likewise having manganese minerals associated. The Bluebird is the principal mine. The manganiferous outcrop was a notable feature in the landscape, and exhibited a broad, rusty-black belt, not rich at the surface, but only showing the silver in depth. Like the veins in the basic granite, these were also formed by replacement of the rock along a shattered strip. Placer mines were early worked near Butte and led to the location of the deep mines. They are still productive and are again referred to under "Auriferous Gravels."[1]

2.10.10. Deer Lodge County. Placer deposits are numerous along the Deer Lodge River, and auriferous quartz veins are known, but the greatest mine is the Granite Mountain, a source of very handsome returns. This is in the southern part of the county, near Phillipsburg, and is a fissure vein in granite, principally with silver ores, although affording considerable gold. On the same vein is the Bimetallic. Farther west sedimentary rocks come in, much metamorphosed by contact with the later irruptive granite. On the edge of the county, and not far from the Drumlummon group of veins, later mentioned, are the veins of the Bald Butte Company, in slates and intrusive diorite. A number of other veins are in the same general region.[2]

2.10.11. Lewis and Clarke County. The placer mines, near Helena (in Last Chance and Prickly Pear gulches), were the first in the county to attract attention. They were found by the prospectors, who spread through the Rocky Mountains as the California gold diggings gave out. Since then many auriferous quartz veins in granite and slates have been developed. Some twenty

---

[1] W. P. Blake, "Silver Mining and Milling at Butte, Mont.," *M. E.*, XVI. 38. "Rainbow Lode, Butte, Mont.," *M. E.*, XVI. 65. Rec. S. F. Emmons, "Notes on the Geology of Butte, Mont.," *M. E.*, XVI. 49. Rec. Richard Pearce, "The Associations of Minerals in the Gagnon Vein, Butte City," *M. E.*, XVI. 62. E. D. Peters, *Mineral Resources of U. S.*, 1883–84, p. 374. E. G. Salisbury, "Placer Mining in Montana," *Engineering and Mining Journal*, Sept. 3, 1887, p. 167. Rec. "Silver Mines of Butte, Mont.," *Ibid.*, April 18, 1885, p. 261. Williams and Peters on Butte, Mont., *Engineering and Mining Journal*, March 28, 1885, p. 208.

[2] H. M. Beadle, "The Condition of the Mining Industry in Montana in 1892," *Engineering and Mining Journal*, Feb. 11, 1893, p. 123. G. W. Goodale and W. A. Ackers, "Concentration, etc., with Notes on the Geology of the Flint Creek Mining District," *M. E.*, 1890. Rec. "The Granite Mountain Mine," *Engineering and Mining Journal*, Dec. 10, 1887; Nov. 23, 1889. E. G. Spilsbury, "Placer Mining in Montana," *Ibid.*, Sept. 3, 1887, p. 167.

miles north of Helena, in the town of Marysville, is the Drumlummon group of veins, which carry refractory silver and gold ores, in a quartz gangue, on the contact between a granite knob and the surrounding metamorphic schists. There are also other veins in the granite. Dikes of intrusive rocks occur associated with the ore bodies.[1]

2.10.12. *Missoula County.* In the northwestern corner of the State is a region of very recent development, and more especially since the Northern Pacific Railroad has been built through it. At Iron Mountain and elsewhere mining districts are growing up, but available descriptions have not yet been received. (See paper by G. E. Culver, cited under 2.10.04.) In Meagher County, in the central part of the State, there are a number of mining districts in the Little Belt Mountains. Neihart is the location of some rich silver mines, and is now connected with Great Falls by rail. Other camps are as yet too remote for profitable working.

IDAHO.

2.10.13. *Geology.*—The southern part of the State extends into the alkaline deserts of the Great Basin and is dry and barren. North of this is the Snake River Valley, which is filled by a great flood of recent basalt which stretches from the Wyoming line nearly across the State. North of the Snake River a large area of granite appears in the western portion and contains many mines. Extensive deposits of gravel also occur. Metamorphic rocks and Paleozoic strata largely constitute the northern portion of the State, and are penetrated by many igneous intrusions. The eastern part lies on the western slopes of the Bitter Root Mountains, whose general geology was outlined under Montana. The geology of Idaho has been but slightly studied, and the few reliable records have resulted from the scattered itineraries of Hayden's survey, isolated mining reports, and the collections of the Tenth Census.[2]

---

[1] J. E. Clayton, "The Drumlummon Group of Veins," etc., *Engineering and Mining Journal*, Aug. 4 and 11, 1888, pp. 85, 106. S. F. Emmons, *Tenth Census*, Vol. XIII., p. 97.

[2] G. F. Becker, *Tenth Census*, Vol. XIII., 52. F. H. Bradley, *Hayden's Survey*, 1872, p. 208. F. V. Hayden, *Ann. Rep.*, 1871, pp. 1, 147; 1872, p. 20. J. S. Newberry, "Notes on the Geology and Botany along the Northern Pacific Railroad," *Annals N. Y. Acad. Sci.*, III. 252. Raymond's *Reports on Mineral Resoures West of the Rocky Mountains*. O. St. John, *Hayden's Survey*, 1877, p. 323; 1878, p. 175.

SILVER AND GOLD, CONTINUED.

2.10.14. Custer County lies north of Lemhi and contains several well-known mines. The Ramshorn is in metamorphic slates on a fissure vein that has rich chutes of high-grade silver ores in a siderite gangue. The Custer and the Charles Dickens are farther west, near Bonanza City, and afford both silver and gold in quartz gangue from veins in porphyry. Smelting ores occur in the region and have been used in some operations based on this treatment. In Boisé and western Alturas counties a granite area forms the greater part of the surface, and in it are numerous productive veins. In the former they are chiefly gold quartz except in the Banner district, where silver predominates.

The placer deposits of Boisé County, which were developed in 1863, were very rich, but are now less productive than in former years. In Alturas County gold quartz veins occur, and also others carrying silver, and the county is a strong producer. The Wood River mines in slates and limestones, southeast of the granite, have already been referred to under Example 32a. Owyhee County is in the southwestern corner of the State. It is probable that the granite of the two last mentioned counties extends under overlying drift and comes up again near Silver City (Becker). Southwest of it quartz porphyry and metamorphic rocks are found, with dikes of basalt. Gold quartz and high-grade silver ores are present. The Poorman Lode is famous for ruby silver ores. W. P. Blake mentions seeing a piece from this mine at the Paris Exposition which weighed about 200 pounds.[1] It was awarded a gold medal. The crystal from which it was broken weighed 500 pounds.[2] In Cassia and Oneida, two other counties in the southern part, placers are being or have been worked, and in Bear Lake County, in the southeast corner, salt and sulphur deposits are recorded.[3]

---

[1] *Amer. Jour. Sci.*, ii., XLV. 97.
[2] Raymond's *Reports on Mineral Resources West of the Rocky Mountains*, 1868, p. 523.
[3] G. F. Becker, *Tenth Census*, **Vol. XIII.**, p. 59. Raymond's *Reports on Mineral Resources West of the Rocky Mountains*, **Rep.** *Director of the Mint*, 1882, p. 227.

CHAPTER XI.

SILVER AND GOLD, CONTINUED.—THE REGION OF THE GREAT BASIN, IN UTAH, ARIZONA, AND NEVADA.

UTAH.

2.11.01. *Geology.*—The eastern half of **Utah**, terminating with the western front of the Wasatch, is in the **Colorado Plateau**, but the western is within the limits of the **Great Basin**. The plateau portion consists largely of Mesozoic strata, quite horizontal and more or less carved by erosion. The east and west arch of the Uintah Mountains, in the northern part, has upheaved them, so that where the **Green River** has cut a channel across, the Paleozoic is exposed in great strength. The Wasatch range rises with a gradual ascent from the east, and then terminates with a great fault line having a steep westerly front. This line of weakness was developed in the Archæan and has been a scene of movement even to recent times. It is a very important structural feature. West of the Wasatch, which is a fine example of block tilting in mountain-making, the mountains belong to the Basin ranges, which are more typically developed in Nevada. The Wasatch section was shown by the Fortieth Parallel Survey to involve 12,000 to 14,000 feet of the Upper Archæan and nearly 30,000 feet of the Paleozoic. In southern Utah the Triassic rocks are important and contain some rich mines.[1]

---

[1] G. F. Becker, *Tenth Census*, Vol. XIII., 38. C. E. Dutton, *Rep. on the High Plateaus of Utah*, Washington, 1880. A. Geikie, "Archæan Rocks of the Wasatch Mountains," *Amer. Jour. Sci.*, iii., XIX. 363. G. K. Gilbert, "Contributions to the History of Lake Bonneville," *Second Ann. Rep. Director U. S. Geol. Survey*, 169–200, and Monograph II. "The Ancient Outlet of Great Salt Lake," *Amer. Jour. Sci.*, iii., XV. 256, XIX. 341; see also A. C. Peale, *Ibid.*, XV. 439. *The Henry Mountains*, Washington, 1877. Hague, King, and Emmons, *Fortieth Parallel Survey*, Vols. I. and II. O. C. Marsh, "On the Geology of the Eastern Uintah Mountains," *Amer. Jour. Sci.*, iii., I. 191. B. Silliman, "Geological and Mineralogical Notes on Some of the Mining Districts of Utah Territory," *Amer. Jour.*

2.11.02. The greater number of the Utah mines are for lead-silver ores and have been mentioned under "Lead Silver." The northwestern county, Box Elder, is in the alkaline desert region of the Great Basin. The mining districts occur in the central part of the State, in the Wasatch and Oquirrh mountains, and are also found in the extreme southwest.

2.11.03. Ontario Mine. Nearly east of Salt Lake City, in Summit County, is the Ontario mine, a vein from four to twenty-three feet wide (averaging eight feet), in quartzite, but extremely persistent, being opened continuously for 6000 feet. In the lower working a porphyry dike has come in as one of the walls. It is extensively altered by fumarole action to clay. The best parts of the mine have quartzite walls. The ores consist of galena, gray copper, silver glance, blende, etc. Other somewhat similar ore bodies are known in the same region but are less developed.[1]

2.11.04. The lead-silver veins of Bingham Cañon, in Salt Lake County, have already been mentioned. Reference may again be made to the great bed-veins of gold quartz associated with them. Ophir Cañon and Dry Cañon, in Tooele County, and the Tintic district, in Juab County, have also been described. In addition to the smelting ores, others have been treated by milling. Quite recently interest has been directed to the mines of the Deep Creek district, on the extreme western border of Utah, in the Ibapah range. Limestones regarded by Blake as Carboniferous, and other sedimentary rocks, have been broken through by great outflows of granite, andesite, hypersthene-andesite, etc. The ore bodies appear to be contact deposits in limestone near igneous rocks, and carry much free gold.[2]

In Beaver County the interesting deposits of the Horn Silver, the Carbonate, and the Cave ore bodies have been mentioned

---

Sci., iii., III. 195. Wheeler, Gilbert, Lockwood and others on Western Utah, *Wheeler's Survey, Rep. Prog.* 1869-71-72. *Idem*, Final Reports, Vol. III.

[1] T. J. Almy, "History of the Ontario Mine, Park City, Utah," *M. E.*, XVI. 35. "The Ontario Mine," *Engineering and Mining Journal*, May 28, 1881, p. 365. D. B. Huntley, *Tenth Census*, Vol. XIII., p. 438.

[2] W. P. Blake, "Age of the Limestone Strata at Deep Creek, Utah, and the Occurrence of Gold," etc., *Amer. Geol.*, January, 1892, p. 47. *Engineering and Mining Journal*, Feb. 23, 1892, p. 253. S. F. Emmons, *Fortieth Parallel Survey*, Vol. II. J. F. Kemp, "Petrographical Notes on a Suite of Rocks collected by E. E. Olcott," *N. Y. Acad. Sci.*, May, 1892.

under Examples 30*g*, 33*a*, and 32*b*. The great iron mines of Iron County will be found under Example 14. In Piute County, near the town of Marysvale, around Mount Baldy, are a number of mines with lead-silver or milling ores in quartz porphyry (copper belt), or between limestone and quartzite (Deer Trail, Greeneyed Monster, etc.). Selenide of mercury is found in the Lucky Boy.[1]

2.11.05. Example 41. Silver Reef, Utah. Native silver, cerargerite, and argentite, impregnating Triassic sandstones, and often replacing organic remains. These deposits were earlier referred to under Example 21, p. 80. They were discovered in 1877. At Silver Reef there are two silver-bearing strata or reefs, with beds of shale between. Above the water line the ore is horn silver; below, it is argentite. At times it replaces plant remains; at other times no visible presence of ore can be noted, although the rock

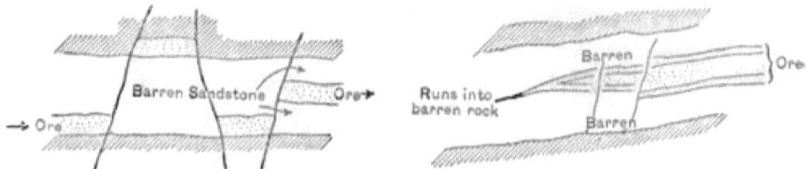

FIG. 56.—*Two sections of the argentiferous sandstone at Silver Reef, Utah. After C. M. Rolker, M. E., IX., p.* 21.

may afford $30 to the ton. The silver always occurs along certain ore channels, distributed through parts of the sandstone. The origin of the deposits has given occasion to a vigorous discussion. J. S. Newberry holds that the silver was deposited in and with the sandstone from the Triassic sea, although it may have been concentrated since in the ore channels. F. M. F. Cazin holds that the organic remains were deposited in and with the sandstone, and that these were the immediate precipitating agents of the ores. R. P. Rothwell explained them much as does Rolker, below. C. M. Rolker, who was for some years in charge of several of the mines, has also written about them, and is probably nearest to the truth. Rolker argues that the impregnation was subsequent to the formation of the sandstone, and was caused by the igneous outbreaks in the neighborhood, and probably runs along old lines of partial weakening or crushing that afterward healed up. Eruptive rocks

---

[1] G. J. Brush, "On the Onofrite, etc.," *Amer. Jour. Sci.*, iii., XXI. 312.

are known in the neighborhood of the ores both in Utah and in the Nacemiento copper district of New Mexico. From what we know of ore deposits in general this seems most probable.[1]

ARIZONA.

2.11.06. *Geology.*—Arizona lies partly in the plateau region and partly in the Great Basin. The Basin ranges converge with the Rocky Mountains, which, however, are chiefly in New Mexico. The uplands of the ranges are well watered and covered with timber, but the low-lying portion of the Great Basin is an arid desert, and in southwestern Arizona is the hottest part of the United States. Cretaceous and Jura-Trias largely form the plateau region. Running southeast to northwest is the great development of Carboniferous limestone so often referred to under "Copper," and underlying this are found Archæan granites, gneisses, etc. A great series of ore deposits is ranged along this contact. In the southwest are mountains of granites and metamorphic rocks. The Territory also contains vast flows of igneous rocks, and in the plateau country between the converging ranges some 20,000 or 25,000 square miles are covered by them. The Grand Cañon of the Colorado has laid bare a magnificent geological section of many thousand feet, from the Archæan to the Tertiary.[2]

---

[1] F. M. F. Cazin, "The Origin of the Copper and Silver Ores in Triassic Sandrock," *Engineering and Mining Journal*, Dec. 11, 1880, p. 381; April 30, 1881, p. 300. "The Silver Sandstone Formation of Silver Reef," *Ibid.*, May 22, 1880, p. 351, Jan. 10, 17, 24, 1880, pp. 25, 48, 79 (Rothwell). A. N. Jackson, "Silver in Sedimentary Sandstone," *Rep. Director of Mint*, 1882, p. 384, reprinted from *Cal. Acad. Sci.* J. S. Newberry, "Report on the Properties of the Stormont Silver Mining Company," etc., *Engineering and Mining Journal*, Oct. 23, 1880, p. 269. "The Silver Reef Mines," *Ibid.*, Jan. 1, 1881, p. 4. C. M. Rolker, "The Silver Sandstone District of Utah," *M. E.*, IX. 21.

[2] "Central Arizona," *Engineering and Mining Journal*, April 23, 1881, p. 285. "Colorado River of the West," review of Ives Expedition, *Amer. Jour. Sci.*, ii., XXXIII. 387. G. F. Becker, *Tenth Census*, Vol. XIII., p. 44. C. E. Dutton, "The Physical Geology of the Grand Cañon District," abstract of *Monograph II., Second Ann. Rep. Director U. S. Geol. Survey*, 49–161; see also the monograph. Patrick Hamilton, *The Resources of Arizona*, A. L. Bancroft & Co., San Francisco, 1884. B. Silliman, "Report on Mining Districts of Arizona, near the Rio Colorado," *Engineering and Mining Journal*, Aug. 11, 1877, p. 111; taken from *Amer. Jour. Sci.*, ii, XLI. 289. C. D. Walcott, "Permian and Other Paleozoic

2.11.07. Apache County is in the northeastern corner. In the southern part of the county gold and silver ores, in veins in limestone, associated with copper ores, are reported, and some small placers.

2.11.08. Yavapai County. Gold and silver ores, in quartz veins, in granite and metamorphic rocks. The Black Range copper district has already been referred to under Example 20e.

Mohave County. Silver sulphides, arsenides, etc., and alteration products in veins in granite, at times showing a gneissoid structure. Only the richest can now be worked.

Yuma County. Quartz veins, with silver ores and lead minerals in metamorphosed rocks (gneiss, slate, etc.), or in granite.

Maricopa County contains both Paleozoic and Archæan exposures. The ore deposits lie mostly along the contact of the two, in granite or highly metamorphosed strata. They are usually quartz veins, with silver ores and copper, lead, and zinc minerals. The Globe district, extending also into Pinal County, is the principal one. Mention has already been made of it under "Copper," Example 20e.

Pinal County adjoins Maricopa on the south and contains a number of important mines. They produce mostly silver ores, with lead and copper associates, and some blende. The gangue minerals are quartz, calcite, etc., occasionally manganese compounds, and sometimes, in the granites, barite. Limestone, slate, sandstone, and quartzite, as well as granite, diabase, and diorite, occur as wall rock.

2.11.09. Silver King Mine. A central mass or chimney of quartz, with innumerable radiating veinlets of the same, carrying rich silver ores and native silver, in a great dike of feldspar porphyry, with associated granite, syenite (Blake), porphyry, gneiss, and slates, all of Archæan age. The veinlets ramify through the strongly altered porphyry, and form a stockwork, which furnishes the principal ores. In the region are also Paleozoic strata, whose upper limestone beds are referred by Blake to the Carboniferous. The minerals at the mine are native silver, stromeyerite, argentite, sphalerite, galenite, tetrahedrite, bornite, chalcopyrite, pyrite, quartz, calcite, siderite, and, as an abundant gangue, barite.

Groups of the Kanab Valley, Arizona," *Amer. Jour. Sci.*, iii., XX. 221. "Pre-Carboniferous Strata in the Grand Cañon of the Colorado, Arizona," *Amer. Jour. Sci.*, December, 1883, 437. *Wheeler's Survey*, Vol. III., and Supplement.

Graham County contains the Clifton-copper district, referred to under Example 20a.

Cochise County is the southeastern county, and contains the Tombstone district, the most productive of the precious metals in the Territory.

2.11.10. Tombstone. A great porphyry dike up to 70 feet wide, cutting folded Palæozoic strata, and itself extensively faulted and altered, and carrying above the water line in numerous vertical joints, or partings, quartz with free gold, horn silver, and a little pyrite, galenite, and lead carbonate. Curiously enough, in the porphyry itself, and far from the quartz veins, flakes and scales of free gold have been found, evidently introduced in solution. Ore also occurs along the side of the dike. There are also other fissures parallel with this principal dike, and still another series crossing these and the axis of the great anticline of the district. Connected with these fissure veins are bedded deposits in the limestone, along the bedding planes or dropping from one to another, appearing to have originated by replacement. Blake offers two explanations of the first-mentioned dike deposit—either that the dike itself held the precious metals, or that they came from the pyrite of the adjoining strata. Several other mining districts of less note occur in the county. The important copper deposits of the Bisbee region have already been mentioned under Example 20f.

2.11.11. Pima County is the central county of the southern tier and has Tucson as its principal city. There are numbers of mines of the precious metals, and a few less important copper deposits.

Yuma County, in the southwestern corner, has some mines along the Colorado River, on quartz veins in metamorphosed rocks, containing silver and lead minerals.[1]

---

[1] G. F. Becker, *Tenth Census*, Vol. XIII., p. 44. G. H. Birnie, "Castle Dome District," *Wheeler's Survey*, 1876, p. 6. W. P. Blake, "The Geology of Tombstone, Ariz.," *M. E.*, X. 334, *Engineering and Mining Journal*, June 24, 1882, p. 328; *The Silver King Mine*, a short monograph, New Haven, March, 1883. Rec. See also *Engineering and Mining Journal*, April 28, 1883, p. 238. J. F. Blandy, "The Mining Region around Prescott, Ariz.," *M. E.*, XI. 286, *Engineering and Mining Journal*, July 21, 1883. "On Tombstone, Ariz.," *Ibid.*, May 7, 1881, p. 316; March 18, 1882, p. 145. "Silver in Arizona," General Review, *Engineering and Mining Journal*, Sept. 21 and 25, 1880, pp. 172, 203. "Central Arizona," *Ibid.*, April 23, 1881, p. 285. O. Loew, "Hualapais District," *Wheeler's Survey.*

NEVADA.

2.11.12. *Geology*.—Nevada lies almost entirely in the Great Basin, only the western portion being in the Sierras. The surface is thus largely formed by the dried basins of former great lakes, principally Lakes Lahontan and Bonneville. A large number of ranges extend north and south through the State, known collectively as the Basin ranges. They have been formed by block tilting on a grand scale and present enormously disturbed strata. The geological sections exposed are of surpassing interest (cf. Example 36), and show Archæan and Paleozoic in great thickness. In these mountains are found the mining districts, while between them lie the alkaline plains.[1]

2.11.13. Lincoln County is in the southeastern corner and contains a number of small mining districts. The ores are in general silver-lead ores in limestone, or veins with sulphuret ores in quartzite and granite. Pioche is one of the principal towns, near which is found the once famous and now reopened Raymond & Ely mine. A strong fissure cuts Cambrian quartzite and overlying limestone, where the latter has not been eroded, and is occupied by a great porphyry dike. Along the contact between the porphyry and the wall rock the chutes of ore have been found. Mr. Ernest Wiltsee, at the Montreal meeting of the American Institute of Mining Engineers, February, 1893, described and figured the Half Moon mine, on this same great fissure, where the quartz-

---

1876, p. 55. B. Silliman, "Report on the Mining District of Arizona near the Rio Colorado," *Amer. Jour. Sci.*, ii., XLI. 289; see also *Engineering and Mining Journal*, Aug. 11, 1877, p. 111. Raymond's Reports, and those of the Director of the Mint.

[1] J. Blake, "The Great Basin," *Proc. Cal. Acad. Sci.*, IV. 275, *Amer. Jour. Sci.*, iii., VI. 59. W. P. Blake, "On the Geology and Mines of Nevada" (Washoe silver region), *Quar. Jour. Geol. Sci.*, Vol. XX., p. 317. H. G. Clark, "Aurora, Nev.: a Little of its History, Past and Present," *School of Mines Quarterly*, III. 133. G. K. Gilbert, "A Theory of the Earthquakes of the Great Basin, with a Practical Application," *Amer. Jour. Sci.*, iii., XXVII. 49. I. C. Russell, "Geology and History of Lake Lahontan, a Quaternary Lake of Northwestern Nevada," *Monograph XI.*, *U. S Geol. Survey*; also *Third Ann. Rep. Director U. S. Geol. Survey*, 195. C. D. Walcott, "Paleontology of the Eureka District," *Mononraph VIII., U. S. Geol. Survey*. Gilbert, Wheeler, Lockwood, and others, "Eastern Nevada: Notes on its Economic Geology," *Wheeler's Survey*, *Rep. Prog.*, 1869, 71, 72; also Vol. III. and Supplement. For further literature, see under Example 36.

ite still retained a limestone cap. The ore-bearing solutions, on reaching a shaly streak containing a limestone layer, departed from the fissure and followed under the limestone, so as to form a lateral enlargement, much like those described and figured from Newman Hill, Colorado, under 2.09.10. The Pahranagat and Tem Pahute districts, still farther south, have had some prominence, but the whole region is so far from the lines of transportation that the conditions are hard ones.[1]

2.11.14. Ney County, next west, has an important mining center, in its northern portion, around the town of Belmont. Quartzite and slates rest on granites in the order named, and in them are veins with quartz gangue and silver chlorides, affording very rich ores. Southeast of Belmont is Tybo.[2]

2.11.15. White Pine County lies to the northeast, and contains the White Pine district. The principal town is Hamilton, about 110 miles south of Elko, on the Central Pacific. The Humboldt range is prolonged southward in some broken hills, consisting chiefly of folded Devonian limestone. At Hamilton these are bent into a prominent anticline, and this has a strong fissure crossing the axis. The geological section is Devonian limestone, thin calcareous shale, thin siliceous limestone, argillaceous shale, probably Carboniferous sandstone, and Carboniferous limestone. The ore bodies occur, according to Arnold Hague, in four forms, all in the Devonian limestone: (1) in fissures crossing the anticlinal axis; (2) in contact deposits between the limestones and shales; (3) in beds or chambers in the limestone parallel to the stratification; (4) in irregular vertical and oblique seams across the bedding. The ore is chiefly chloride of silver in quartz gangue. It is thought by Mr. Hague to have probably come up through the main cross fissure, and, meeting the impervious shale, to have spread through the limestone in this way.[3]

Egan Cañon is in the northern part of the county and shows a geological section of granite, quartzite, and slate in the order named. In slates, and perhaps extending into the quartzite, is a quartz vein five to eight feet wide carrying gold and silver ores.

---

[1] E. P. Howell, *Wheeler's Survey*, III. 257. G. M. Wheeler, Report, *Wheeler's Survey*, 1869, p. 14.

[2] S. F. Emmons, *Survey of the Fortieth Parallel*, Vol. III., p. 393. G. K. Gilbert, "On Belmont and Neighborhood," *Wheeler's Survey*, III. 36.

[3] J. E. Clayton, "Section of the Rocks at Hamilton, Nev.," *Cal. Acad. Sci.* A. Hague, *Fortieth Parallel Survey*, Vol. III., p. 409.

Eureka County is the next county west of White Pine. The deposits at Eureka have already been described under "Lead-silver" (Example 36).

2.11.16. Lander County lies next west of Eureka. The Toyabe range runs through it from north to south and in its southern portion, in Ney County, contains the Belmont deposits. (See above, 2.11.14.) At Austin, which is 80 or 90 miles south of the Central Pacific Railroad, now connected with it by a branch, are the mines of the Reese River district, named from the principal stream near by. From Mount Prometheus, which consists of biotite granite or granitite, and which is pierced by a great dike of rhyolite, a western granite spur runs out known as Lander Hill. The ore bodies are in this hill, and are narrow fissure veins with a general northwest and southeast trend, carrying rich ruby silver ores, with gray copper, galena, and blende, in a quartz gangue with associated rhodochrosite and calcite. They are also often faulted. At times they show excellent banded structure. Antimony has recently been found in this region.[1] (See under "Antimony.")

2.11.17. Elko County lies north of White Pine and Eureka counties and contains the Tuscarora mining district. The deposits are high-grade silver ores in veins, in a decomposed hornblende andesite.[2]

Humboldt County is the middle county of the northern tier, and contains a number of mining districts, which produce both silver and gold from quartz veins in the Mesozoic slate. Small amounts of the precious metals come also from Washoe County, in the northwest corner of the State.[3]

Churchill County adjoins Lander on the west and possesses a few silver mines.

Esmeralda County, in the southwest, has a considerable number of rich silver and gold mines, which produce high-grade ores from veins, with a quartz gangue in metamorphic rocks, slates, schists, etc. (See also under "Nickel.")

2.11.18. Storey and Lyon are two small counties in the western central portion of the State, but the former contains the most important and interesting ore deposit in Nevada, if indeed it is not the largest and richest single vein yet discovered.

---

[1] S. F. Emmons, *Fortieth Parallel Survey*, Vol. III., p. 349.
[2] G. F. Becker, *Tenth Census*, Vol. XIII., p. 34.
[3] *Ibid.*, p. 33.

2.11.19. Comstock Lode. A great fissure vein, four miles long, forked into two branches above, along a line of faulting in eruptive rocks of the Tertiary age and chiefly andesites. In the central part of the vein the displacement has been about 3000 feet, shading out, however, at the ends. The ores are high-grade silver ores in quartz, and occur in great bodies, called "bonanzas," along the east vein. Over $325,000,000 in gold and silver has been extracted, in the ratio of two of the former to three of the

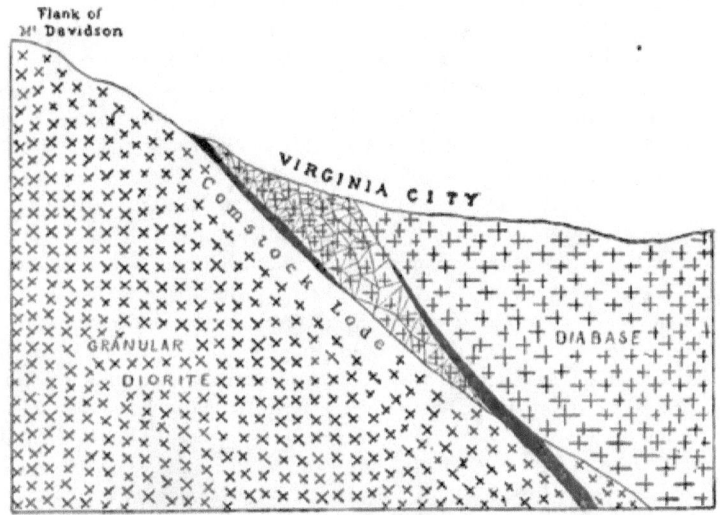

FIG. 57.—*Section of Comstock Lode on line of Sutro Tunnel. After G. F. Becker, Monograph III., U. S. Geol. Survey. The colors of the original are here indicated by line-work.*

latter. The vein lies on the easterly slope of a northeasterly spur of the Sierras. West of it is Mount Davidson. The outcroppings lie on the flank of the latter, about 6500 feet above the sea and 1500 feet below the summit. The general strike of the vein is east of south and it dips east. Views regarding the geology of the Comstock have changed in the course of years, as they have been influenced by the successive writings of Von Richthofen, King, Church, Becker, and Hague and Iddings, the points in especial controversy being the determinations of the rock species.

2.11.20. It may be remarked that the whole scheme of the

classification of our volcanic (effusive) rocks rests largely on Von Richthofen's early studies, and that perhaps the most important generalization of late years is due to the work of Hague and Iddings on the same. Von Richthofen (1885) described the ore body as filling a fissure on the contact of a so-called syenite and an eruptive rock that he called "propylite." The ore and gangue are thought to have been brought up from below by solfataric action, in which fluorine, chlorine, and sulphur were the principal dissolving agents. Clarence King (1867–68, published in 1870) brings out forcibly the fact that the footwall of the vein approximates closely the natural continuation of Mount Davidson, and contends that the vein filled a fissure between the syenite of which Mount Davidson consists and the late Tertiary eruptive rocks poured out against its flanks. He traced the geological succession of these and explained the filling of the vein by solfataric action, attendant on a thin dike of andesite, which forced its way into the contact. J. A. Church (1877) thought that the diorite (called syenite above) of Mount Davidson had been poured out originally in thin horizontal sheets, which were folded in east and west folds. This was to account for the bedding of the rocks of the lode as now seen. On the diorite was poured out next the propylite, likewise in successive horizontal sheets. Then they were all tilted along north and south axes, and eruptions of andesite penetrated between their sheets in very large amount. Further movements forced the convexities of the first-formed folds against the andesites and crowded their substance sidewise, to some extent, into the synclinals. This movement slightly parted the beds, affording watercourses through which rose siliceous waters. These dissolved away the neighboring beds, leaving extensive quartz bodies in their places. They also removed the andesite caps. No ore was formed as yet. Now followed great trachyte eruptions on the east, and they loaded the hanging wall of the lode so heavily as to cause a downward movement of it on the foot, making a new series of openings, and into these poured the ore-bearing solutions which brought the precious metals. No one who has intelligently followed this explanation will doubt that Mr. Church has shown great ingenuity, and yet few would be inclined to have very much confidence in this long, unnatural hypothesis when a simple course will lead to the same results. At the time of Mr. Church's visit the workings were becoming very deep and the great heat which has been since such an obstacle was manifesting itself. Flooded drifts, it was thought, had been observed to grow hotter, and from this the hypothesis of kaolinization was conceived. It was that the kaolinization of the feldspar in the deeply buried rock occasioned the heat of the lode.

2.11.21. G. F. Becker (1879–82) comments on the excessive alteration which the rocks have undergone, as it figures largely in his hypothesis of origin. He then traces the results of faulting, and shows that under conditions like those present the surface would tend to assume a logarithmic curve, which coincides surprisingly well with the present outline of the country. After describing the lode itself, the origin of its metalliferous contents is traced as follows. Waters under hydrostatic pressure from the heights to the west are supposed to have percolated toward the lode, passing through deeply buried regions of heat. They were probably diverted from rising directly through the lode by an impervious clay seam, and were thus forced to soak through the diabase hanging, relieving it in passage of the metals, which were afterward deposited in the higher portions of the lode. The metals themselves were probably largely derived from the augite of the rock. Mr. Becker had as an associate Dr. Carl Barus, who studied the heat phenomena (especially the hypothesis of kaolinization) and the electrical manifestations of the lode. The result of Dr. Barus's careful experiments threw great doubt on kaolinization as a source of heat. The electrical experiments were not satisfactory. They were carried on also at Eureka, Nev., but no very definite results were reached.

2.11.22. The correct determination of the eruptive rocks neighboring to the Comstock has been of great importance, not alone because of their scientific interest, but as bearing on the fact as to whether the lode itself was a contact fissure between two different rocks, or whether it was simply a fissure vein. It is worthy of note that in connection with it Von Richthofen developed one of the first important attempts to classify the volcanic rocks, and that Hague and Iddings have finally urged that the peculiar crystalline structures of all eruptive rocks depend primarily on the heat and pressure (i.e., depth below the surface) under which they have solidified, destroying thus the time element in classification. This is, to be sure, an old idea, but it gains its best confirmation from the Comstock. Von Richthofen, in his report to the Sutro Tunnel Company, and in his later memoir on "The Natural System of the Volcanic Rocks" (*Cal. Acad. Sci.*,

1867; also *Zeitschrift d. d. geol. Gesell.*, 1868, 663), distinguished in the Washoe district syenite, metamorphic rocks, quartz-porphyry, propylite, sanidine-trachyte, and very subordinate andesite. Mr. King referred much of the propylite of Von Richthofen to andesite, but retained the propylite as a distinct species, although remarking the close affinities of the two. The quartz-porphyry he called quartz-propylite. In other respects no changes are introduced. Zirkel (*Fortieth Parallel Survey*, Vol. VI.) determined the syenite as granular diorite, and while accepting hornblende-propylite and quartz-propylite as separate species, the greater part of the quartzose rock he called dacite. He introduced for the first time augite-andesite, rhyolite, and basalt. Mr. Church paid less attention to lithology, and used the terms of his predecessors somewhat loosely. Mr. Becker makes the following classification: granular diorite, porphyritic diorite, micaceous diorite-porphyry, quartz-porphyry, earlier diabase, later diabase, earlier hornblende-andesite, augite-andesite, later hornblende-andesite, and basalt. In this it will be seen that several new varieties are introduced, but the main mass of Mount Davidson was still considered diorite, and the vein was thought to lie between this and some of the other species mentioned, especially diabase. In 1885, Arnold Hague and J. P. Iddings completed new microscopical studies upon the materials collected by Mr. Becker, and the results were published as Bulletin 17 of the United States Geological Survey ("On the Development of Crystallization in the Igneous Rocks of Washoe," etc.). These two writers had had more to do with the eruptive rocks of the Great Basin and the Pacific slope than any other geologists, and hence brought to the review an exceptional experience. Nowhere else in the world are such exposures and thorough sections afforded, alike in depth and in horizontal extent. They proved that the diabase and augite-andesite shade into each other, the differences in crystallization being due to depth; that the hornblende of the so-called diorite was largely secondary from original augite, being derived by paramorphic change (uralitization), and that the diorite was a structural variety of the diabase; that the porphyritic diorites shade into the earlier hornblende-andesites and are structural varieties of them; that the mica-diorites and hornblende-andesite are identical in the same way; that the assumed Pre-Tertiary age of the quartz-porphyry was unwarranted, and that it was partly dacite and partly rhyolite, the two shading into each other; that the younger diabase, so called, of the sub-surface dike was identical with the rock elsewhere occurring on the surface and called basalt, and was really a basalt, owing its holocrystalline character to its depth; and finally,—the most important conclusion of all in this connection, although the other conclusions are among the most important advances made in late years,—"that the Comstock Lode occupies a line of faulting in rock of Tertiary age, and cannot be considered as a contact vein between two different rock masses." The crystalline structure of the Washoe rocks has been subsequently treated by Mr. Becker. ("The Washoe Rocks," *Bull. Cal. Acad. Sci.*, Vol. II., p. 93, January, 1887; "Texture of Massive Rocks," *Amer. Jour. Sci.*, ii., Vol. XXXIII., p. 50, 1887.) The various structures—granular, porphyritic, and glassy—are referred more to differences in composition and fluidity than to circumstances of solidification.[1]

---

[1] G. F. Becker, "Geology of the Comstock Lode and the Washoe District," *Monograph III., U. S. Geol. Survey.* Rec. See also *Engineering and Mining Journal*, March 1, 1884, p. 162; *Second Ann. Rep. Director U. S. Geol. Survey.* Rec. J. A. Church, *The Comstock Lode: Its Formation and History.* New York, John Wiley & Sons. Reviewed in *Engineering and Mining Journal*, Feb. 21, 1880, p. 397. See also shorter papers in the *Engineering and Mining Journal*, Dec. 28, 1878, p. 456; July 19, 1879, p. 35; Dec. 12, 1885, p. 397; Jan. 23, 1886, p. 52. "On the Changes in the Comstock Vein," *Engineering and Mining Journal*, Dec. 18, 1886, p. 434; "The Discovery of the Comstock Lode," *Ibid.*, Dec. 5 and 19, 1891, and other papers in 1892 by Dan Dequille. Hague and Iddings, "On the Development of Crystallization in the Igneous Rocks of Washoe, Nev.," etc., *Bull.* 17, *U. S. Geol. Survey.* See also *Bull.* 6, *Cal. Acad. Sci.*, and *Engineering and Mining Journal*, Dec. 11, 1886, p. 415.

CHAPTER XII.

THE PACIFIC SLOPE: WASHINGTON, OREGON, AND CALIFORNIA

WASHINGTON.

2.12.01. *Geology.*—Little is available in the way of systematic descriptions of the geology of Washington, and an attractive field remains to be developed. The rocks of the Rocky Mountains extend across the panhandle of Idaho and show in northeastern Washington, affording considerable amounts of ores. They are prevailingly granite and gneiss, which have escaped being covered by the enormous volcanic outbreaks of Tertiary and later time. West of the granites a great plateau country of somewhat diversified surface is met. It seems to have been an ancient lake basin, but is now covered by igneous rocks and deposits of volcanic tuff. Still farther west the Cascade chain forms the central divide of the State. The rocks are granites, flanked by Paleozoic, Mesozoic, and metamorphic strata, much like the Sierras of California. They were upheaved in large part before the Cretaceous, and since then other movements have occurred. There are vast developments of igneous rocks, forming, as at Mount Tacoma (Rainier), some of the highest American peaks. West of the Cascade range is a great valley formerly marking a drainage system, but now covered partly by glacial drift and partly by the waters of Puget Sound. The glacial deposits are enormous, and render the problem of working out the geology very difficult. Some glaciers remain on the heights even to the present day. West of the Puget Sound Basin is the northerly extension of the Coast range, locally called the Olympics, and largely Cretaceous and Tertiary strata.[1]

---

[1] G. F. Becker, *Tenth Census*, Vol. XIII., p. 27. G. A. Bethune, *First Ann. Rep. State Geologist*, 1891. A. Bowman, "Mining Developments on the Northwest Pacific Coast and their Wider Bearing," *M. E.*, XV. 707. J. MacFarlane, *Geol. Railway Guide*, second edition, p. 262; notes by Pumpelly, Willis, and others. Rec. J. S. Newberry, "Geology and Bot-

2.12.02. Good descriptions of the ore deposits of Washington are greatly needed. The First Annual Report of the State Geologist has little of scientific value, and the other accounts are ancient history. There are gold placers in Yakima, Stevens, and Kittitas counties, largely worked by Chinese. But in Okanogan and Stevens counties, in the northeast, the developments of deep mining for silver ores, although recent, are considerable. The veins are largely in metamorphic rocks and contain the usual sulphides in quartz gangue.[1]

OREGON.

2.12.03. *Geology.*—Northeastern and northern central Oregon are formed by a prolongation southward of the igneous plateaus of Washington. Slates and granite appear in Baker County on the east, in the Blue Mountains, and the geology seems to resemble the Sierras. All southeastern Oregon belongs in the Great Basin, which comes north from Nevada, but is better watered than the southern portion. It is traversed by several subordinate ranges of the block-tilted basin type. Of these the Stein Mountains are the most prominent. The general surface is formed by Quaternary lake deposits and great outbreaks of igneous rocks. West of the basin and the plateau the Cascade range traverses the State, and is cut by the Columbia River on the north and the Klamath on the south. The range consists of granite and metamorphic rocks, etc., the latter chiefly Mesozoic. In northern Oregon a broad valley intervenes between the Cascade and the Coast ranges, but in the southern part the two ranges run together, and their distinction has been only partly worked out. (See *Bull.* 33, *U. S. Geol. Survey.*) In the Coast range Cretaceous and Tertiary strata predominate.[2]

---

any of the Northern Pacific Railroad," *Trans. N. Y. Acad. Sci.*, III., 1884, p. 253. C. A. White, "Puget Group of Washington," *Amer. Jour. Sci.*, iii., XXXVI. 443. B. Willis, "Our Grandest Mountain and Deepest Forest," *School of Mines Quarterly*, VIII. 152. "Report on the Coal Fields of Washington Territory," *Tenth Census*, Vol. XV., p. 759. "Changes of River Courses in Washington Territory due to Glaciation," *Bull.* 40, *U. S. Geol. Survey.*

[1] G. A. Bethune, *First Ann. Rep. State Geol*, 1891. C. B. Fenner, "The Monte Cristo District, Snohomish County," *School of Mines Quarterly*, November, 1892. "The Mines of Kittitas County," *Engineering and Mining Journal*, Dec. 24, 1892, p. 608.

[2] G. F. Becker, *Tenth Census*, Vol. XIII., p. 27. T. Condon, "On

2.12.04. Oregon is an important producer of gold both from placers and from veins. Baker County, on the east, presents the characteristic placers and gold quartz of the California Sierras, and is the most productive section of the State. Grant and Josephine counties also have placers, and smaller amounts come from a few others. In the extreme southeast, near the California line, is Curry County, containing:

2.12.05. Example 44a. Port Orford. Auriferous beach sands at the foot of gravel cliffs, and shifting with the winds and tides. At Port Orford the ocean has access to great sea cliffs of gravel which it breaks down by the force of the waves. A sorting action ensues, performed by the undertow and the littoral current. The heavier gold dust is concentrated and gathered up by the miners at low tide. Some submarine work has also been attempted. The product is not great, and the deposit is chiefly interesting in its scientific bearing. It runs along into California as well. Auriferous sands occur at Yakutat Bay, Alaska. (See J. Stanley-Browne, *Nat. Geog. Mag.*, Vol. III., 1891.) The gold of the Potsdam sandstones of the Black Hills has been concentrated in a similar way in early geologic time, and the magnetite sands which were referred to under 2.03.13 furnish something of a parallel.[1]

Some Points Connected with the Igneous Eruptions along the Cascade Mountains of Oregon," *Amer. Jour. Sci.*, iii. XVIII. 406. J. S. Diller, "Notes on the Geology of Northern California," *Bull.* 33, *U. S. Geol. Survey*. J. C. Frémont, "Observations on the Rocky Mountains and Oregon," *Amer. Jour. Sci.*, ii., III. 192. George Gibbs, "Notes on the Geology of the Country East of the Cascade Mountains, Oregon," *Amer. Jour. Sci.*, ii., XX. 275. J. Leconte, "On the Great Lava Flood of the West, and on the Structure and Age of the Cascade Mountains," *Amer. Jour. Sci.*, iii., VIII. 167, 259. C. King, *Fortieth Parallel Survey*, Vol. I. J. MacFarlane, *Geol. Railway Guide*, p. 316. J. S. Newberry, *Pacific R. R. Reports*, Vol. VI., pp. 1-73. I. C. Russell, "A Geological Reconnoissance in Southern Oregon," *Fourth Ann. Rep. Director U. S. Geol. Survey*, pp. 435-462.

[1] G. F. Becker, *Tenth Census*, Vol. XIII., p. 27, general account of Oregon. W. P. Blake, "Gold and Platinum from Cape Blanco (Port Orford)," *Amer. Jour. Sci.*, ii., XVIII. 156. "Remarks on the Extent of the Gold Regions of California and Oregon," etc., *Amer. Jour. Sci.*, ii., XX. 72. A. W. Chase, "The Auriferous Gravel Deposits of Gold Bluffs, California," *Cal. Acad. Sci.*, 1874; *Amer. Jour. Sci.*, iii , VII. 367. "Dredging for Gold," *Engineering and Mining Journal*, June 23, 1883, p. 360. B. Silliman, "Cherokee Gold Washings," *Amer. Jour. Sci.*, iii., VI. 132. W. P. Watts, "Sands in Santa Cruz County, California," *Rep. Cal. State Mineralogist*, 1890, p. 622.

CALIFORNIA.

2.12.06. Geology.—The topography and geology of northern California have been but recently made clear. Diller considers that the southern end of the Cascade range is Mount Shasta; that the Sierras proper terminate near the north fork of the Feather River, but the line is continued about fifty miles farther north, in the Lassens Peak volcanic ridge, and that all else west and south of Mount Shasta belong to the Coast range. Central California, as is well known, has the Sierras on the east, the great Sacramento Valley in the middle, with the Coast range on the west. The arid regions of the Great Basin just touch the northeastern corner, but on the southern extremity they swing around and form a large part of the State. The Great Basin portion is formed by Quaternary lake deposits. The Sierras consist of central granite and gneiss, with great developments of slates and eruptives on their flanks. The excessive metamorphism has largely destroyed the fossils, but enough have been found to prove that while in large part Jurassic, yet Carboniferous and Cretaceous representatives are also present. The western slopes have the mantles of gravel, which have furnished so much gold, and with these are large outflows of basalt. The upheaval of the Sierras occurred before the middle Cretaceous time. The Coast range consists of rocks of late Cretaceous and early Tertiary age, extending into the Miocene. They were upheaved in post-Miocene time. Great outbreaks of andesite also occurred, and later basalts. The principal product of California is gold, but recently a district which furnishes considerable silver has been developed. This will first be described, in order to lead up to gold. The copper and iron resources have already been mentioned, and the mercury, antimony, and chromium deposits remain for description after the precious metals.[1]

---

[1] G. F. Becker, "Notes on the Early Cretaceous of California," *Amer. Jour. Sci.*, iii., II. 201. "Antiquities from under Tuolumne Table Mountain, California," *G. S. A.*, II. 189. "Cretaceous Metamorphic Rocks of California," *Amer. Jour. Sci.*, iii., XXXI. 348. "Structure of a Portion of the Sierra Nevada of California," *G. S. A.*, II. 50. "Notes on the Stratigraphy of California," *Bull.* 19, *U. S. Geol. Survey.* W. P. Blake, "Notes on California," *Amer. Jour. Sci.*, ii., XVIII. 441. W. H. Brewer epitomizes Whitney's report, *Amer. Jour. Sci.*, ii., XLI. 231; also 351. J. D. Dana, "Notes on Upper California," *Amer. Jour. Sci.*, ii., VII. 376. J. S. Diller, "Geology of the Lassen Peak District," *Eighth Ann. Rep. Di-*

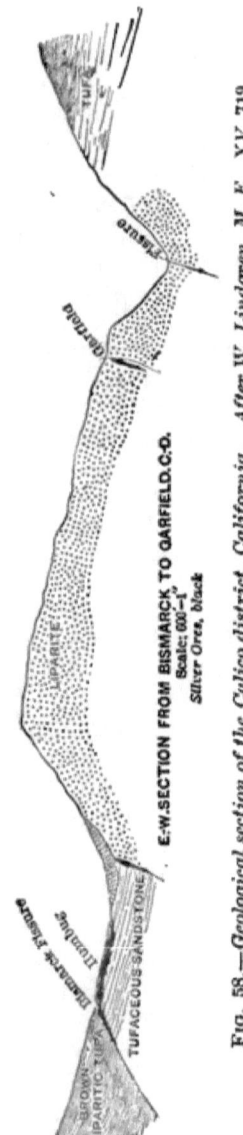

Fig. 58.—Geological section of the Calico district, California. After W. Lindgren, M. E., XV. 719.

2.12.07. Calico District. Deposits of silver chloride in fissure veins, small fractures and pockets in liparites and tufaceous sandstones, probably of the Pliocene series. They occur in southwestern California, in that portion of the State belonging rather to the Great Basin than to the Pacific slope. An immense outbreak of liparite has formed a series of elevations, and the attendant tufas are extensively developed.

rector U. S. Geol. Survey, pp. 401, 435. "On the Cretaceous Rocks of Northern California," Amer. Jour. Sci., iii., XL. 476. "On the Geology of Northern California," Proc. Phil. Soc. of Wash., Jan. 16, 1886; Abstract. Amer. Jour. Sci., iii., XXXIII. 152. "Geology of the Taylorville Region, Plumas County," Bull. Geol. Soc. Amer., III. 369. G. K. Gilbert, "The Recency of Certain Volcanoes of the Western United States," A. A. A. S., XXIII. 29. A. Hyatt, "Jura and Trias of Taylorville, California," Bull. Geol. Soc. Amer., III. 395. William Irelan, State Mineralogist, Ann. Rep., 1880, and following, especially 1890, geology by counties. J. Leconte, "Post-tertiary Elevation of the Sierra Nevadas, shown by the River Beds," Amer. Jour. Sci., iii., XXXII. 167. "Old River Beds of California," Ibid., iii., XIX. 190; iii, XXXVIII. 261. "Extinct Volcanoes about Lake Mono, and their Relations to the Glacial Drift," Ibid., iii., XVIII. 35. Jules Marcou, "Report on the Geology of a Portion of Southern California," Wheeler's Survey, Ann. Rep., 1876, App., p 158. J. E. Mills, "Stratigraphy and Succession of the Rocks of the Sierra Nevada of California," Bull. Geol. Soc. Amer., III. 413. E. Reyer, Theoretische Geologie, p. 537, 1888. I. C. Russell, "The Quaternary History of Mono Valley, California," Eighth Ann. Rep. Director U. S. Geol. Survey, pp. 267-400. H. W. Turner, "The Geology of Mount Diablo, with the Chemistry of the Rocks, by W. H. Melville," G. S. A., II.

The ore is thought by Lindgren to have come in heated solution from below and to have filled the fissures and overflowed, forming the surface deposits in the tufas. (Cf. Silver Cliff, Colorado.) There are deposits of gold ores in the same region.[1]

2.12.08. Example 44. Auriferous Gravels. (1) River gravels, or placers in the beds of running streams. These have been often referred to in other States, but the type is placed in Cali-

FIG. 59.—*View of the Union diggings, Columbia Hill, Nevada County, California. From a photograph.*

fornia, as they are there best known. They were the first gravels washed in 1849, and, although substantially exhausted by 1860, were very productive. Eastward from the great Sacramento

---

383. J. A. Veatch, "Notes on a Visit to the Mud Volcanoes of the Colorado Desert," etc., *Amer. Jour. Sci.*, ii., XXVI. 288. J. D. Whitney and others, reports of the California Geological Survey, issued at Cambridge, Mass. L. G. Yates, "Notes on the Geology and Scenery of the Islands forming the Southern Line of the Santa Barbara Channel," *A. G.*, V. 43.

[1] W. Lindgren, "The Silver Mines of Calico District, California," *M. E.*, XV. 717.

Valley the surface rises with a quite gentle gradient to the summit of the Sierras. The country consists chiefly of metamorphic rocks, which have yielded a very few well-determined fossils of Carboniferous, Jurassic, and Cretaceous ages; but the identity of

FIG. 60.—*View of the Timbuctoo diggings, Yuba County, California. From a photograph.*

the strata in all the stretch is difficult to make out, because where the fossils were originally present they are almost entirely destroyed by metamorphism. Down the slopes of the range the modern streams have flowed and cut deep cañons in which gravels have gathered. Out in the more open country the gravels have also accumulated and have furnished some productive bars. The gold has been derived principally from the quartz veins of the slates, which are later described, and has been mechanically concentrated in the streams. Before coming to its final rest it may have lodged in the high or deep gravels, of which mention will next be made.

It is accompanied by magnetite as a general thing, by zircon, garnet, and rarely by other heavy metals, such as platinum and iridosmine. The greatest amount is usually near the bed rock, and when this is at all porous the gold may work into it to a small distance from the top. The gold is usually in flattened pellets of all sizes, from the finest dust to nuggets of considerable weight. They show evidence of being water worn. The interesting phenomena connected with the possible circulation of the precious metal in solution through the gravels are discussed under the deep gravels. Important deposits of the same general character as these have also been dug over near Santa Fé, N. M.; in California Gulch, near Leadville, Colo.; at Fairplay, Colo.; in San Miguel County, Colorado; in the Sweetwater district, Wyoming; near Butte, Mont.; in Last Chance and Prickly Pear gulches, near Helena, Mont.; in the Black Hills; in southern Idaho, especially along the Snake River; and at various points in Washington and Oregon. Placers of this type have also been found on the slopes of the Green Mountains and in the Southern States, but they never have proved of serious importance.

2.12.09. (2) High or Deep Gravels. With the exhaustion of the river gravels the gold seekers of California were driven to prospect on the higher slopes, where auriferous gravels much less accessible had been long noted. Increasing observation and development have shown that these are the relics of a former and very extensive drainage system, which was more or less parallel with the present streams, but of greater volume. The beds lie in deep gulches in the slates, and are capped in most cases by basaltic lava flows or by consolidated volcanic tuffs, called cement. They extend some 250 miles along the Sierras and up to 5000 feet above the sea. They have at times great thickness, reaching 600 feet at Columbia Hill, but drop elsewhere to 1 or 2 feet. They vary from a maximum width in workable material of 1000 feet to a minimum of 150. The inclosing slates on the sides of the old river valley are called "the rims," and on them are sometimes found other gravels. In some districts channels, belonging to two or three periods of flow, have been traced. They tend to follow the softer strata, breaking at times across the harder rocks. The channel filling consists of gravel, sand, and clays, volcanic tuffs, and firm basalt. With these are great quantities of silicified trees, and even standing trunks project through some beds. The gravel is oftenest formed of white quartz boulders, but may contain all the metamorphic rocks of the neighborhood, and even boulders brought from a great distance. The gravel at times is cemented together by siliceous and calcareous matter, and then requires blasting; but loose gravel also occurs. The clays are locally called "pipe clays," and are often interbedded with sand layers. They are blue when unoxidized, giving rise to the term "blue lead," but red oxidized clays are not infrequent. The clays

FIG. 61.—*Generalized section of a deep gravel bed, with technical terms. After R. E. Browne, Rep. Cal. State Mineralogist, 1890, p. 437.*

contain many leaf impressions of species thought by Lesquereux to be late Tertiary. The gravels also contain bones of extinct vertebrates, and have afforded some authentic human remains and stone implements of good workmanship. The volcanic tuffs have been strong factors in modifying the original drainage lines. They have flowed into the ancient valleys in a state of mud and have then consolidated.

2.12.10. The richest gravels are those nearest the bed rock. In these the distribution of the gold is governed more or less by the character of the ancient channels. It favors the inside of bends and the tops of steeper runs. The gradients of the old channels were fairly high, often running 100 to 200 feet per mile. Gold has also been found by assay in pyrite that has been formed in the gravels since their deposition, and from this it is evident that the precious metal does circulate in solution with sulphate of iron, but on this slender foundation some quite unwarranted chemical hypotheses for the origin of nuggets have been based. Substantially all the gold has been derived by the mechanical degradation of the quartz veins in the slate.

2.12.11. The depths to which the modern streams have cut out their channels below the old drainage lines have received considerable attention. Whitney avers that no disturbance has taken place since the old gravels were laid down, but Leconte thinks that there has been a tilting or elevation of the higher parts of the range, all moving as a block. Becker has recently shown in the

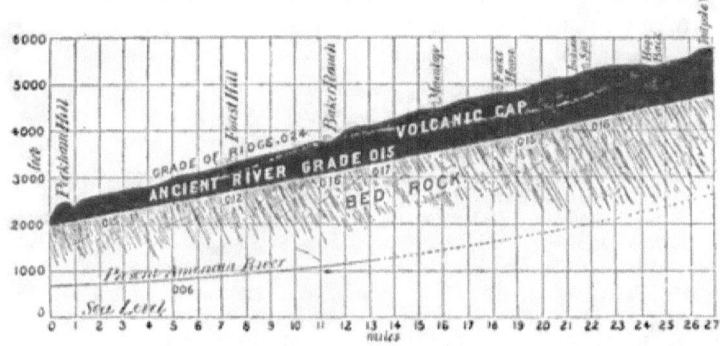

FIG. 62.—*Section of Forest Hill Divide, Placer County, California, to illustrate the relations of old and modern lines of drainage. After R. E. Browne, Rep. Cal. State Mineralogist*, 1890, p. 444.

high portions a great series of small north and south faults with uniform downthrow on the western side or upthrow on the eastern. (See paper below, cited from Geological Society of America.) This is of varied intensity in different portions and is limited to the strip just west of the summit. It occurred in the Pliocene and increased the gradient of the streams where the present deep cañons occur, but had no effect near the plains, where the old and new channels are nearly on the same level.

2.12.12. After the formation of the deep gravels and after the volcanic flows, glaciation took place in great extent over the mountain sides, but it was doubtless later in time than the glacial period of the East. References to the similar great development of the ice in Washington have already been made. Many hypotheses were early advanced to explain the deep gravels. They have been referred to the ocean, to ocean currents, and to glaciers; but it is now well established that they are river gravels, formed when the rainfall was probably in excess of what it is to-day, and when the attitude of the land toward the ocean may have been different.[1]

2.12.13. Example 45. Gold Quartz Veins. Veins of gold-bearing quartz, usually described as segregated veins, in slates and other metamorphic rocks, and more or less parallel with the bedding. The quartz contains auriferous pyrite, free gold, arsenopyrite, chalcopyrite, tetrahedrite, galena, and blende, but pyrites is far the most abundant. Some tellurides have been noted by Silliman at Carson Hill, Calaveras County. The veins approximate at times a lenticular shape, which is less marked in California than in some other regions, and which shows analogies of shape with pyrites lenses (Example 16) and magnetite lenses (Example 12). In such cases the fissure-vein character is some-

---

[1] G. F. Becker, "Notes on the Stratigraphy of California," *Bull.* 19, *U S. Geol. Survey.* "Structure of the Sierra Nevadas," *G. S. A.*, II. 43. W. P. Blake, "The Various Forms in which Gold Occurs," *Rep. Director of the Mint*, 1884, p. 573. A. J. Bowie, Jr., "Hydraulic Mining in California," *M. E.*, VI. 27. R. E. Browne, "The Ancient River Beds of the Forest Hill Divide," *Rep. Cal. State Mineralogist*, 1890, p. 435. Rec. T. Egleston, "Formation of Gold Nuggets and Placer Deposits," *M. E.*, IX. 63. "Working Placer Deposits in the United States," *School of Mines Quarterly*, VII., p. 101. J. H. Hammond, "Auriferous Gravels of California," *Rep. Director of the Mint*, 1881, p. 616. Rec. *Rep. Cal. State Mineralogist*, 1889, p. 105. H. G. Hanks, "Placer Gold," *Rep. Director of the Mint*, 1882, p. 728. H. G. Hanks and William Irelan, *Rep. Cal. State Mineralogist*, Annual. T. S. Hunt, "On a Recent Formation of Quartz, and on Silicification in California," *Engineering and Mining Journal*, May 29, 1880, 369. J. Leconte, "The Old River Beds of California," *Amer. Jour. Sci.*, iii., XIX. 80, p. 176. J. J. McGillivray, "The Old River Beds of the Sierra Nevada of California," *Rep. Director of the Mint*, 1881, p. 630. R. I. Murchison, "Siluria," etc. Contains a sketch of the distribution of gold over the earth. J. S. Newberry, "On the Genesis and Distribution of Gold," *School of Mines Quarterly*, Vol. III.; *Engineering and Mining Journal*, Dec. 24 and 31, 1881. J. A. Phillips, "Notes on the Chemical Geology of the California Gold Fields," *Philos. Mag.*, Vol. XXXVI., p. 321; *Proc. Roy. Soc.*, XVI. 294; *Amer. Jour. Sci.*, ii., XLVII. 134. B. Silliman, "On the Deep Placers of the South and Middle Yuba, Nevada County, California," *Amer. Jour. Sci.*, ii., XL. 1. J. D. Whitney, "Auriferous Gravels of the Sierras," Cambridge, 1880. "Climatic Changes in Later Geological Times," Cambridge.

what obscure. In California the veins occupy undoubted fissures in the slates. The largest and best known is the so-called Mother Lode, which is a lineal succession of innumerable larger and smaller quartz veins that run parallel with the strike, but which cut the steep dip of the slates at an angle of 10°. It was doubtless formed by faulting in steeply dipping strata. The wall rocks of the California veins are serpentine, diabase, diorite, and granite, as well as slate, for all these enter into the western slopes of the Sierras. The serpentine is probably a metamorphosed igneous rock, while the diabase and diorite form great dikes. Considerable calcite, dolomite, and ankerite occur with the quartz, and very often it is penetrated by seams of a green chloritic silicate, which is provisionally called mariposite, as it is probably not a definite mineral, but rather an infiltration of decomposition products. The quartz veins vary somewhat in appearance, being at times milk white and massive (locally called "hungry," from its general barrenness), at times greasy and darker, and again manifesting other differences, which are difficult to describe, although more or less evident in specimens. The richer quartz in many mines is somewhat banded, and is called ribbon quartz. The quartz has been studied in thin sections, especially in rich specimens, by W. M. Courtis, who shows that fluid or gaseous inclusions of what is probably carbonic acid are abundant. In rich specimens the cavities tend to be more numerous than in poor, but more data are needed to form the basis of any reliable deductions. Some quartz showed evidence of dynamic disturbances. The walls of the veins are themselves impregnated with the precious metal and the attendant sulphides. The rich portions of the veins occur in chutes to a large degree.

2.12.14. The great Mother Lode is the largest group of veins in California. It extends 112 miles in a general northwest direction. Beginning in Mariposa County, in the south, it crosses Tuolumne, Calaveras, Amador, and El Dorado counties in succession. It is not strictly continuous nor is it one single lode, but rather a succession of related ones, which branch, pinch out, run off in stringers, and are thus complex in their general grouping. Over 500 patented locations have been made on it. Whitney has thought it may have originated from the silicification of beds of dolomite, but others regard it, with greater reason, as a great series of veins along a fissured strip. The veins are often left in strong relief by the erosion of the wall rock, and thus are called ledges, or reefs. Some discussion has arisen over the condition of the gold in the pyrite, but in most cases it is the native metal mechanically mixed, and not as an isomorphous sulphide. It has been detected in the metallic state, in a thin section of a pyrite crystal from Douglass Island, Alaska, as later set forth (2.13.04), and the fact that it remains as the metal when the pyrite is dissolved in nitric acid makes this undoubtedly the general condition. The association of gold with bismuth, which has been shown by R. Pearce to occur in the sulphurets of Gilpin County, Colorado (referred to on p. 212), and the difficulty experienced in amalgamating some ores, indicate the possibility of other combinations. When crystallized, gold has shown, in one specimen and another, nearly all the holohedral forms of the isometric system, but the octahedron and rhombic dodecahedron are commonest. The veins are younger than any of the igneous dikes with them. They may have been filled, as thought by Whitney, during the metamorphism of the rocks attendant upon their upheaval in post-Jurassic time. Certain it is that a very extensive circulation of siliceous solutions was in progress. For the gold in the similar veins of Australia a precipitation by organic matter has been urged. (See William Nicholas, "The Origin of Gold in Certain Victorian Reefs," *Engineering and Mining Journal*, Dec. 15, 1883.) In the development of explanations of origin, however, a wide field for study yet remains.[1]

---

[1] M. Attwood, "On the Wall Rocks of California Gold Quartz and the Source of the Gold," *Rep. Cal. Mineralogist*, 1888, p. 771 (thought to be due to igneous injection in diabase). W. P. Blake, "On the Parallelism between the Deposits of Auriferous Drift of the Appalachian Gold Field and those of California," *Amer. Jour. Sci.*, ii., XXVI. 128. "Remarks on the Extent of the Gold Region of California and Oregon," etc., *Ibid.* ii., XX. 72. "The Carboniferous Age of a Portion of the Gold-bearing Rocks of California," *Ibid.*, ii., XLV. 264. W. H. Brewer, reply to above, *Ibid.*, ii., XLV. 397. A. Bowman, "Geology of the Sierra Nevada in Relation to Vein Mining," *Min. Resources West Rocky Mountains*, 1875, p. 441. W. H. Brewer, "On the Age of the Gold bearing Rocks of the Pacific Coast," *Amer. Jour. Sci.*, ii., XLII. 114. F. G. Corning, "The Gold Quartz Mines of Grass Valley, California," *Engineering and Mining Journal*, Dec. 11, 1886, p. 418. W. M. Courtis, "Gold Quartz," *M. E.*, 1889, Ottawa meeting. H. W. Fairbanks, "Geology of the Mother Lode," *Tenth Ann. Rep. Cal. Mineralogist;* also in briefer form in *Amer. Geol.*, April, 1891, p. 201. Rec. "On the Pre-Cretaceous Rocks of the California Coast Ranges," *Amer. Geol.*, March, 1892, February, 1893. J. H. Hammond,

"Mining of Gold Ores in California," *Tenth Ann. Rep. State Mineralogist*, p. 852. Rec. P. Laur, "Du Gisement et de l'Exploration de l'Or en Californie," *Ann. des Mines*, Vol. III., 1863, p. 412. G. W. Maynard, "Remarks on Gold Specimens from California," *M. E.*, VI. 451. J. S. Newberry, "On the Genesis and Distribution of Gold," *School of Mines Quarterly*, III., p. 16. A. Remond, "Mining Statistics," *No. 1, Cal. Geol. Survey* (tabular statement of quartz mining and mills between the Merced and Stanislaus rivers). J. A. Phillips, "Mining and Metallurgy of Gold and Silver," Wiley, N. Y.; also treatise on Ore Deposits, p. 524. Rec. C. M. Rolker, "The Late Operations in the Mariposa Estate," *M. E.*, VI. 145. B. Silliman, "Notice of a Peculiar Mode of Occurrence of Gold and Silver in the Foothills of the Sierra Nevada, California," *Amer. Jour. Sci.*, ii., XLV. 92; *Cal. Acad. Sci.*, Vol. III., p. 353. H. M. Turner, review of recent papers by H. W. Fairbanks and others on California geology, in *Amer. Geol.*, June, 1893. Rec. J. D. Whitney, *Cal. Geol. Survey*, Geology, Vol. I., p. 212. J. S. Wilson, "On the Gold Regions of California," *Quar. Jour. Geol. Sci.*, Vol. X., p. 308, 1854.

CHAPTER XIII.

GOLD ELSEWHERE IN THE UNITED STATES AND IN CANADA.

2:13.01. Example 45a. Southern States. (1) Gold quartz veins (segregated veins) in metamorphic slates, talcose schists, etc., of late Archæan or early Palæozoic age, with numerous associated trap (diabase) dikes. (2) Beds of auriferous slates, gneiss, feldspathic and hydromica schists, and even limestone. The general geology of the southern Atlantic States has been outlined in the introduction. Reference may again be made to the Coastal Plain of Quaternary, Tertiary, and Mesozoic rocks, and to the Archæan strip back of this. In the latter are found the gold deposits. At times they resemble the Western quartz veins, but they are also extremely diverse in character, and involve almost every sort of rock. Gold has even been found in a trap dike by Genth. It is generally in pyrite, and the rock, where productive, is heavily charged with this mineral. The trap dikes have also exerted an important influence, and in some localities, as at the Haile mines, South Carolina, the rock is rich only near them. They have probably stimulated the ore-bearing solutions. The belt of auriferous rocks begins in Maryland, although gold is known in the States farther north. It runs with varying width into Alabama, where it terminates. It reaches a maximum of 70 miles in North Carolina. The country rock in these unglaciated regions is often covered to a great depth by the residual clays and other products of its alteration. These are as much as 100 feet in places. This material is sometimes called laterite. Where the original rocks have been auriferous it has furnished loose material for panning and washing, which is essentially different from the Western placers. It gradually works down hill, and has been called by Kerr "frost drift." The ores of the Southern States are generally low grade, and need careful management to be made profitable.[1]

2.13.02. *Example 45b.* Ishpeming, Mich. In the metamorphic rocks of the iron-bearing areas in the Lake Superior region gold-bearing quartz veins have been discovered and several mines have been opened. The Ropes mine is the best known and has yielded some good ore.[2]

ALASKA.

2.13.03. *Geology.*—Our knowledge of the geology of Alaska is very fragmentary, and is chiefly obtained from the reports of exploring expeditions along the coast and up the Yukon River. The Cretaceous and Tertiary rocks of Washington and of British America extend north into the southern portion. Igneous rocks

---

[1] W. P. Blake, A. Eaton, D. Olmstead, C. U. Shepard, the elder Silliman, and others have made many references to gold in the Southern States in the early numbers of the *American Journal of Science.* H. M. Chance, "Auriferous Gravels of North Carolina," *Amer. Philos. Soc.*, July 15, 1881, p. 477. H. Credner, "Report of Explorations in the Gold Fields of Virginia and North Carolina," *Amer. Jour. of Mining*, 1868, pp. 361, 377, 393, 407. W. B. Devereux, "Gold and its Associated Minerals at Kings Mountain, North Carolina," *Engineering and Mining Journal*, Jan. 15, 1881, p. 39. S. F. Emmons, "Notes on the Gold Deposits of Montgomery County, Maryland," June 11, 1881, p. 397. F. A. Genth, "Contributions to Mineralogy," *Amer. Jour. Sci.*, ii., XXVIII. 246. F. C. Hand, "Southern Gold Fields," *Engineering and Mining Journal*, Dec. 7, 1889, p. 495. G. B. Hanna, "Fineness of Native Gold in the Carolinas and Georgia," *Ibid.*, Sept. 18, 1886, p. 201. W. C. Kerr, "Gold Gravels of North Carolina," *M. E.*, VIII. 462. "Some Peculiarities in the Occurrence of Gold in North Carolina," *M. E.*, X. 475; also Geological Report on North Carolina, 1875. O. M. Lieber, "Ueber das Gold-vorkommen in North Carolina," *Gangstudien*, Vol. III., p. 253. Lieber states that Whitney's *Metallic Wealth* was written to "boom" certain mines! "Gold in South Carolina," *Gangstudien*, Vol. III., pp. 253, 481. P. H. Mell, "Auriferous Slate Deposits in Southern Mining Regions," *M. E.*, IX. 399. W. B. Phillips, "The Lower Gold Belt of Alabama," *Bull. No. 3, Geol. Survey Ala.*, 1892. E. G. Spillsbury, "Gold Mining in South Carolina," *M. E.*, XII. 99; *Engineering and Mining Journal*, June 23, 1883, p. 362. A. Thies and W. B. Phillips, "The Thies Process, etc., at the Haile Mine, South Carolina," *M. E.*, September, 1890. A. Thies and A. Mezger, "Geology of the Haile Mine," *M. E.*, September, 1890.

[2] C. D. Lawton, *Rep. Mich. Com. of Mineral Statistics*, 1887, p. 167. "The New Michigan Gold Field," *Engineering and Mining Journal*, Sept. 22, 1888, p. 238. M. E. Wadsworth, *Am. Rep. Mich. State Geologist*, issued January, 1892.

often pierce them. Tertiary coals are recorded from several localities in the archipelago in the southern part. The Aleutian Islands mark a long stretch of volcanic eruptions. W. H. Dall has given a general section of the banks of the Yukon River from the British boundary to the delta. There are, in rough order from east to west, granite, talcose slates, and Azoic rocks, 150 miles; sandstone and conglomerate, 250 miles; shales with one small coal seam, 75 miles; blue slate, conglomerate, eruptive rocks, slate, eruptive rock, blue slate, and black sandstone, 325 miles; blue sandstone, slate, trap, blue and black slate, volcanic rock, to the northwest mouth. The interior is largely composed of tundras, or great plains of moss and other plants, frozen into perpetual ice at a depth of a foot or so. These hide the geology. Doubtless the slates of the upper waters have supplied the gold, which is rich in some of the river gravels. On the Aleutian Islands some brown Miocene sandstones are seen (Shumagin), and granite and metamorphic rocks.[1]

2.13.04. Example 46. Douglass Island. A dike or boss of granite 400 feet wide, piercing slates regarded as Triassic by G. M. Dawson, and impregnated (i.e., the granite) with auriferous pyrites. This enigmatical ore body is thought by Dawson to be the upper part of a granite dike. It consists in great part of a mass of quartz, feldspar, calcite, and pyrite, in which are buried the so-

---

[1] "Alaska as a Mining Territory," *Engineering and Mining Journal*, June 27, 1885, p. 444. "Mineral and Agricultural Wealth of Alaska," *Engineering and Mining Journal*, Aug. 24, 1887, p. 134. T. A. Blake, Rep. on the Geol. of Alaska, *Ex. Doc.* No. 177, Fortieth Congress, New Series, p. 314, Washington, 1868. W. H. Dall, "Explorations in Alaska," *Amer. Jour. Sci.*, ii., XLV. 96. Rec. "Notes on Alaska and the Vicinity of Bering Straits," *Ibid.*, iii., XXI. 104. "Notes on Alaska Tertiary Deposits, Geological Section of the Shumagin Islands," *Ibid.*, iii., XXIV. 67. "Alaska and its Resources," Washington, 1870. Rec. "Glaciation in Alaska," *Bull. Phil. Soc.*, Vol. VI., p. 33, Washington, 1884. G. H. Dawson, "Report on the Yukon District in 1887," *Geol. Survey of Canada*, 1887-88, Vol. III., Part B, pp. 14B-18B, 154B-156B. H. W. Elliot, "Our Arctic Provinces," p. 163, New York, 1887. E. J. Glave, "Pioneer Packhorses in Alaska," *The Century*, September and October, 1892. R. G. McConnell, "Glacial Features of Parts of the Yukon and Mackenzie Basins," *Geol. Soc. of Amer.*, I., p. 540. I. C. Russell, "The Surface Geology of Alaska," *Geol. Soc. of Amer.*, I., p. 99. E. R. Skidmore, "Alaska," *Rep. Director of the Mint*, 1883, p. 17, and 1884, p. 17. J. Stanley-Brown, "Auriferous Sands at Yakutat Bay, Alaska," *Nat. Geog. Mag.*, Vol. III., 1891.

called kernels, which have been shown by F. D. Adams to be masses of less altered granite, almost without pyrite. Both varieties show abundant cataclastic or crushed structure, as an evidence of having suffered from dynamic movements. Adams concludes that the mass was originally a hornblende granite that has been subjected to solfataric action, which has brought in the gold. The gold in itself is in irregular masses in the pyrite. The Treadwell mine, located here, is very extensive, and the chief source of Alaska bullion.[1]

2.13.05. Example 45c. Nova Scotia. The southeastern portion of Nova Scotia is composed of Cambrian slates. They stretch from Canso to Yarmouth, and, together with associated granites, cover from 6000 to 7000 square miles. There are two well-marked divisions. The upper, 3000 feet thick, consists of dark pyritous slates, with beds of quartzite and small irregular veins; the lower, 8000 feet thick, has quartzites, sandstones, and slates, which in parts contain the veins. The slates are folded along east and west axes. The veins are not large, averaging from 4 to 8 inches, while 20 inches is very exceptional. The gold is both free and associated with the usual sulphides, among them often mispickel. The assays are not high, but with careful working the mines pay good returns.[2]

---

[1] F. D. Adams, "On the Microscopical Character of the Ore of the Treadwell Mine, Alaska," *Amer. Geol.*, August, 1889, p. 88. G. M. Dawson, "Notes on the Ore Deposits of the Treadwell Mine, Alaska," *Amer. Geol.*, August, 1889, p. 84. *Mining and Scientific Press*, San Francisco, Sept. 27, Oct. 4, 1884.

[2] J. W. Dawson, "On Recent Discoveries of Gold in Nova Scotia," *Canadian Naturalist and Geologist*, December, 1861. E. Gilpin, Jr., "The Nova Scotia Gold Mines," *M. E.*, XIV. 674. Rec. H. Y. Hind, "Report on the Mount Uniacke, Oldham, and Renfrew Gold Mining District," Halifax, 1872; *Amer Jour. Sci.*, iii., IV. 497. D. Honeyman, "On the Geology of the Gold Fields of Nova Scotia," *Quar. Jour. Geol. Sci.*, Vol. XVIII., p. 342, 1862. T. S. Hunt, "On the Gold Region of Nova Scotia," *Can. Geol. Survey*, 1868; *Canadian Naturalist*, February, 1868. W. E. Logan, "Notes on the Gold of Eastern Canada," *Can. Geol. Survey*, 1864. O. C. Marsh, "The Gold of Nova Scotia," *Amer. Jour. Sci.*, ii., XXXII. 395. A. Michel and T. S. Hunt, "Report on the Gold Region of Canada," *Can. Geol. Survey*, 1866. H. S. Poole, "The Gold Leads of Nova Scotia," *Quar. Jour. Geol. Sci.*, Vol. XXXVI., p. 307. A. R. C. Selwyn, "On the Gold Fields of Quebec and Nova Scotia," *Can. Geol. Survey*, 1870-71, pp. 252-289. B. Symons, "The Gold Fields of Nova Scotia," *Trans. Min. Asso. and Inst. Cornwall*, III. 80, 1892.

2.13.06. Example 45d. Gold elsewhere in Canada. At the headwaters of the Chaudière River, in eastern Quebec, auriferous gravels have been located, and also quartz veins in the metamorphic rocks. They have been worked to a small extent. Gold occurs in many places north and west of Lake Superior in the region of the Lake of the Woods. Some small mining has been carried on. An interesting and important vein, carrying auriferous mispickel in quartz, occurs in Marmora, Hastings County, just north of Lake Ontario. It is more fully described under "Arsenic," as it is the one American source of that metal. Auriferous gravels occur in British Columbia in not a few places and are being worked. They are also found along the disputed boundary of Alaska, and scattered reports of their existence have reached civilization from the Mackenzie River and the remote Northwest.[1]

2.13.07. The following table gives an idea of the relative importance of the several States. Full details of the United States and other countries are given in the Annual Reports of the Director of the Mint, the *Mineral Resources* of the United States Geological Survey, and the annual statistical number of the *Engineering and Mining Journal*.

	1881.		1890.	
	Silver.	Gold.	Silver.	Gold.
Alaska.................		$15,000	$9,697	$762,500
Arizona................	$7,300,000	1,060,000	1,292,929	1,000,000
California	750,000	18,200,000	1,163,636	12,500,000
Colorado...............	17,160,000	3,300,000	24,307.070	4,150,000
Dakota.................	70,000	4,000,000	129 292	3,200,000
Georgia................		125,000	517	100,000
Idaho..................	1,300,000	1,700,000	4,783,838	1,850,000

---

[1] "Descriptive Catalogue of the Economic Minerals of Canada at the Colonial and Indian Exhibition," London, 1886, p. 54. "The Marmora Gold Mine," *Engineering and Mining Journal*, Oct. 23, 1880, p. 266. R. Bell, "Mineral Resources of the Hudson Bay Territories," *M. E.*, February, 1886. E. Coste, "Report on the Gold Mines of the Lake of the Woods," *Rep. Prog. Can. Survey*, 1882-84, K, 3-22. W. M. Courtis, "Animikie Rocks and their Vein Phenomena as shown at the Duncan Mine, Lake Superior," *M. E.*, XV. 671. R. W. Ells, "Mining Industries of Eastern Quebec," *M. E.*, October, 1889. T. S. Hunt, "On Gold in the Laurentian Rocks of Canada," *A. A. A. S.*, XVIIth meeting. A. Michel and T. S. Hunt, "Report on the Gold Regions of Canada," *Can. Geol. Survey*, 1866-68. R. P. Rothwell, "The Gold-bearing Mispickel Vein of Marmora, Ontario," *M. E.*, IX. 409.

GOLD ELSEWHERE IN UNITED STATES AND CANADA.

	1881.		1890.	
	Silver.	Gold.	Silver.	Gold.
Montana	$2,630,000	$2,330,000	$20,363,636	$3,300,000
Nevada	7,060,000	2,250,000	5,753,535	2,800,000
New Mexico	275,000	185,000	1,680,808	850,000
North Carolina		115,000	7,757	118,500
Oregon	50,000	1,100,000	96,969	1,100,000
South Carolina		35,000	517	100,000
Utah	6,400,000	145,000	9,050,505	680,000
Washington		120,000	90,505	204,000
Other States	5,000	20,000	461,574	130,000
Total	$43,000,000	$34,700,000	$70,485,714	$32,845,000

A glance at the table will indicate where the heaviest producers and most important districts are situated. In the next few years a heavy falling off in silver and a relative increase in gold seems inevitable.

CHAPTER XIV.

THE LESSER METALS: ALUMINIUM, ANTIMONY, ARSENIC, BISMUTH, CHROMIUM, MANGANESE.

ALUMINIUM.

2.14.01. The importance of aluminium grows with improved and cheaper methods of production. Its sources have been alums, either natural or artificial, corundum, cryolite, and bauxite. The first of these is formed in nature by the decay of pyrite in shales and slates, and is little, if at all, used at present. The second is now more valuable as an abrasive. Cryolite ($Al_2F_6.6NaF$), a peculiar mineral, occurring in quantity only in Greenland, has been most largely employed until the recent discoveries of bauxite have made it less necessary. The cryolite forms an immense bed or vein in gneiss at Evigtok, on the Arksut Fjord, Greenland.[1] Bauxite ($Al_2O_3.3H_2O$) has long been valuable as a refractory material, but at present it is also used as a source of aluminium. Bauxite occurs in quantity in several of the Southern States. In Floyd County, Georgia, it covers about half an acre. Near Little Rock, Ark., it is in greatest quantity, and forms an interbedded mass in ferruginous Tertiary sandstone. Small amounts of oxide of iron partially replace the alumina at times, but the average grade is better than the foreign. It is considered by J. F. Williams as probably a hot-spring deposit, but the origin is obscure. Discovered deposits aggregate over a square mile and range up to 40 feet thick.[2]

---

[1] G. Hagermann, "On Some Minerals associated with the Cryolite in Greenland," *Amer. Jour. Sci.*, ii., XLII. 93. J. W. Taylor, "On the Cryolite of Evigtok, Greenland," *Quar. Jour. Geol. Sci.*, XII. 140.

[2] J. C. Branner, "Bauxite in Arkansas," *Amer. Geol.*, Vol. VII., p. 131, 1891. *Ann. Rep. Geol. Survey Arkansas*, 1888, Vol. I. J. F. Williams, *Ann. Rep. Geol. Survey Arkansas*, 1880, Vol. II., p. 124. E. Nichols, "An Aluminium Ore, Bauxite," *M. E.*, XVI. 905.

ANTIMONY.

Senarmontite, Sb_2O_3; Sb. 83.56; O. 16.44.

Stibnite (Antimonite, Antimony Glance), Sb_2S_3; Sb. 71.8; S. 28.2.

2.14.02. Antimony occurs in composition with several silver ores, but almost its sole commercial source is stibnite. The oxide, senarmontite, is rarely abundant enough to be an ore. Stibnite was one of the minerals formerly cited as having originated in veins by volatilization from lower sources. But it has probably, in all cases, been derived from solutions of alkaline sulphides.

2.14.03. Example 47. Veins containing stibnite, usually in quartz gangue. California, Kern County. At San Emigdio a vein of workable size has been found. It has a quartz gangue and is in granite. The vein varies from a few inches to several feet across, and has afforded some metal. Several others are known in San Benito and Inyo counties.

2.14.04. Nevada, Humboldt County. Stibnite has been known for some years in veins with quartz gangue. The Thies-Hutchens mines, about 15 miles from Lovelock station, were productive in 1891. Lander County. The most important of the American mines are the Beulah and Genesee, at Big Creek, near Austin. The vein is reported as showing three feet of nearly pure stibnite. It produced 700 tons of sulphide in 1891, and was operated in 1892.

2.14.05. Arkansas, Sevier County. Stibnite occurs in veins with quartz gangue in southwestern Arkansas. Some attempts have been made to develop them, but the ore is reported to be too remote for profitable working. The veins appear to be generally interbedded in Trenton shales and to lie along anticlinal axes, which trend northeast. They are all controlled by the United States Antimony Company of Philadelphia.

2.14.06. New Brunswick, York County. Veins of quartz or quartz and calcite, carrying stibnite, occur over several square miles. The wall rocks are clay slates and sandstones of Cambro-Silurian age. The mines have been commercially productive. The veins vary from a few inches to six feet.

2.14.07. Example 48. Utah, Iron County. Disseminations of stibnite in sandstone and conglomerate, following the stratification. In Iron County, southwestern Utah, masses of radiating needles occur in sandstones and between the boulders of an associated conglomerate. Very large individual pieces have been obtained, but not enough for profitable mining. Blake thinks that the ore has crystallized from descending solutions. Eruptive rocks are present above the sandstones.

2.14.08. An interesting deposit of senarmontite was worked for a time in Sonora, just south of the Arizona line, but it was soon exhausted.[1]

ARSENIC.

2.14.09. This metal occurs with many silver ores in the West and in arsenopyrite, or mispickel, a not uncommon arseno-sulphide in the gold quartz veins, east and west. At the Gatling mines, in the town of Marmora (more lately called Deloro), in Hastings County, Ontario, auriferous mispickel occurs in great quantity in granite, in veins with quartz gangue. Considerable oxide of arsenic has been obtained in the past from the roasters, but for five years or more the mines have not been productive. For reference to the printed descriptions see under "Gold in Canada" (2.13.07).

BISMUTH.

2.14.10. Bismuth occurs with certain silver ores in the San Juan district, Colorado, and is referred to in describing the country under "Silver and Gold" (2.09.10). Lane's mine, at Monroe, Conn., has furnished museum specimens of native bismuth in quartz. Some neighboring parts of Connecticut have afforded

---

[1] General references: W. P. Blake, "General Distribution of Ores of Antimony," *Mineral Resources of the U. S.*, 1883–84, p. 641. Arkansas: T. B. Comstock, *Geological Survey of Arkansas*, 1888, I., p. 136. F. P. Dunnington, "Minerals of a Deposit of Antimony Ores in Sevier County, Arkansas," *A. A. A. S.*, 1877. Rec. J. W. Mallet, *Chemical News*, No. 533. C. E. Waite, "Antimony Deposits of Arkansas," *M. E.*, VII. 42. C. P. Williams, "Notes on the Occurrence of Antimony in Arkansas," *M. E.*, III. 150. California: W. P. Blake, "Kern County," *U. S. Pac. R. R. Explorations and Survey*, Vol. V., p. 291. H. G. Hanks, *Rep. Cal. State Mineralogist*, 1884. See also subsequent reports by William Irelan, Jr. Mexico: E. T. Cox, "Discovery of Oxide of Antimony in Sonora," *Amer. Jour. Sci.*, XX. 421. J. Douglass, "The Antimony Deposit of Sonora," *Engineering and Mining Journal*, May 21, 1881, p. 350. Nevada: *Engineering and Mining Journal*, 1892, p. 6. New Brunswick: L. W. Bailey, "Discovery of Stibnite in New Brunswick," *Amer. Jour. Sci.*, ii., XXXV. 150, and in *Rep. on the Geol. of New Brunswick*, 1865; also H. Y. Hind, in the same. Utah: D. B. Huntley, *Tenth Census*, Vol. XIII., p. 463.

bismuth minerals, and not a few other places in the country contain traces, but the San Juan is the only serious one as yet.[1]

CHROMIUM.

2.14.11. Chromite, whose theoretical composition is $FeO.Cr_2O_3$, with Cr_2O_3 68%, often has MgO and Fe_2O_3 replacing its normal oxides. The percentage of Cr_2O_3 is thus reduced. It is always found in association with serpentine, which has resulted from the alteration of basic rocks consisting of olivine, hornblende, and pyroxene. These minerals contain the chromic oxide probably as a base in their fresh condition, but lose it on alteration. A chrome spinel, picotite, which is an original mineral in these rocks, likewise affords it. The chromite is scattered through the serpentine, often forming masses of large size. Traces of nickel minerals are frequently noted associated with the chromite.

2.14.12. Example 49. Disseminations of chromite in serpentine. Pennsylvania and Maryland. Great areas of this rock are found in southeastern Pennsylvania and in the adjacent parts of Delaware and Maryland. Considerable mining has been done in the past. Woods mine, in Lancaster County, Pennsylvania, has furnished great quantities, and other large producers are situated in the Bare Hills, near Baltimore. This section is now no longer commercially productive. Chromite has also been announced from several places in the South, no one of which has yet sent notable quantities to the market.

2.14.13. California. As already mentioned under the precious metals, great areas of serpentine occur on the western flanks of the Sierras and in the Coast range. In Del Norte, San Luis Obispo, Placer, and Shasta counties, California, they furnish commercial amounts of chromite. In some places the ore is followed by underground mining, and in others it is gathered as float material. The irregular distribution, always characteristic of the mineral, renders underground work uncertain. Good ore should afford 50% Cr_2O_3, and in California no ore less than 47% is accepted. It brings in the East $22 to $35 per ton. Considerable quantities are imported.[2]

---

[1] *Mineral Resources of the U. S.*, 1885, p. 399. B. Silliman, "Bismuthinite from the Granite District, Utah," *Amer. Jour. Sci.*, iii., VI. 123. H. L. Wells, "Bismuthosphærite from Willimantic and Portland, Conn.," *Amer. Jour. Sci.*, iii., XXXIV. 271.

[2] F. D. Chester, *Ann. Rep. Penn. Survey*, 1887, p. 93, describes the

COBALT (SEE UNDER "NICKEL").

MANGANESE.

2.14.14. Ores: Pyrolusite MnO_2, Mn. 63.2, braunite, Mn_2O_3, Mn 69.68. Some SiO_2, which may be chemically combined, is usually present, and small amounts of MgO, CaO, etc. Psilomelane has no definite composition, but usually contains barium or other impurities. An Arkansas variety has afforded Brackett MnO, 77.85.

There are various other oxides and hydroxides, which are rarely abundant enough to be ores. The carbonate, rhodochrosite, and the silicate, rhodonite, are rather common gangue minerals with ores of the precious metals. Franklinite is also an important source (2.07.04). Pyrolusite and psilomelane are the commonest ores the country over, but braunite is the one in the Batesville (Ark.) region. Manganese is widely distributed, and yet is commercially important in but few localities. It imitates limonite very closely in its occurrence and is often associated with this ore of iron. To make a manganese ore valuable, at least 40% metallic manganese should be present, and this is a lower limit than was formerly admissible when the ores were chiefly used in chemical manufactures. Under present conditions, if iron is present, the ore may be suited to spiegel, although even lower in manganese than 40%. Further, there should be low phosphorus; Penrose says not over 0.2 to 0.25% in Arkansas, and not over 12% SiO_2. High-grade ores run 50 to 60% manganese.

2.14.15. Example 50. Manganese ores, chiefly psilomelane

---

serpentine along the State line near Delaware. D. T. Day, *Mineral Resources of the U. S.*, 1882, p. 428; 1883–84, p. 567 (Rec.); 1885, p. 357; 1886, p. 176; 1887, p. 132. J. Eyerman, "On Woods Mine, Pennsylvania," *Mineralogy of Penn.*, Easton, 1889. P. Fraser, "The Northern Serpentine Belt in Chester County, Pennsylvania," *M. E.*, XII. 349. Rep. C3 (Lancaster County), *Penn. Geol. Survey.* T. H. Garrett, "Chemical Examination of Minerals associated with Serpentine," *Amer. Jour. Sci.*, ii., XIII. 45, and XV. 332. Also F. A. Genth, *Amer. Jour. Sci.*, ii., XLI. 120. E. Goldsmith, "Chromite from Monterey County, California," *Phil. Acad. Sci.*, 1873, p. 365. William Irelan, Jr., *Reps. Cal. State Mineralogist*, especially 1890, pp. 167, 189, 313, 582, 583, 638. G. H. Williams, "The Gabbros and Associated Hornblende Rocks near Baltimore," *Bull.* 23, *U. S. Geol. Survey*, pp. 50–59. "The Geology of the Crystalline Rocks near Baltimore," distributed at the Baltimore meeting of the Institute of Mining Engineers, February, 1892. Rec.

and pyrolusite, often in concretionary masses, disseminated through residual clay, which with the ores has formed by the alteration of limestones and shales. The deposits are entirely analogous to Examples 2 and 2a, under "Iron." Along the Appalachians the favorite horizon is just over the Cambrian (Potsdam) quartzite. Such is the case at Brandon and South Wallingford, Vt., where the ores occur in a great bed of clay between quartzite and limestone. They are referred to under Example 2a, where mention is made of the associated limonites and interesting lignite. They have never been important producers of manganese. Crimora, in

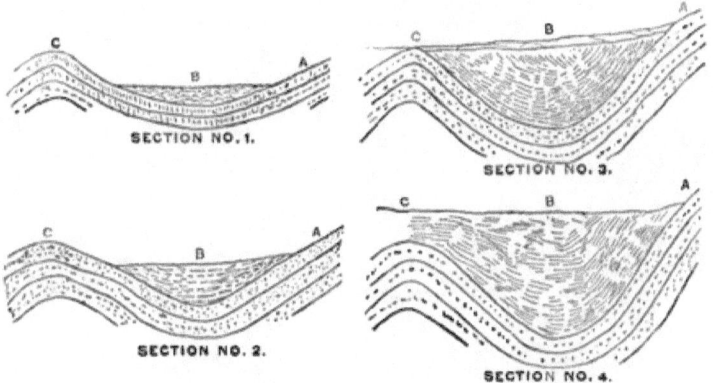

FIG. 63.—*Sections of the Crimora manganese mine, Virginia. The trough is formed by Potsdam sandstone and is filled with clay carrying nodules of ore. After C. E. Hall, M. E., June,* 1891.

Augusta County, Virginia, is the largest mine in the country. The containing clay bed is very thick, as a drill hole of 276 feet failed to strike rock. The ores occur in pockets, which as a maximum are 5 to 6 feet thick and 20 to 30 feet long, and of lenticular shape. Other irregular stringers and smaller masses run through the clay, which preserves the structure of the original rock. Potsdam quartzite underlies it. Other similar bodies occur at Lyndhurst and elsewhere in the Great Valley of Virginia. Less important deposits are found at higher horizons. Cartersville, Ga., is second to Crimora in production. The ores again occur in pockets in a stiff clay and are associated with quartzite, which is not sharply identified as yet. It may be Cambrian (Potsdam) or Upper Silurian (Medina). West of Cartersville is the Cave Spring

IDEAL SECTIONS SHOWING THE FORMATION OF MANGANESE-BEARING CLAY FROM THE DECAY OF THE ST. CLAIR LIMESTONE.

BOONE CHERT   MANGANESE-BEARING CLAY   IZARD LIMESTONE
ST. CLAIR LIMESTONE   SACCHAROIDAL SANDSTONE

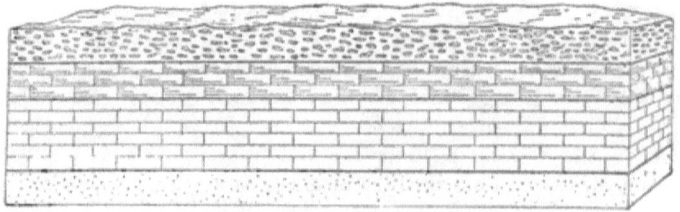

FIG. 1.—ORIGINAL CONDITION OF THE ROCKS.

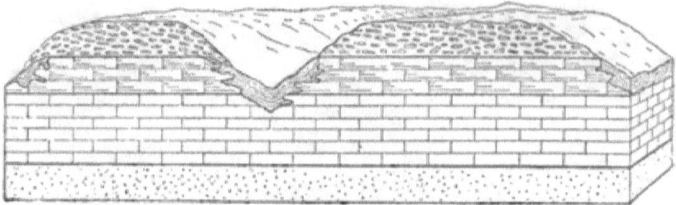

FIG. 2.—FIRST STAGE OF DECOMPOSITION.

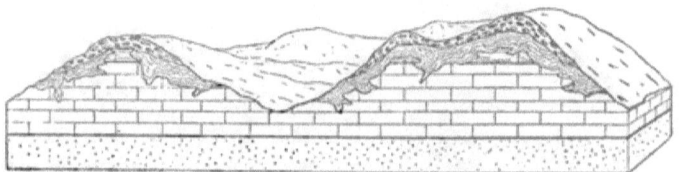

FIG. 3.—SECOND STAGE OF DECOMPOSITION.

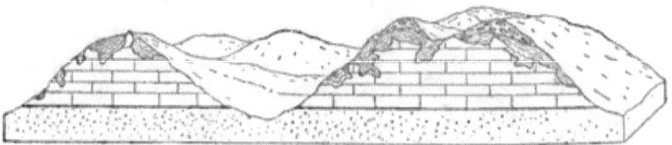

FIG. 4.—THIRD STAGE OF DECOMPOSITION.

FIG. 64.—*Geological sections illustrating the formation of the manganese ores in Arkansas. After R. A. F. Penrose, Geol. Survey of Ark., 1890, Vol. I., p. 177.*

region, where the ores occur with Lower Silurian cherts. There are numerous other localities not yet of commercial importance along the Appalachians, in Tennessee and elsewhere. Full descriptions will be found in Penrose's report, cited below.

2.14.16. Batesville, Ark. The ore is braunite, and is found in masses disseminated in a residual clay left by the alteration of a

FIG. 65.—*The Turner mine,* Batesville region, Arkansas. *After R. A. F. Penrose, Geol. Survey Ark.,* 1890, *Vol. I., p.* 272.

limestone locally called the St. Clair. It is of geologic age between the Trenton and Niagara periods, and is underlain by another limestone called the Izard, which is later than the Calciferous. On the St. Clair a series of cherts called the Boone cherts is found, which is of Subcarboniferous (Mississippian) age. The clays are sometimes in valleys, sometimes on hillsides, according to the unequal decay of the limestone. South of the Batesville district are the Boston Mountains, a range of low hills 500 feet high, and from these the manganiferous rocks form a low monocline to the north. This district is in northern central Arkansas. Southwestern Arkansas contains a second district in which the ore occurs in a great stratum of novaculite of probable Lower Silurian age. The ores are of no practical importance, being too lean and too disseminated. Small amounts of manganese ore have been obtained in California, in San Joaquin County, and from Red Rock, in San Francisco harbor. The former, and perhaps others in the State, may prove important hereafter.

2.14.17. Quite productive deposits are found in pockets at Markhamville, Kings County, N. B., in Lower Carboniferous limestone. Some thousands of tons have been shipped. Other mines are situated at Quaco Head. At Tenny Cape, in the Bay of Minas, Nova Scotia, is another deposit in Lower Carboniferous limestone which has furnished several thousand tons of ore. Others less important occur on Cape Breton.

The production of manganese ores in the United States seems to be falling off. In 1887 it reached its maximum, 34,524 long tons. In 1889 it was 23,927 tons, divided as follows: The Virginias, 14,616; Arkansas, 2528; Georgia, 5208; other States, 1575.[1]

---

[1] "Manganese Mines near Santiago, Cuba," *Engineering and Mining Journal*, Nov. 24, 1888, p. 439. H. P. Brumell, "Notes on Manganese in Canada," *Amer. Geol.*, August, 1892, p. 80. D. T. Day, *Mineral Resources*, 1882, p. 424; 1883-84, p. 550. F. P. Dunnington, "On the Formation of the Deposits of Oxides of Manganese," *Amer. Jour. Sci.*, iii., XXXVI. 175. Rec. W. M. Fontaine, "Crimora Manganese Deposits," *The Virginias*, March, 1883, pp. 44-46. Rec. C. E. Hall, "Geological Notes on the Manganese Ore Deposits of Crimora, Va.," *M. E.*, June, 1891. E. Halse, "Notes on the Occurrence of Manganese Ore, near Mulegé, Baja California, Mexico," *Trans. N. of Eng. Min. and Mech. Eng.*, XLI. 302, 1892. H. Hoy, "Ores of Manganese and their Uses," *Proc. and Trans. N. S. Inst. Nat. Sci.*, Halifax, II., 1864-65, p. 139. "Manganese Mining in Merionethshire, England," *Engineering and Mining Journal*, Dec. 18, 1886, p. 438. R. A. F. Penrose, *Ann. Rep. Ark. Geol. Survey*, 1890, Vol. I. The best work published. Rec. "Origin of the Manganese Ores of Northern Arkansas," etc., *A. A. A. S.*, XXXIX. 250. J. D. Weeks, *Mineral Resources of the U. S.*, 1885, p. 303 (Rec.); 1886, p. 180; 1887, p. 144. D. A. Wells, "On the Distribution of Manganese," *A. A. A. S.*, VI. 275. C. L. Whittle, "Genesis of the Manganese Deposits at Quaco, N. B.," *Proc. Bost. Soc. Nat. Hist.*, XXV., p. 253.

CHAPTER XV.

THE LESSER METALS, CONTINUED—MERCURY, NICKEL AND COBALT, PLATINUM, TIN.

MERCURY.

2.15.01. Ores: Cinnabar, HgS. Hg. 86.2, S. 13.8. Metacinnabarite is a black sulphide of mercury. Native mercury also occurs.

Mercury deposits are found in workable quantities in the United States only in California, in Oregon, and in one locality in Nevada. In all cases cinnabar is the principal ore. The California deposits are limited to the Coast range and in their formation seem to have followed great basaltic eruptions of post-Pliocene age.

2.15.02. Example 50. New Almaden. Cinnabar with subordinate native mercury, in a gangue of crystallized and chalcedonic quartz, calcite, dolomite, and magnesite, forming a stockwork, or "chambered vein," in shattered metamorphic rocks (pseudo-diabase, pseudo-diorite, serpentine, and sandstone). There are two main fissures, making a sort of V, with a wedge of country rock between. The ore bodies are in the fissures and also in the intervening wedge, and have associated with them much attrition clay. A great dike of rhyolite runs nearly parallel to the fissures, and to this Becker attributes the activity of circulations which filled the vein. New Idria is farther south, high up toward the summit of the Coast range. The ore is deposited in shattered metamorphic rocks of Neocomian (Lower Cretaceous) age, and in overlying Chico beds. The ore is accompanied by bitumen. Basalt is abundant ten miles away. North of San Francisco other mines have been opened, among which are the Oat Hill, Great Eastern, and Great Western. The mines are in a region pierced by eruptions of basalt and andesite, which doubtless gave impetus to the ore-bearing solutions. The ores are deposited in both metamorphic and unaltered sedimentary rocks.

2.15.03. Example 50a. Sulphur Bank. This is in the same general region as the last, but from its peculiar character has been one of the best known of ore deposits. A great flow of basalt has come down to the shores of Clear Lake from the west. Waters charged with alkaline (including ammonia) carbonates, chlorides, borates, and sulphides, and with CO_2, H_2S, SO_2, and marsh gas, have circulated through it. Sulphur and sulphuric acid have formed at the surface, and the latter has dissolved the bases of the rock, leaving pure white silica behind. Lower down, cinnabar

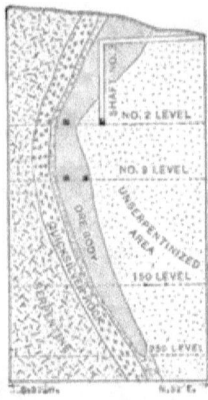

FIG. 66.—*Section of the Great Western cinnabar mine. After G. F. Becker, Monograph XIII., U. S. Geol. Survey, p. 360.*

is found, both in the basalt and in the underlying sedimentary rocks, with other sulphides and chalcedony. Leconte attributed its precipitation to cold surface waters, charged with sulphuric acid, which trickled down and met the hot alkaline solutions. Becker refers the same to the ammonia set free toward the surface by diminished heat and pressure. The California cinnabar deposits have been often, but wrongly, referred to vapors of the sulphide volatilized by internal heat and condensed above.

2.15.04. Example 50b. Steamboat Springs, Nevada. These springs are in Nevada, only six miles from the Comstock Lode. Granite is the principal rock, while on it lie metamorphic rocks of the Jura-Trias, and much andesite and basalt. The hot springs, coming up through small fissures, deposit chalcedony in some places, carbonates in others, with cinnabar as well as gold. The following minerals have been noted : "Sulphides of arsenic, antimony, sulphides or sulphosalts of silver, lead, copper, and zinc, oxide, and possibly sulphide of iron; manganese, nickel and cobalt compounds, and a variety of the earthy minerals" (Becker). Becker thinks the source of the cinnabar is in all classes in the underlying granite, and that it has come up in solution with sodium sulphide, and been precipitated toward the surface by the other compounds in the hot alkaline waters, with which it would remain in solution at greater depths, temperatures, and pressures. The Steamboat Springs are often and properly cited as metalliferous veins in active process of formation.[1]

NICKEL AND COBALT.

2.15.05. These two metals almost always occur together. Their ores are the following :

Millerite	NiS,	Ni. 64.4	S. 35.6.
Niccolite	NiAs,	Ni. 44.0	As. 56.0.
Linnaeite	Co_3S_4	Co. 58.0	S. 42.0.

Also in small percentages in pyrrhotite, and a few oxidized compounds. They may be in pyrrhotite as the doubtful mineral polydymite, a sulphide of Co, Ni, and Fe.

2.15.06. Example 16c. (See 2.03.16 and 2.04.02.) Pyrrhotite Beds or Veins. Lenticular masses of pyrrhotite interbedded in

---

[1] W. P. Blake, "Quicksilver Mine at Almaden, Cal.," *Amer. Jour. Sci.*, ii , XVII. 438. G. F. Becker, "Quicksilver Deposits of the Pacific Slope," *Monograph XIII., U. S. Geol. Survey*, Chap. 17. Rec. "On New Almaden," *Cal. Geol. Survey*, I., p. 68. S. D. Christy, "On the Genesis of Cinnabar Deposits," *Amer. Jour. Sci.*, June, 1878, p. 453 ; *Engineering and Mining Journal*, Aug. 2, 1879, p. 65. D. de Cortazar, "General Review of Occurrence, etc., of Mercury," *Reps. and Awards Group I.*, *Centennial Exposition*, p. 196. William Irelan, *Ann. Reps. Cal. State Mineralogist*. Laur, "On Steamboat Springs," *Annales des Mines*, 1863, 423. J. Leconte and Rising, "Metalliferous Vein Formation at Sulphur Bank," *Amer. Jour. Sci.*, July, 1882 ; *Engineering and Mining Journal*, Aug. 26, 1882, p. 109. J. Leconte, "On Steamboat Springs," *Amer. Jour. Sci.*, June, 1883, p. 424. "Genesis of Metalliferous Veins," *Amer. Jour. Sci.*, July, 1883. J. A. Phillips, "On Sulphur Bank, California," *Phil. Mag.*, 1871, p. 401 ; *Quar. Jour. Geol. Sci.*, XXXV., 1879, p. 390. Rolland, *Annales des Mines*, XIV., 384, 1878. B. Silliman, "Notes on the New Almaden Quicksilver Mines," *Amer. Jour. Sci.*, ii., XXXVII. 190. Siveking, *B. und H. Zeitung*, 1876, p. 45.

gneisses and schists as described for pyrite. They are known at various places in the East. Openings have been made at Lowell, Mass., Chatham and Torrington, Conn., and Anthony's Nose, N. Y., the last in search of pyrite for sulphuric acid. Nickel appears up to 3% of the ore, but these mines have never amounted to much.[1]

2.15.07. Example 16d. Gap Mine, Lancaster County, Pennsylvania. A great wedge or lense of hornblende rock appears to be inclosed in mica schists. For a space of from 6 to 30 feet of its outer portion it is impregnated with millerite, chalcopyrite, siderite, etc. The millerite occurs as a coating, lining cracks. Nickeliferous pyrrhotite is also found. A great trap dike is near. The mine is not at present worked. It presents some important analogies with the Sudbury deposits.[2]

2.15.08. Example 16e. Sudbury District, Ontario. Breccia of diorite, cemented by nickeliferous pyrrhotite and chalcopyrite, forming large but more or less irregular deposits; also deposits of purer sulphides, apparently in great veins. An extensive area of Huronian schistose rocks runs northwesterly from the juncture of Lakes Superior and Huron. It contains some Archæan inliers, and two great belts of diorite whose strike is northeast, parallel with the schists. These latter include a great variety of rocks and dip at a high angle. The diorites are crossed in places by diabase dikes, which seem to exert an enriching influence. Also, where the diorite belt pinches in, ore bodies are found near gneiss and quartz syenite. Where the breccia structure is developed they clearly lie in lines of dynamic disturbances which are parallel with the general strike. The purer deposits, as in the Stobie mine, present a great thickness of pyrrhotite. In the instance cited it is 160 feet across; but even in this, horses of diorite and more or less angular inclusions occur. The chalcopyrite is in pockety masses in the pyrrhotite. The deposits extend seventy miles in a northwest direction, and over a maximum breadth of fifty miles. They produce far more nickel than any other region. The interesting plati-

---

[1] H. Credner, "Anthony's Nose," *B. und H. Zeit.*, 1866, p. 17. W. E. C. Eustis, "The Nickel Ores of Orford, Quebec," *M. E.*, VI. 208. *Engineering and Mining Journal*, March 16, 1878.

[2] W. P. Blake, *Mineral Resources*, 1882, p. 399. J. Eyerman, *Mineralogy of Penn.* P. Fraser, Rep. CCC, *Second Penn. Survey*, p. 163. Wharton, "Analyses of Nickel Ore from the Gap Mine," *Phil. Acad. Sci.*, 1870, p. 6.

num compound, sperrylite, from this region, is mentioned under "Platinum."[1]

2.15.09. Example 23a. Mine la Motte. Considerable pyrite occurs with the lead ores mentioned under Example 23, and this is separated in the ore dressing and treated by itself, as it contains nickel and cobalt. Such pyrite is most abundant at Mine la Motte, and considerable matte is made and shipped abroad. The siegenite in Potsdam sandstone is interesting, but not abundant enough to be practically available. See under Example 23 for additional literature.[2]

2.15.10. Numerous other localities of nickel ores, chiefly of oxidized character, have been reported, as at Webster, Jackson County, N. C., in the olivine rock, dunite; in Nevada at the Lovelock mines, Churchill County, and near Riddle Station, Douglass County, Ore. At the last place the ores are hydrated silicates of magnesium and nickel associated with serpentine, which is derived from an altered olivine rock. Nickeliferous pyrrhotite is reported in the neighboring county, Jackson, at Rock Point. Nickel ores have also been reported from Salina County, Arkansas. Millerite occurs in a vein with quartz gangue in black shales. It is not practically productive. Nickel is also reported in a rather fine conglomerate from Logan County, Kansas. It occurs with manganese and limonite in the cementing material of the rock.

In 1887 there were produced 183,125 lbs. of metallic nickel in the United States, and 144,841 lbs. in 1891. Canada produced 1,336,627 lbs. in 1891.[3]

2.15.11. The principal foreign source of nickel is New Cale-

---

[1] A. E. Barlow, "On Sudbury," *Ottawa Naturalist*, June, 1891. R. Bell, "The Nickel and Copper Deposits of the Sudbury District, Canada," *Bull. Geol. Soc. Amer.*, Vol. II., p. 125. T. G. Bonney, "Notes on a Part of the Huronian Series near Sudbury, Canada," *Quar. Jour. Geol. Sci.*, XLIV. 32. F. W. Clarke and Ch. A. Catlett, "Platiniferous Nickel Ore from Canada," *Amer. Jour. Sci.*, iii., XXXVII. 372. J. H. Collins, "Notes on the Sudbury Ore Deposits," *M. E.*, October, 1889, *Engineering and Mining Journal*, Oct. 26, 1890; *B. und H. Zeit.*, Vol. L., p. 148. E. D. Peters, "On Sudbury Ore Deposits," *M. E.*, October, 1889; *Engineering and Mining Journal*, Oct. 26, 1890; *B. und H. Zeit.*, Vol. L., p. 148.

[2] J. M. Neill, "Notes on the Treatment of Nickel and Cobalt Mattes at Mine la Motte," *M. E.*, XIII. 634.

[3] *Ark. Geol. Survey*, 1888, Vol. I., pp. 34, 35. F. P. Dewey, "On the Nickel Ores of Russell Springs, Logan County, Kansas," *M. E.*, XVII. S. H. Emmons, "The Nickel Deposits of North Carolina," *Engineering and Mining Journal*, April 30, 1892. H. B. v. Foullon, "On Riddle, Ore-

donia, in the South Pacific. The ores are garnierite and other hydrated silicates of nickel and magnesium in serpentine. In 1890 they furnished 885,300 lbs. Norway comes next, and in 1889 produced 149,872 lbs. Sweden in the same year afforded 17,632 lbs. These latter are from sulphide ores.[1]

PLATINUM.

2.15.12. Some hundreds of ounces of platinum are annually gathered from placer washings in northern California, and two or three times as much more from British Columbia. Much iridium and osmium are associated with it. In October, 1889, F. L. Sperry, the chemist of the Canadian Copper Company, of Sudbury, discovered a heavy crystalline powder in the concentrates of a small gold-quartz mine in the district of Algoma. He detected the presence of platinum, and sent the material to Professors Wells and Penfield of Yale, by whom it was analyzed, and crystallographically determined to be the arsenide of platinum, $PtAs_2$, the first compound of platinum, other than an alloy, detected in nature. It has been appropriately named sperrylite by Wells, and although not at present a source of platinum, it may merit attention, as the price of the metal has recently approximated that of gold. The chief reliance of the world for platinum is Russia, whose deposits are in the Urals. More or less comes also from Colombia, South America, and from placer washings elsewhere. Serpentine is very generally its mother-rock.[2]

---

gon," *Jahrbuch d. k. k. geol. Reichsanstalt*, Vienna, XLII., 223, 1892. Rec. A. D. Hodges, "Notes on the Occurrence of Nickel and Cobalt in Nevada," *M. E.*, X. 657. W. R. Ingalls, General paper, *Engineering and Mining Journal*, Jan. 2, 1892, p. 40. Rec. *Mineral Resources*, 1887, p. 126. S. B. Newberry, "Nickel Ores from Nevada," *Amer. Jour. Sci.*, iii., II. 122. H. Wurtz, "On the Occurrence of Cobalt and Nickel in Gaston County, North Carolina," *Amer. Jour. Sci.*, ii., XXVII. 34.

[1] J. Heard, Jr., "Caledonia Nickel and Cobalt," *Engineering and Mining Journal*, Aug. 11, 1888, p. 103. J. H. L. Vogt, "On Nickel Production and Occurrence." Good paper in Swedish: *Geologiska Foreningens, i.* Stockholm, Band, XIV., p. 433.

[2] California: *Engineering and Mining Journal*, June 29, 1889, 587. B. Silliman, "Cherokee Gold Washings, California," *Amer. Jour. Sci.*, iii., VI. 132. Canada: **F. W. Clarke and Ch.** Catlett, "Platiniferous Nickel Ore from Canada," *Amer. Jour. Sci.*, iii., XXXVII. 372. H. L. Wells and S. L. Penfield, "Sperrylite, a New Mineral," *Amer. Jour. Sci.*, iii., XXXVII. 67. Russia: A. Daubree, "On the Platiniferous Rocks of the Urals," *Trans. French Acad. Sci.*, March, 1875; *Amer. Jour. Sci.*, iii., IX. 470. General paper by C. Bullman, *The Mineral Industry for 1892*, p. 373. Rec.

THE LESSER METALS, CONTINUED.

TIN.

2.15.13. Ores: Cassiterite, SnO_2, Sn. 78.67, O. 21.33. The sulphide stannite is a rather rare mineral.

Cassiterite occurs in small stringers and veins on the borders of granite knobs or bosses, either in the granite itself or in the adjacent rocks, in such relations that it is doubtless the result of fumarole action consequent on the intrusion of the granite. It appears that the tin oxide has probably been formed from the fluoride. A favorite rock for the ore is the so-called greisen, a mixture of quartz and muscovite or lithia mica, and probably an

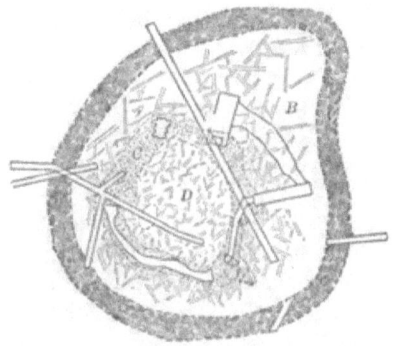

FIG. 67.—*Horizontal section of the Etta knob.* After W. P. Blake, *Mineral Resources*, 1884, *p.* 602.

original granite altered by fumarole action. Topaz, tourmaline, and fluorite are found with the cassiterite, indicating fluoric and boracic fumaroles. Cassiterite seems also to crystallize out of a granite magma with the other component minerals. Cassiterite, being a very heavy mineral, accumulates in stream gravels, like placer gold, affording thus the stream tin. When of concentric character it is called wood tin. It is not yet demonstrated that the United States have workable tin mines.

2.15.14. Example 51. Black Hills. Knobs of granite rock containing cassiterite, disseminated in a mass of albite and mica, and associated with immense crystals of spodumene. Columbite, tantalite, and beryl are also found. There are two granite knobs which are best known, the Etta and the Ingersoll. The former is a conical hill, 250 feet high by 150 feet by 200 feet, piercing mica and garnetiferous slates. Tunnels show it to have a concentric structure—first, a zone of mica; second, a zone of great spodumene crystals, with an albitic, so-called greisen and cassiterite in the interstices; lastly, a mixture of quartz and feldspar as a core. Other tin-bearing granites occur as dikes, or veins, as much as 80 feet wide, and bearing the so-called greisen and tin ore in quartz. They are called segregated veins by Carpenter, who doubts their igneous character, probably on good ground. No tin is yet commercially produced. The tin deposits extend also into Wyoming.[1]

2.15.15. Tin ores as stream tin have been found in gold washings in Montana and Idaho. Tin is also known in the Temescal Mountains, southern California, and according to Blake is in various small veinlets in a granite region. This locality attracted much interest years ago, but has never yielded any practical results until lately. Operations after being carried on for a time have, however, proved a failure.[2]

2.15.16. Narrow veins carrying cassiterite are being exploited in the granite and schistose rocks of Rockbridge and Nelson counties, Virginia, in North Carolina, and in Alabama. Companies have been formed to work the two former, but as yet without a notable output.[3]

2.15.17. Cassiterite has been discovered in narrow veins in mica schists with lepidolite and fluorite at Winslow, Me., and is known at other places in Maine and New Hampshire. A salted tin prospect several years ago spread the impression that tin was to be found in southwest Missouri.[4]

---

[1] W. P. Blake, *Mineral Resources*, 1884-85, p. 602. Rec. *Amer. Jour. Sci.*, September, 1883, p. 235; *Engineering and Mining Journal*, Sept. 8, 1886. "Tin Ore Deposits of the Black Hills," *M. E.*, XIII. 691. F. R. Carpenter, *Prelim. Rep. Dak. School of Mines*, 1888; also *M. E.*, XVII. 570. "Tin in the Black Hills," *Engineering and Mining Journal*, Nov. 28, 1884, p. 353. *Mineral Resources of the U. S.*, annually under "Tin."

[2] W. P. Blake, "Occurrence of Wood Tin in California, Idaho, and Montana," *Mining and Scientific Press*, San Francisco, Aug. 5, 1882. H. G. Hanks, *Rep. Cal. State Mineralogist*, 1884, p. 121.

[3] H. D. Campbell, "Tin Ore, Cassiterite, in the Blue Ridge in Virginia," *The Virginias*, October, 1883. A. R. Ledoux, "Tin in North Carolina," *Engineering and Mining Journal*, Dec. 14, 1889, p. 521; see also February, 1887, p. 111. McCreath and Platt, *Bull. Iron and Steel Asso.*, Nov. 7, 1883, p. 209. W. Robertson, *London Mining Journal*, Oct. 18, 1884. A. Winslow, "Tin Ore in Virginia," *Engineering and Mining Journal*, November, 1885. Rec.

[4] W. P. Blake, *Mineral Resources*, 1884, p. 538. C. H. Hitchcock, "Discovery of Tin Ore and Emery at Winslow, Me.," *Engineering and Mining Journal*, Oct. 2, 1880, p. 218. T. S. Hunt, "Remarks on the Occurrence of Tin Ore at Winslow, Me.," *M. E.*, I. 573. C. T. Jackson, "Tin Ore at Winslow, Me.," *Proc. Bost. Soc. Nat. Hist.*, XII. 267.

CHAPTER XVI.

CONCLUDING REMARKS.

2.16.01. In review of the western border of the country, we note the elevated plateau rising from the Mississippi to the Rocky Mountain range, which consists of various ranges of general north and south or northwest and southeast trend, with broad valleys between. Next comes the Colorado Plateau, and then the Wasatch Mountains and the Great Basin, with its various subordinate north and south ranges. These are succeeded by the Sierra Nevada, and the great valley of California, the Coast range, and finally the Pacific Ocean.

From the Archæan to the close of the Carboniferous there were granite islands around which active sedimentation proceeded. At the close of the Carboniferous the elevation of the Wasatch and the region of eastern Nevada occurred. At the close of the Jurassic the elevation of the Sierra Nevada took place. The chief upheaval of the Rocky Mountain system came at the close of the Cretaceous and that of the Coast range at the close of the Miocene Tertiary. Smaller and less important oscillations have occurred before and since. Each elevation was accompanied by foldings, faultings, and extensive outpourings of eruptive rocks. The resultant fractures and the solfataric action, occasioned by the dying volcanic activity, constitute the primary cause of the formation of the ore deposits, which in some cases lie in ranges along the lines of faulting or of disturbances, and in others are irregularly scattered. We can recognize the Coast range belt with mercury and chromium; the California gold belt in the western Sierras; the silver belt of Utah on the western flank of the Wasatch; a belt in Arizona from southeast to northwest, along the contact between Palæozoic limestone, mostly Carboniferous, and the Archæan; and the great stretch of lead-silver mines in the Carboniferous limestones of Colorado. The other areas are scattered, and apparently exhibit no such grand general relations to these geographical and geological phenomena.[1]

2.16.02. In the Mississippi Valley, W. P. Jenney has remarked the connection of the antimony and silver deposits of Arkansas with the Onachita uplift that traverses that State and Indian Territory; also, the location of the Missouri lead and zinc ores along the Ozark uplift; and he has referred the Wisconsin lead and zinc mines, as well as those in the neighboring part of Iowa and Illinois, to an uplift south of the Archæan area of Wisconsin. The limitation of the Lake Superior copper deposits to the Keweenawan system may be mentioned, and such parallelism as prevails among the Lake Superior iron ores. In the East the great belt of Clinton Ores; the long succession of Siluro-Cambrian limonites in the Great Valley; the black-band ores and clayironstones of the Carboniferous; the closely similar geological relations of nontitaniferous, magnetite lenses in the Archæan gneisses; and the general association of titaniferous magnetites with rocks of the gabbro family the country over—all are striking illustrations of broad, general geological features that may characterize extended areas. To these may be appended the great series of pyritous beds or veins in the slates and schists of the East, the gold belt of the southeastern States, and the small copper deposits associated with the Triassic traps and sandstones. Aside from these, while there are important mines not included in the list, the others do not exhibit the same widespread uniformities of structure or associations. Yet, from the list cited, it forcibly appears that similar conditions have brought about related ore bodies over great stretches of country; and while in the opening schemes of classification points of difference were emphasized, in the closing pages points of resemblance may be with equal right brought to the foreground.

2.16.03. A few general conclusions suggest themselves from the preceding pages.

(1) The extreme irregularity in the shape of metalliferous de-

---

[1] G. F. Becker, *Amer. Jour. Sci.*, 3d Series, Vol. XXIII, 1884 p. 209. W. P. Blake, *Rep. Cal. State Board of Agriculture*, 1866. S. F. Emmons, "The Structural Relations of Ore Deposits," *M. E.*, XVI. 804. R. W. Raymond, "Geographical Distribution of Mining Districts in the United States," *M. E.*, I. 33. *Fortieth Parallel Survey*, Vol. III., Chap. I. "Precious Metals," *Tenth Census*, Vol. XII.

posits, and from this the unwisdom of the United States law in the West, which is based on well-defined fissure veins. The only practicable method is that a man should own all that is embraced in his property lines, whether the ore body outcrops outside or not. "A square location is the square thing" (Raymond).

(2) The very general proximity of eruptive rocks in some form to the ore bodies. Except in the case of iron, there are only a few where these are not present, and apparently strong factors in the circulations which formed the ore. The lead and zinc deposits of eastern and western Missouri and the neighboring States, and of New York and Virginia, are almost the only ones, and we are justified in concluding that eruptive rocks are of great importance.

(3) We know from the investigations of Sandberger and others that the dark silicates of many rocks contain percentages of the common metals. The choice is open whether to refer the ore to original dissemination in these, and derive it by gradual concentration, probably at great depths, or to some indefinite unknown source, which can only be described as "below."

ADDENDA.

Page 65. In the report of the State Geologist of Michigan for 1891–92 (issued January, 1893), pp. 144, 145, Dr. M. E. Wadsworth has published a "Preliminary Classification of Metalliferous or Ore Deposits." The main outline is as follows :

I. Eruptive Deposits (*a*) Non-Fragmental.
   (*b*) Fragmental.
II. Mechanical Deposits (*a*) Unconsolidated.
   (*b*) Consolidated.
III. Chemical Deposits (*a*) Sublimations.
   (*b*) Water Deposits.
   (*c*) Impregnations or Replacements.
   (*d*) Segregations or Cavity Deposits.

Each of the above except III. (*d*) is then subdivided so that the table becomes practically a classification of rocks. Indeed, a moment's consideration will show that the scheme in its main divisions is closely modeled after the prevailing classification of rocks. III. (*d*) Segregations or Cavity Deposits contains the following : 1. Pockets. 2. Chambers. 3. Contact Deposits. 4. Veins, including Gash Veins, Segregated Veins, Reticulated Veins or Stockwork, Contact Veins, Fissure or Fault Veins.

The author states in some appended comments that the table is not limited to those deposits now practically worked (which we ordinarily understand the expression *ore* deposits to mean), but is intended to include all that have been or may be of value. But in this respect there is good ground for preferring to make our classifications in ore deposits, as in mineralogy, zoölogy, etc., embrace only the authenticated varieties, expecting additions to be incorporated as discovered and suitably described.

Page 73. In connection with the precipitation of iron ores by limestones, an added note on the chemistry of the reaction is desirable. Dr. J. P. Kimball, in a paper on the "Genesis of Iron Ores by Isomorphous and Pseudomorphous Replacement of Limestone," etc., *Amer. Jour. Sci.*, September, 1891, p. 231 (the paper is concluded in the *American Geologist*, December, 1891), brings out the fact that alkaline carbonates, of which calcium carbonate is one, precipitate from solutions of ferrous sulphate or ferrous carbonate, under ordinary conditions, *hydrous* ferrous carbonate, an unstable salt that quickly oxidizes to a hydrous oxide. The argument is then adduced that bodies of siderite or *anhydrous* ferrous carbonate could not have originated by direct precipitation, but have probably done so by pseudomorphous replacement of calcium carbonate. The author then follows the possible metamorphism or changes of these bodies to other forms of iron ore, citing, however, some as possible examples that are clearly unwarranted. The chemical distinctions thus brought out undoubtedly have their weight and importance; but siderite as a frequent veinstone crystallizes as the anhydrous carbonate, and in surroundings giving no reason to infer that it has replaced calcium carbonate. It has also been obtained artificially by several investigators (as cited by Fouqué and Levy in *Synthèse des Minéraux et des Roches*), and there is still reason for believing that it does not necessarily always form in nature as a pseudomorph after calcium carbonate.

Page 74. Attention has been lately directed to the great deposits of bog ore in the Three Rivers district of Quebec. (P. H. Griffin, "The Manufacture of Charcoal-Iron from the Bog and Lake Ores of Three Rivers District, Province of Quebec, Canada," *Trans. Inst. Min. Eng.*, Montreal meeting, February, 1893.) Although the deposits have been long known and have been the basis of a small industry, they are now utilized on a larger scale for car-wheel iron. They occur in the small lakes and swamps that receive the drainage of the old Laurentian highlands on the north. This water, more or less charged with iron, drops its burden as bog ore wherever it stands. By collecting supplies from a fairly extended district, quite large amounts of ore are obtained. The deposits furnish ideal illustrations of the general origin of bog ores, as outlined in preceding pages. They occur also on the bottoms of lakes and are obtained by dredging. The lake ore seems to run somewhat richer than that gathered in the bogs. Both are low in sulphur but have about 0.3% phosphorus.

Page 78. Volume I. of the Annual Report of the Geological Survey of Arkansas consists of a report by R. A. F. Penrose on the "Iron Deposits of Arkansas." It at once appears that there is little prospect of Arkansas producing any notable amounts of iron ore. Such as have been found are practically all limonite or brown hematite, and are generally very low in iron. The ores occur in five districts, viz.: Northeastern Arkansas, northwestern Arkansas, the valley of the Arkansas River, the Ouachita Mountains, and southern Arkansas. They are generally associated with sandstones or cherty limestones. The first-named district makes the best showing. In it the ores are in Lower Silurian (Calciferous or lower) sandstones, cherts, and limestones. In the second district they are in Lower Silurian cherts and Lower Carboniferous sandstones. In the third, they occur with Carboniferous and Lower Carboniferous strata, but are also in the form of recent spring deposits. In the Ouachita Mountains they are with Lower Silurian shales and novaculites. In this district the magnetite of Magnet Cove occurs, but it is only an interesting mineral and not in any practical quantity. The last district has the ores in sands and clays of the Eocene. Its continuation in Texas and Louisiana has been already mentioned in the main text.

Page 122. Too little attention was given to titaniferous magnetite in the text; for although these ores are not now of value, they are exciting considerable attention and are of great scientific interest. They are almost invariably in wall rock that consists of plagioclase, with augite, hypersthene, and hornblende, one or all. The rock may thus be gabbro, norite, or diorite, and is of igneous (plutonic) character. The ores appear to be excessively basic developments of the wall rock, which were formed during its cooling and crystallization. Subsequent metamorphism, mountain-making processes and the like, sometimes give them a gneissic structure, and stretch out the ore into apparent beds. In the great series of these labradorite rocks which is so extensive in Canada, and which was called by T. Sterry Hunt the Norian, these ores are very abundant. The same rocks form the central Adirondacks and contain some enormous bodies of titaniferous ore. Prof. E. Emmons of the New York Geological Survey in 1835-40 gives much space to them as occurring near Lake Sanford and Lake Henderson.[1] The later Survey of Uno Sebenius has indicated an even greater extent than was known to Emmons, although

---

[1] E. Emmons, *Rep. on Second District, N. Y. State Survey*, pp. 244-255, 1842. A. J. Rossi, "Titaniferous Ores in the Blast Furnace," *Trans. Amer. Inst. Min. Eng.*, February, 1893. J. C. Smock, *Bulletin of N. Y. State Museum*, p. 37.

that was over 500 feet wide and 1600 feet long. Numerous other less extensive ore bodies also occur in the neighborhood, and very many elsewhere in the mountains. They are all titaniferous, although, as often happens with such ores, they are low in phosphorus and sulphur and at times quite high in alumina. At present they are not utilized, but it is to be hoped that in time processes will be developed to treat them. Fairly high titaniferous ores occur in New Jersey on Schooley's Mountain and to the southwest, and small amounts, say up to 1% of TiO_2, are known in many others.[1] The wall rock of these ores should receive microscopic examination to determine if it affords a mineralogical parallel with those of the Adirondacks.

Titaniferous ores are also known in Virginia and North Carolina,[2] in the gabbros of Minnesota, in Wyoming, in Colorado—the last three of which have already been mentioned in the text. Microscopic determinations of the wall rock, where not already made, would be of great interest. Their geological relations have long been known in Sweden and Norway, but an igneous form of origin has been but recently advocated. (Cf. p. 56.)

Page 130. In the review of the methods of formation of magnetite-lenses, the method by segregation or as segregated veins was omitted. As it is viewed with favor by many reliable observers, it should have its place. By this method the iron oxide is conceived to concentrate from a state of dissemination in the walls by slow secretion in solution, to form the ore bodies along certain favorable beds. The action is regarded as analogous to the formation of concretions, and is illustrated on a small scale by the well-known disks of pyrite and calcite that form in clays and shales. It is a curious fact, however, that some magnetites are in wall rock that hardly shows a trace of even a dark silicate. Thus the lenses at Hammondville, in the Lake Champlain district, are in a pure, white gneiss that only has a little garnet near the ore, but it is possible that elsewhere the explanation may be the most reasonable one. Where it applies we would expect hornblende and other ferruginous minerals in the walls.

---

[1] B. F. Fackenthal, "Titaniferous Ores in the Blast Furnace," *Trans. Amer. Inst. Min. Eng.*, February, 1892. R. W. Raymond, *Ibid.*

[2] H. B. C. Nitze, "Notes on some of the Magnetites of Southwest Virginia and the Contiguous Territory of North Carolina," *Trans. Amer. Inst. Min. Eng.*, June, 1891. Reprint p. 13. See also discussion of the paper at the same meeting.

Page 143. The limestones called the Second Magnesian, and stated to be of Lower Silurian age in 2.04.17, have been lately shown to be Cambrian. (G. C. Broadhead, "The Correct Succession of the Ozark Series," *Amer. Geol.*, April, 1893, p. 260. F. L. Nason, "The Magnesian Series of the Ozark Uplift," *Ibid.*, February, 1893, p. 91. A. Winslow, "Notes on the Cambrian in Missouri," etc., *Amer. Jour. Sci.*, March, 1893, p. 221.)

Page 172. C. R. Boyd has published in the *Engineering and Mining Journal*, June 17 and 24, 1893, a quite complete description of the lead and zinc mines and works of the Wythe Company, near Austinville, Va. The geological horizon is stated to be in the Knox group of Safford, which is, as stated in the text, near the Calciferous of New York. While giving a thorough historical sketch and description of the works, little additional to the brief outline on p. 172 is given on the geology.

Page 247. Mr. Waldemar Lindgren has published (June, 1893) in the *Bulletin of the Geological Society of America*, Vol. IV., pp. 257-298, a most valuable paper on "Two Neocene Rivers of California." He traces out the location and geological history of the deep gravels along the drainage line of the American and Yuba rivers, and adds much to our knowledge of their former location and gradients. He concludes that the old divide in general coincided with the present one, but that the slope of the Sierra has been considerably increased since the time when the Neocene (*i.e.*, Miocene and Pliocene), ante-volcanic rivers flowed over its surface. "It finally appears probable . . . that the surface of the Sierra Nevada has been deformed during this uplift, and that the most noticeable deformation has been caused by a subsidence of the portion adjoining the great valley, relatively to the middle part of the range" (p. 298).

General Addenda. I. Through the courtesy of Dr. R. W. Raymond, secretary of the American Institute of Mining Engineers, the author has been permitted to read, in advance of its publication, the great essay on the origin of ore deposits, that Prof. Franz Posepny of Vienna has sent to the Institute for the July meeting, 1893. The paper is a theoretical discussion of the origin of ores, with illustrations selected from all parts of the world, but especially from Europe and America. It forms one of the most important contributions to the literature that has yet been made. Posepny distinguishes at the outset between rocks and mineral deposits; *i.e.*, between original materials, such as wall rock, and secondary introductions, such as veins, etc. The former he calls "idiogenites," the latter "xenogenites," basing the names on the familiar Greek terms that run through all our literature. The latter are especially characterized by "crustification," by which term is indicated what has been termed "banded structure," on p. 35. The subject of cavities is then taken up, and, while minute pores are stated to be in all rocks, a distinction is made between the larger openings, which originate in a rock mass as a part of its own structure, such as contraction joints in igneous rocks, amygdaloids, and the like, and those induced by outside causes, such as fault fissures.

The circulation of water through these is next treated: first, surface waters or "vadose" circulations, which descend; second, ascending waters from great depths, such as springs in deep mines, hot springs, etc. The common salts in solution in these latter are tabulated, being of course mostly alkaline carbonates, sulphates, chlorides, etc. The "exotic" metallic admixtures which would bear on the origin of ores are next discussed, so far as possible with analyses of actual cases. The alterations produced by mineral springs in rocks and the structural relations of the deposits of mineral springs, especially as expressed by "crustification," are then described. This preliminary material clears the way for the general discussion of the origin of ore bodies. The argument running all through the paper is that ore bodies, even when apparently interbedded with sedimentary rocks, are of secondary introduction and, in general for veins, are from deep-seated sources. Precipitation from descending solutions and filling by lateral secretion are strongly controverted.

The discussion of origin follows in its arrangement the following classification of ore deposits:

I. Filling of spaces of discission (fissures).
II. Filling of spaces of dissolution in soluble rocks.
III. Metamorphic deposits in soluble rocks; in simple sediments; in crystalline and eruptive rocks.
IV. Hysteromorphic (*i.e.*, later or last formed) deposits. Secondary deposits due to surface action (*i.e.*, placers, etc.).

The treatment, both in the introductory pages and in the later discussions, is often strikingly similar to that of this book, and the underlying argument is much the same. The standpoint in both essays is essentially a genetic one, and the main difference lies in the fact that the one is an exposition of an individual's views, fortified by examples from all parts of the world; the other endeavors to be a judicial statement with a complete description of the ore bodies of the United States alone.

II. An extended treatise on the useful minerals, earthy as well as metallic, by M. M. E. Fuchs and L. De Launay, has recently appeared (*Traité des Gîtes Minéraux et Métallifères,* Paris, 1893). The book is based on the lectures in economic geology delivered in the École Supérieure des Mines, at Paris, in the last fifteen years by the two authors. (Professor Fuchs died in 1889, and was succeeded by Professor De Launay.) A vast amount of valuable information is brought together and discussed from various points of view, useful applications and methods of treatment being set forth as well as geological occurrence. So far, however, as the ore deposits and other useful minerals of the United States are concerned, the authors have suffered from the unavoidable limitations of those not native and conversant in a discriminating way with our literature. In many cases, if not in most, the descriptions used by them are from German or other foreign sources, and are often antiquated and curiously mixed. While placing at command much that is not easily accessible on foreign ore deposits and those in outlying parts of the world, and also much that is valuable in the way of general discussion, nevertheless the book cannot be considered in its American relations as adding much to our literature.

INDEX.

Abiquin, N. M., copper ores, 154.
Adams, F. D., on granite at Douglass Island, 255.
Adirondack titaniferous ores, 280.
Agglomerates, 200.
Alabama, Clinton ore, **95, 96.**
    Gold mines, 252.
    Limonite, 78, 84.
    Tin, 274.
Alaska, **geology, 253, 254.**
Aleutian **Islands,** 254.
Algonia, **platinum,** 272.
Alice mines, Butte, Mont., 220.
Alleghany Mountains, geology, 8.
Alturas Co., Idaho, 223.
Aluminium, 258.
Amador Co., Cal., 249.
American River, Cal., **282.**
Ancram lead mine, N. Y., **157.**
Anthony's Nose, nickel **mine, 270.**
    Mine of pyrrhotite, **131.**
Anticlines, 11.
    Cause of **cavities,** 11.
Antimony, **259, 260.**
    Nevada, **232.**
Apache Co., **Ariz., 228.**
Appalachians, general description, 6, 7.
    Topography, 6, 7.
    Geology, **7, 8.**
Argentine, **Clear Creek Co., Colo.,** 211.
Arizona **copper mines, 144-152.**
    Geology, 227.
    Lead-silver ores, 198.
    Prince copper mine, **Arizona,** 150.
    Silver mines, **275.**
Arkansas, antimony, 259.
    Bauxite, 258.
    Iron ores, 279.
    Limonite, Addenda.
    Manganese, 264-266.
    Nickel, 271.
    Silver mines, 201.
Arksut Fjord, Greenland, 258.
Arlington, N. J., copper mines, **153.**
Arnold iron mines, N. Y., 122.
Arsenic, 260.

Ascension by infiltration, **29, 30-32.**
Ashcroft iron mines, Colo., **126.**
Aspen, Colo., 186-190, 211.
    Iron mines near, 126.
Atlantic copper mine, Mich., 141.
Augusta, Va., manganese, 263.
Auriferous beach sands, 240.
    Gravels, 243-248.
Austin, Nev., 232.
    Antimony, 259.
Austinville, Va., zinc mines, 282.
Bachelor Mountain, near Creede, Colo., 209.
Baker Co., Ore., 239, 240.
Bald Butte Co.'s mines, Mont., **221.**
Baltimore, Md., chromite, 261.
Banded structure of veins, 35.
Bannack City, Mont., 219.
Banner district, Boisé **Co., Idaho,** 223.
Bare Hills, Md., chromite mine, 261.
Barton Hill iron mines, N. Y., 121.
Barus, C., experiments on electrical activity of veins, 40, 41.
    Experiments **on** the Comstock Lode, 235.
Bassick mine, Colo., **35, 36, 213.**
Batesville, Ark., manganese, 265, 266.
Bauxite, 258.
Beach sands, form of **ore body, 58.**
Bear Lake Co., Idaho, **223.**
Beaver Co., Utah, 225.
Beaverhead Co., Mont., 219.
Becker, G. F., precious metals in diabase, 25.
    Cited on Steamboat Springs, Nev., 32.
    On the Comstock Lode, 233-237.
    On California gravels, 247.
    On California mercury, 267-269.
    Various metals in granite, 26.
Beds, in sedimentary rocks, 10.
Belmont, Nev., 231, 232.
Belts of ore deposits in the West, 275, 276.
Berks Co., Penn., iron mines, 125.
Bernalillo Co., N. M., 204.
Bertha zinc mines, Wythe Co., Va., 172.

288   INDEX.

Bethlehem, Penn., zinc mines near, 174.
Beulah antimony mine, Nev., 259.
Big Cottonwood Cañon, Utah, 192.
Big Creek, Nev., 259.
Bimetallic mine, Butte, Mont., 221.
Bingham Cañon, Utah, 192-225.
    Cited as illustration of gossan, 39.
Bisbee copper district, Ariz., 148.
Bismuth, 260, 261.
Bitter Root Mountains, Idaho, 222.
Black-band iron ore, 87.
Black Copper groups of copper mines, Arizona, 150.
Black Hills, geology, 9.
    Geology and mines, 216-218.
    Gold in Potsdam, 240.
    Placers, 245.
    Tin, 273.
Black Range copper mines, Arizona, 151.
    Copper district, 228.
Blake, W. P., on the Deep Creek mines, 225.
    On the Silver King mine, Arizona, 228.
    On Tombstone, Ariz., 229.
    On Utah antimony, 260.
    Ruby silver ore, Poorman lode, 223.
Block iron ore, 87.
Block Island, R. I., magnetite sands, 128.
Blow, A. A., illustration of fault by, 17.
    Cited on Leadville, Colo., 185.
Bluebird mine, Butte, Mont., 221.
Blue lead in California gravels, 246.
Blue Mountains, Oregon, 239.
Bog iron ore, 73.
    Precipitated by algæ, 75.
Bog ore at Three Rivers, 279.
Boisé Co., Idaho, 223.
Bonanza City, Idaho, 223.
Bonanza of ore, 37.
Bonne Terre, Mo., lead mines, 158.
Bonneville, Lake, 230.
Bonsacks, Va., illustration of gossan, 39.
    Zinc mines, 172.
Boone cherts, Arkansas, 265.
Boss of igneous rock defined, 11.
Boston Mountains, Ark., 265.
Boulder Co., Colo., 214.
    Iron ores, 126.
Box Elder Co., Utah, 225.
Boyd, C. R., on Virginia zinc mines, 282.
Brandon, Vt., manganese, 263.
Bristol, Conn., copper mine, 153.

British Columbia gold gravels, 256.
    Platinum, 272.
Britton, N. L., cited on the geology of the Highlands, N. J., 123.
Broadhead, G. C., cited on Missouri geology, 282.
Brooks, T. B., on the Marquette district, 101, 102.
    Origin of ores, 106.
Browne, D. H., on phosphorus in iron mines, 130.
Browne, R. E., on California gravels, 246, 247.
Brown hematite, Arkansas, 279.
Brunton, D. W., cited on Aspen, Colo., 188.
Bucks Co., Penn., iron mines, 125.
Buckwheat mine, Franklin Furnace, N. J., 177.
Buena Vista, Colo., 212.
Buford Mountain, Mo., 116.
Buhrstone ore, 88.
Bull Domingo mine, Colo., 36, 213.
Burden spathic ore mines, Hudson, N. Y., 89.
Burro Mountains, N. M., 203.
Butte, Mont., 220.
    Copper mines, 136, 137, 142.
    Illustrates infiltration by ascension, 32.
    Placers near, 245.
Calaveras Co., Cal., 249.
Caldwell Co., Ky., lead and zinc mines, 166.
Calico district, Cal., 242.
California, Chromite, 261.
    Copper mines, 136.
        Literature, 136.
    Geology, 241.
    Gulch, near Leadville, Colo., 211, 245.
    Kern Co., antimony, 259.
    Lead-silver ores, 198.
    Magnetite, 126.
    Mercury, 267, 268.
    Platinum, 272.
    Tin, 274.
    Yuba and American rivers, 282.
Callon scheme of classification of ore deposits, 47.
Calumet and Hecla copper mines, Michigan, 141.
    Iron mine, Colorado, 126.
Campbell Mountain near Creede, Colo., 209.
Campo Seco, Cal., copper mines, 136.
Canada bog ores, 279.
    Gold, 255, 256.
    Magnetite ore mines, 122.
    Titaniferous magnetite, 280.

Canso, N. S., 255.
Cape Breton, N. S., manganese, 266.
Capelton, Quebec, mines of pyrite, 131; chalcopyrite. 134.
Carbonate Hill, Leadville, Colo., 183.
Carbonate mine, Utah, 195.
Carpenter, F. R., on the Black Hills, S. D., 217.
On Black Hills tin, 274.
Carson Hill, Calaveras Co., Cal., 248.
Cartersville, Ga., manganese, 263.
Cascade Mountains, 238.
Cascade Range, Cal., 241.
Cassia Co., Idaho, 223.
Cassiterite, 273.
   In granite veins, 60.
Cave mine, Utah, 195.
Caves, origin of, 21.
   As a form of ore body, 57.
Cave Spring manganese mines, Georgia, 263.
Cazin, F. M. F., on Silver Reef, Utah, 226.
Cedar Mountain, Mo., 116.
Central City, Colo., copper mines, 138.
Chaffee Co., Colo., 212.
   Magnetite, 126.
Chalcopyrite with pyrite, 134–136.
   Literature, 135.
Chamberlain, S. C., on Wisconsin lead and zinc mines, 161–163.
Chambered vein, 267.
Charlemont, Mass., mines of pyrite, 131.
Charles Dickens mine, Idaho, 223.
Chatham, Conn., nickel mine, 270.
Chateaugay iron mines, New York, 72, 120.
Chaudière River, Canada, gold, 256.
Chauvenet, R., cited on Colorado iron ores, 126.
Chester, A. H., on the average yield of certain well-known iron ores, 71.
Chicago copper mine, Arizona, 150.
Chico beds, California, 267.
Chimney of ore, 37.
Chromite distribution, 261.
Chromium, 261.
   Mines, California, 275.
Chugwater Creek, Wyo., iron ores, 126
Church, J. A., on the Comstock Lode, 233, 234.
Churchill Co., Nev., 232.
Chute of ore, 37.
Cincinnati uplift, 9.
Cinnabar, 267.
Classification of ore deposits, 42–62.

Classification of ore deposits, underlying principles, 42.
   Wadsworth, 278.
Clay ironstone defined, 86.
   Deposits of, 86–88.
Clay selvage, seam, or parting, 37.
   Attrition clay, 37.
   Residual clay, 37.
Clayton, J. E., on Butte, Mont., 220.
Clayton's law, 37.
Clear Creek Co., Colo., 211, 212, 214.
Clear Lake, Cal., mercury, 267.
Clerc, F. L., cited on zinc mines of southwestern Missouri, 167–170.
Cliff copper nine, Michigan, 141.
Clifton, Ariz., copper district, 145.
Clinton ore, 92–98.
   Distribution, 92–97.
   Origin, 97.
Clinton ores, general relations, 276.
Coal measures, classification of in Pennsylvania, 87.
Coastal Plain, 252.
   Geology, 8.
   Topography, 6.
Coast Range belt of mines, 275.
   Chromite, 261.
   Oregon, 39.
   Washington, 238.
Cobalt, 262.
Cochise Co., Ariz., 229.
Cœur d'Alène, Idaho, lead-silver ores, 192.
Colfax Co., N. M., 204.
Colombia, platinum, 272.
Colorado, geology, 204.
   Iron (bog) ore, 74.
   Lead-silver mines, 182, 191, 275.
   Limonite, 79.
   Magnetite, 126.
      Literature, 126.
   Plateau, 7–9.
   Silver and gold, 205–216.
   Spathic iron ore, 89.
   Titaniferous ores, 281.
Columbia Co., N. Y., lead mines, 157.
Columbia Hill, gravels, California, 245.
Comb-in-comb structure in a vein, 35.
Comstock Lode, Nev., 233–237.
Comstock, S B., on Red Mountain, Colo., 190.
   On vein systems of the San Juan, 206.
Conejos Co., Colo., 212.
Connecticut, bismuth, 260.
   Lead mines, 157.
   Limonite, 80, 85.
   Magnetite sands, 128.
   Roxbury siderite, 91.

Contact deposits, 58.
Copper Basin, Ariz., 151.
Copper Creek, near Gothic, Colo., **211**.
Copper Falls copper mine, Michigan, 141.
Copper Mountain, Ariz., copper district, 145.
Copperopolis, Cal., copper mines, 136.
Copperopolis copper mine, Utah, 152.
Copper ores, analyses of ore minerals, **134**.
Copper Queen mine, Arizona, 150.
Copper, statistics, 155.
Cordillera region, general geology, 275.
Cornwall, Penn., iron mines, 125, 127.
 Literature, 127.
Costillo Co., Colo., iron mines, 126.
Cotta, B. von, **on filling of mineral veins**, 28.
 Schemes of classification **of ore deposits**, 44–46.
Courtis, W. M., on the microscopic structure of gold quartz, 249.
Cranberry magnetite mines, North Carolina, 125.
Credner, H , on origin **of Marquette** ores, 106.
Creede, Colo., 207–209.
Crimora, Va., manganese, 263.
Crismon-Mammoth copper **mine**, Utah, 152, 194.
Crittenden Co., Ky., lead and zinc veins, 166.
Crosby, W. O., cited on joints, 13.
Cross, W., on the **mines near Rosita, Colo., 213.**
Crustification, **283.**
Cryolite, 258.
Cumberland **iron** mine, Colorado, 126.
Cumberland River region, Kentucky, 78.
Curry Co. ore, 240.
Curtis, J. S., cited on caves, 21.
 On **Eureka, Nev., 196, 197.**
 On **growth of aragonite crystals, 34.**
 On replacement, 33.
 On silver in quartz porphyry, **25.**
Custer Co., Colo., 212, 213.
 Idaho, 223.
Custer mine, Idaho, 223.
Dakyns and Teall cited **on origin** of magnetite, 129.
Dall, W. H., on Alaska, 253.
Dana, divisions of geological time, 4.
Davison Co., N. C., lead mines, 157.
Dawson, G. M., on Douglass Island, 254.
Deadwood, S. D., 217.

Deep Creek mines, Utah, 225.
Deep gravels, California, 245.
Deer Lodge Co., Mont., 221.
Deer Trail mine, Utah, 226.
De la Biche, electrical **activity of veins**, 41.
De Launay, treatise of, 285.
Delaware, chromite, 261.
Del Norte Co., Cal., chromite, 261.
Deloro, Can., arsenic mine, 260.
Deloro (Marmora), Can., gold mine, 256.
Devereux, W. B., cited on Colorado iron ores, 126.
Devonian system at Hamilton, Nev., 231.
Diaclases. See under "Joints."
Dickerson iron mine, New Jersey, **130.**
Dike defined, 11.
 Distinction from vein, 11.
 Cause of earthquakes, 17.
Diller, J. S., cited on Kentucky peridotite, 166.
 On California, 241.
Dillsburg iron mines, Pennsylvania, 127.
Doe Run, Mo., lead mines, 158.
Dolomitization, 21, 22.
Dolores Co., Colo., 205.
Doña Ana Co., N. M., 181.
Douglass Co., Ore., nickel mines, **271.**
Douglass Island, Alaska, 250, 254.
Drinker, H. S., cited on the Saucon Valley zinc mines, 175.
Drumlummon mines, Montana, 221, 222.
Dry Cañon, Utah, 194, 225.
Ducktown, Tenn., illustration of gossan, 39.
 Mines of pyrite and chalcopyrite, 131, 135.
Dutton, Capt. C. E., cited on the geology of New Mexico, 203.
Dyestone iron **ore**, 92.
"Dyestone ranges,' 94.
Eagle Co., **Colo., 211.**
Eagle River district, Colo., 186.
Earth, average composition of crust, 72.
Eastern sandstone, Keweenaw Point, Mich., 139.
Egan Cañon, Nev., 231.
El Dorado Co., Cal., 249.
Electrical activity of ore bodies, 235; of veins, 40.
Elk Mountains, Colo., 211.
 Iron mines, 126.
Elko, Nev., 231.
Elko **Co., Nev., 232.**

Ely copper mines, Vermont, **134, 135.**
Emma mine, Utah, 194.
Emmons, E., on Adirondack titaniferous ores, 280.
Emmons, S. F., cited **on replacement,** 33.
    On Butte, **Mont., 137.**
    On country **rock of Boulder Co.,** Colo., 214.
    On Leadville, Colo., 183-185.
    On Red Mountain, Colo., 190.
    On the mines near Rosita, Colo., 213.
Endlich, F. M., on gold mines near Ouray, Colo., 206.
Enterprise, Miss., spathic ore, 89.
Esmeralda Co., Nev., 232.
Etta granite knob, with **tin ore, 273.**
Eureka Co., Nev., 232.
    Electrical activity of ore, **235.**
    Illustration gossan, 39.
    Lead-silver, **196.**
Evigtok, Greenland, **cryolite, 258.**
Fahlbands defined, 61.
Fairplay, Colo., 212, 245.
Farish, J. B., illustration by, 20.
    On the veins at Newman Hill, **near** Rico, Colo., 206, 207.
Faults, cause of, 13.
    Defined and described, **17-20.**
    Hade, 17.
    Normal, 18.
    Reversed, 18.
    Schmidt's law **of, 18.**
    Step faults, 20.
Feather River, Cal., 241.
Felch Mountain area **and** iron ores, Michigan, 104.
Fisher Hill iron **mines, New** York, 121.
Flagstaff mine, **Utah, 194.**
Flagstone iron **ore, 87.**
Flaxseed iron **ore, 92.**
Flötz defined, 43.
Floyd **Co.** (Ga.) **bauxite, 258.**
Flucan **of a** vein, 37.
Foerste, A. F., on Clinton ore, 97.
Folds, various kinds defined, 11.
    Axes of, 11.
    Cause of cavities, **13.**
    "Pitch" of, 11.
Forest of Dean iron mine, **New York,** 123.
Forest Queen mine, Ruby **district,** Colo., 211.
Formation defined, 5.
Fort Laramie, Wyo., iron ore near, 114.
Fossil iron ore, 92.
Foster and Whitney, cited **on Mar**quette district, 101.

Foster and Whitney, Marquette district, origin of its ores, 104.
    On Keweenaw Point copper, **142.**
Fouqué and Levy cited, 279.
Franklin copper mines, Michigan, 141.
Franklin Co. (Mo.) **lead and zinc** mines, 165.
Franklin Co. (Va.) magnetite, 125.
Franklin, N. J., 123.
Franklin Furnace, N. J., 174, 175-179.
    Iron mines, 124.
Franklinite, 262.
Freeland, F. T., cited on faults, 19.
Fremont Co. (Colo.) magnetite, 126.
Friedensville zinc mines, Penn., 174.
Frisco, Utah, 195.
Frost drift, North Carolina, 252.
Fryer Hill, Leadville, Colo., 183.
Fuchs and Le Launay, treatise **of,** 284.
Fulton, J., on the Menominee district, 107, 108.
Gagnon vein, Butte, Mont., 137.
Galena (town), S. D., lead-silver mines, 191.
Gangue minerals forming, 23.
    Source of, 26.
Gap mine, Lancaster Co., Penn., 270.
Gap mine of pyrrhotite, Pennsylvania, 131.
Gatling arsenic mines, **Deloro, Can.,** 260.
Genesee antimony mine, Nevada, 259.
Genth, F. **A., on** Boulder Co., Colo., 214.
    On gold in Southern States, 252.
Geology, modern standpoints, 3.
    Tabulation of geological subdivisions, 4, 5.
Georgetown, Colo., 214.
Georgia, bauxite, 258.
    Clinton ore, 95.
    Limonite, **84.**
    Manganese, 263.
Geyser mine, Custer Co., Colo., 213.
Gilbert, G. **K.,** cited on joints, 13.
    Illustration loaned by, 15.
Gilpin Co., Colo., 214, 250.
    Banded veins, 35.
    Copper mines, 138.
Glendale, Mont., 219.
    Lead-silver ores, 191.
Glenmore estate, West Virginia, **98.**
Globe copper district, Arizona, **150,** 228.
Gogebic district of Lake Superior, 108.
Gold, Alaska, 254, 255.
    Black Hills, 217.
    California, 243-250.

Gold in sea beaches, 240.
  Nova Scotia, 255.
  Occurrence and ores, 200.
  Oregon, 240.
  Quartz veins, 248.
  Southern States, 252.
  Statistics, 256.
  Washington, 239.
Gold Cup mine, Tin Cup, Colo., 211.
Gossan, 38.
Gothic, Colo., 211.
Gouge of clay in a vein, 37.
Graham Co., Ariz., 229.
Granby, Mo., zinc and lead mines near, 166–171.
Grand Cañon of the Colorado, 227.
Granite, Chaffee Co., Colo., 212.
Granite Mountain mine, Butte, Mont., 221.
Granite veins with cassiterite, 60.
Grant Co., N. M., 203.
Grant Co., Ore., 240.
Great Basin, 3–7, 10.
  In California, 241.
  In Oregon, 239.
Great Eastern mercury mine, 267.
Great Lakes, geology, 8, 9.
Great Valley, 8, 276.
  Iron ores of, 79.
Great Western mercury mine, 267.
Greeneyed Monster mine, Utah, 226.
Greenland cryolite, 258.
Green Mountain placers, 245.
Green River, Utah, 224.
Greisen with tin ores, 273.
Griffin, P. H., on bog ores of Quebec, 279.
Grimm, J., scheme of classification of ore deposits, 50.
Groddeck, A. v., scheme of classification of ore deposits, 51.
Gulf region, geology, 9.
Gunnison Co. (Colo.) iron mines, 126.
Gunnison region, Colorado, 211.
Hade of a fault, 17.
Hague, A., cited on Wyoming iron ore, 126.
  On Hamilton, Nev., 231.
Hague, A., and Iddings, J. P., on the Comstock Lode, 233, 236, 237.
Haile gold mine, South Carolina, 252.
Hall, C. E., cited on the iron mines of the Adirondacks, 122.
Hamilton, Nev., 231.
Hammondville (N. Y.) iron mines, 122.
Hampton and Eureka copper mines, Arizona, 151.
Hancock, Mich., 139.
Hanging Rock region of Kentucky, 77.

Hanging Rock region of limonite ores, 77.
  Spathic ores, 88.
Harrington, B. J., cited on the origin of magnetite, 130.
Hartman zinc mine, Pennsylvania, 174.
Hastings Co., Can., 260.
Haworth, E., cited on zinc and lead ores of southwestern Missouri, 170.
Hayes, C. W., illustration by, 18.
Hecla iron mine, Colorado, 126.
Hecla mines, Glendale, Mont., 191.
Hecla mines, Montana, 219.
Helena, Mont., 221.
Hematites, red and specular, table of analyses, 118.
Henderson Lake, N. Y., titaniferous ores, 280.
Henrich, C., cited on Aspen, Colo., 189.
  On Clifton (Ariz.) copper mines, 146.
  On the Slayback lode, New Mexico, 204.
Henry Co., Va., magnetite, 125.
Hesse, Germany, bog ore, 75.
Hibernia iron mines, New Jersey, 124.
High or deep gravels, California, 245.
Hills, R. C., cited on the chemistry of replacement, 34.
  On the Little Annie mine, Colo., 212.
  On the vein systems of the San Juan, 206.
  Zone of enrichment in the mines of the Summit district, Colo., 39, 40.
Hinsdale Co., 205.
Hoefer, H., cited on faults, 19.
Hogan Mountain, Mo., 116.
Homestake mine, Eagle Co., Colo., 211.
Honorine mine, Utah, 194.
Horizon defined, 5.
Horn Silver mine, Utah, 195.
Horses of barren rock in a vein, 36.
Hot Springs iron mines, Colorado, 126.
Huerfano Co., Colo., 212.
Humboldt Co., Nev., 232.
  Antimony, 259.
Humboldt-Pocahontas mine, Custer Co., Colo., 213.
Humboldt range, Nev., 231.
Hunt, T. S., on Canadian titaniferous ores, 280.
  On the iron mines of the Adirondacks, 122.
Hurd iron mines, New Jersey, 124.

Hysteromorphic deposits, 284.
Ibapah range, Utah, 225.
Ice, veins of, on Mount McClellan, Colo., 211.
Idaho, geology, 222.
    Lead-silver ores, 191, 192.
    Tin, 274.
Idaho Springs, Colo., 214.
Iddings, J. P., on basaltic columns, 12.
    On the Comstock Lode, 233, 236, 237.
Idiogenites of Posepny, 284.
Igneous rocks, defined and roughly classified, 6.
    Forms assumed by, 11.
Illinois lead and zinc mines, 161.
Independence, Colo., 211.
Ingersoll granitic knob, with tin ore, 273.
International Geological Congress, 1885, classification recommended by, 3, 4.
Inyo Co., Cal., antimony, 259.
Iowa lead and zinc mines, 161.
    Limonite, 79.
Ireland, bog ore, 75.
Iridium, with platinum, 272.
Iron Co., Utah, 226.
    Magnetite, 128.
        Literature, 128.
Iron hat, 38.
Iron Hill, Leadville, Colo., 183.
Iron in rocks, 72.
Iron King iron mine, Colorado, 126.
Iron Mountain, Colo., 126.
Iron Mountain, Missoula Co., Mont., 222.
Iron Mountain, Mo., 116, 130.
    Bibliography, 118, 119.
    Illustration of subaërial decay, 59.
    Porphyries, 117.
Iron ores, bibliography of general papers, 69.
    Brown hematite ore, 73–85.
    General remarks on, 70–73.
    Impurities, 71.
Iron ore, magnetite, 119–130.
Iron ore localities :
    Adirondack Mountains, 120–122.
    Alabama, 78, 84.
        Clinton ore, 95, 96.
    Arkansas, Addenda.
    Colorado, 79, 89.
        Hall's Valley and Handcart Gulch, Park Co., 74.
    Connecticut, 80–85, 91.
    Georgia, 84.
        Clinton ore, 95.
    Hesse, Germany, 75.

Iron ore localities :
    Iowa, 79.
    Ireland, 75.
    Kentucky, 77, 78, 88.
        Clinton ore, 93.
    Louisiana, 79.
    Maryland, 82.
        Clinton ore, 94.
    Massachusetts, 80–84, 89.
    Michigan 78.
        General outline, 100.
        Marquette district, 102–105.
        Menominee district, 107, 108.
        Penokee-Gogebic district, 108.
    Minnesota, 78.
        Vermilion district, 110, 111.
        Mesabi range, 111.
    Missouri, 78.
        Cambrian, 99.
        Iron Mountain, 116.
        Lower Carboniferous, 98.
        Pilot Knob, 114.
    Mississippi, 89.
    Montana, 89.
        Great Falls, 74.
    New Jersey, 81.
    New York, 80, 85,
        Burden Mines, 89.
        Clinton ore, 93.
        Jefferson Co., 99.
        Rye, 74.
        Staten Island, 74.
        Wawarsing, 91.
    North Carolina, 74, 84, 89.
        Specular, 114.
    Nova Scotia, Clinton ore, 97.
    Ohio, 78, 88.
        Clinton ore, 92, 93.
    Oregon, Port Townsend Bay, Portland, 74, 75.
    Pennsylvania, 87.
        Adams Co., 85.
        Blair Co., 77.
        Carbon Co., 77.
        Franklin Co., 76, 77.
        Fulton Co., 77.
        Huntingdon Co., 76, 77.
        Juniata Co., 77.
        Lehigh Co., 81, 84.
        Mifflin Co., 77.
        Perry Co., 77.
        York Co., 81, 84, 85.
        Clinton ore, 93.
        Mansfield ores, 98.
    Quebec, Three Rivers, Addenda.
    Tennessee, 78, 82.
        Clinton ore, 94.
    Texas, 78.
    Utah, 79.
    Vermont, 80.

294 INDEX.

Iron ore localities:
  Virginia, 77-82.
    Clinton ore, 94.
    Specular, 113.
  West Virginia, 88.
    Clinton ore, 94.
    Oriskany ore, 98.
  Wisconsin, Clinton ore, 92.
    Menominee district, 107, 108.
    Penokee - Gogebic district, 108, 109.
  Wyoming, 89.
Iron ore, pyrite, 131, 132.
  Red and specular hematite, 92-119.
Iron ores, origin of, 278.
  Siluro-Cambrian limonites, 79-85.
  Spathic, 86-91.
  Statistics, 133.
  Swedish lakes, 75.
  Table of and compositions, 70.
Irving and Van Hise on the Marquette district, 101, 102.
  Origin of the ores, 106.
Irving, R. D., cited on Silver Islet, 202.
  On replacement, 33.
  Penokee-Gogebic district, 109.
Ishpeming, Mich., gold, 253.
Isle Royale, Lake Superior, 139, 141, 143.
Izard limestone, 265.
Jackson Co., N. C., nickel, 271.
Jacque Mountain, Summit Co., Colo., 211.
Jackson, Ore., nickel, 271.
James River, Va., specular ores near, 113.
Jasper Co. (Mo.) zinc and lead mines, 167.
Jayville iron mines, New York, 122.
Jefferson Co., Mont., 219.
Jefferson Co. (Mo.) lead and zinc mines, 165.
Jenney, W. P., cited on Arkansas silver mines, 201.
  On lateral enrichments, 37.
  On lead ores of southeastern Missouri, 159.
  On lead and zinc mines of the Mississippi Valley, 164.
  On mines of Mississippi Valley, 276.
  On the Head Center mine, Tombstone, Ariz., 31.
  On zinc and lead mines of southwestern Missouri, 170.
Joints defined, 12.
Joplin (Mo.) zinc and lead mines, 166-171.

Josephine Co., Oregon, 240.
Juab Co., Utah, 225.
Julien, A. A., action of organic acids on rocks, 21.
  Cited on the origin of magnetite, 130.
Kansas, nickel, 271.
Kaolinization at the Comstock Lode, 235.
Kelley Lode, New Mexico, 181.
Kemp, J. F., scheme for the classification of ore deposits, 53-55.
  On Missouri lead deposits, 159.
Kentucky, lead and zinc ores of Livingston Co., 165.
  Clinton ore, 93.
  Limonite, 77, 78.
  Limonite or brown hematite ores, 77, 78.
  Spathic ores, 88.
Kern Co. (Cal.) antimony, 259.
Kerr, W. C., on North Carolina gold, 252.
Keweenaw Point (Mich.) copper, 139-143.
Kimball, J. P., cited on the Burden mines, New York, 89.
  On origin of iron ores, 278.
King, C., on the Comstock Lode, 233, 234.
Kinahan, J. H., cited on joints, 13.
Kittitas Co., Washington, gold, 238.
Klausen in the Austrian Tyrol, veins illustrating lateral secretion, 31, 32.
Knob of igneous rock defined, 11.
Koehler, G., cited on faults, 19.
  Scheme of classification of ore deposits, 47.
Lahontan, Lake, 230.
Laccolite defined, 11.
Lac la Belle, Keweenaw Point, 142.
Lager defined as contrasted with "Flötz" and "Gang," 43.
Lakes, A., cited on Aspen, Colo., 189.
Lake Champlain iron region, 120.
  Magnetite sands, 128.
Lake Co., Colo., 211.
Lake of the Woods, gold district, 256.
Lake Superior gold region, 256.
  Iron ore districts, 100-113.
  Mines, 276.
Lake Valley (N. M.) lead-silver ores, 181.
Lancaster Co. (Penn.) chromite, 261.
Lander Co., Nev., 232.
  Antimony mines, 259.
Lander Hill, Nev., 232.
Lane's Mine, Monroe, Conn., 260.
La Plata Co., Colo., 205.

Lassens Peak, Cal., 241.
Last Chance Gulch, near Helena, Mont., 221, 245.
Lateral enrichments in a vein, 37.
  secretion, 29.
Lead City, Black Hills, 217.
Leadville, Colo., 182–185.
  Figure of Moyer Fault, 17.
Lead minerals, 156.
  Statistics, 160.
Lead-silver ores, 181.
Leconte, J., on California gravels, 247.
  On joints, 13.
  On mercury deposits, California, 267.
  Scheme of classification of veins, 45.
Lehigh Co. (Penn.) iron mines, magnetites, 125.
Lenses, or lenticular beds of ore, origin of, 59.
Lesser metals, 258–277.
Lesquereux, cited on California gravels, 246.
Lewis and Clarke Co., Mont., 221.
Lewis Mountain, Mo., 116.
Lime Creek (Colo.) Lower Carboniferous fossils, 189.
Limonite, Arkansas, 289.
Limonite, deposits of, 73–86. See also under iron ore for geographical distribution.
  Analyses, 86.
Lincoln Co., Nev., 230.
Lincoln Co., N. M., 204.
Lindgren on Wickes, Mont., 192.
Lindgren, W., on Calico district, Cal., 243.
  On Neocene rivers of California, 282.
Linnaeite, 269.
Little Annie mine, Rio Grande Co., Colo., 212.
Little Belt Mountains, Mont., 222.
Little Cottonwood Cañon, Utah, 192.
Little River iron mines, New York, 121.
Little Rock (Ark.) bauxite, 258.
Livingston Co. (Ky.) lead and zinc mines, 165.
Llano Co. (Texas) copper mines, 139.
Logan Co. (Kan.) nickel, 271.
Lottner-Serlo scheme of classification of ore deposits, 47.
Longdale iron mines, Virginia, 77.
Louisa Co. (Va.) mines of pyrite, 131.
Louisiana limonite, 79.
Lovelock mines (Churchill Co., Nev.) nickel, 271.

Lovelock, Nev., 259.
Lowell (Mass.) nickel mine, 270.
Low Moor iron mines, Virginia, 78.
Lowville (Lewis Co., N. Y.) lead mine, 157.
Lubeck (Me.) lead mine, 157.
Lucky Boy mine, 226.
Lyndhurst (Va.) manganese, 263.
Lynchburg, Va., iron ores near, 113.
Lyon Co., Nev., 232.
Lyon Mountain (N. Y.) iron ores, 120.
Mackenzie River, gold gravels, 256.
Madison Co., Mont., 219.
Magdalena Mountains, N. M., 181.
Magna Charta mine, Butte, Mont., 220.
Magnet Cove, Ark., magnetite, 280.
Magnetite iron ore, 120–131.
  Analyses, 131.
  General description, 120.
  Origin of, 129.
  Origin in igneous magmas, 56.
  Ore bodies, general relations, 276.
  Sands, 128.
Maine tin, 274.
Manganese, 262–266.
  Statistics, 266.
Mansfield ores, Pennsylvania, 98.
Margerie and Heim, cited on faults, 18.
Maricopa Co., Ariz., 228.
Mariposa Co., Cal., 249.
Mariposite, 249.
Markhamville, N. B., manganese, 266.
Marmora, Can., arsenic mine, 260.
  Gold mines, 256.
Marquette district, 100.
Marshall tunnel, Georgetown, Colo., 38.
Maryland, chromite, 261.
  Clinton ore, 94.
  Gold mines, 252.
  Limonite, 82.
Marysvale, Utah, 226.
Marysville, Mont., 222.
Massachusetts, lead mines, 157.
  Limonite, 80–84.
  Spathic ore at Gay Head, 89.
McGee, W. J., Appomattox and Columbia formations, 5.
  On joints, 13.
Meagher Co., Mont., 222.
Melaconite, Keweenaw Point, 142.
Mendota mines, Keweenaw Point, 142.
Menominee district, Lake Superior, 107.
Mercury, 267–269.
  Mines, California, 275.

Mesabi range, Minnesota, 111, 112.
Metacinnabarite, 267.
Metamorphic rocks, defined and roughly classified, 6.
Michigan, copper, 139.
    Gold, 253.
    Limonite, 78.
    Marquette district, 102–105.
    Menominee, 107, 108.
    Penokee-Gogebic, 108.
Middletown, Conn., lead mine, 157.
Milan, N. H., mines of pyrite, 131.
    Chalcopyrite, 134.
Millerite, 269.
Mine la Motte, Mo., lead mines, 158, 159.
    Nickel, 271.
Mineral Point, Wis., copper ores, 164.
Mineville, N. Y., iron mines, 121, 122.
Mine waters with dissolved metals, 40.
Minnesota copper mines, 143.
    Limonite, 78.
    Mesabi range, 111.
    Titaniferous ores, 281.
    Vermilion district, 110, 111.
Missabe. See Mesabi.
Mississippi, spathic ores at Enterprise, 89.
Mississippi Valley, geology, 9.
    Lead and zinc mines, 161.
    Relations of ore bodies, 276.
Missoula Co., Mont., 222.
Missouri, Cambrian bed hematite in Crawford, Dent, and Phelps Counties, 99.
    Copper at St. Genevieve, 143.
    Iron Mountain, 116.
    Lead mines of southeastern Missouri, 158.
    Limonite, 78.
    Lower Carboniferous red hematite, 98.
    Pilot Knob, 114.
    Tin, 274.
    Zinc and lead in the southwest, 166.
Mitchell Co., N. C., iron mines, 125.
Mohave Co., Ariz., 228.
Moisie, Can., magnetite sands of, 128.
Monarch district, Chaffee Co., Colo., 186, 212.
Monocline defined, 11.
Montana, Butte, 136 137.
Montana, geology of, 218.
    Iron ore at Great Falls, 74.
    Lead-silver ores, 191, 192.
    Silver and gold mines, 219–222.

Montana, spathic iron ore at Sand Coulée, 89.
    Tin, 274.
Montville, N. J., 123.
Morenci, Ariz., copper district, 145.
Möricke, on gold in Chilean volcanic rocks, 25.
Mosquito range, Colo., 182.
    In Park Co., Colo., 212.
Mother Lode, Cal., 249.
Mount Baldy, Utah, 226.
Mount Davidson, Nev., 233.
Mount Hope, N. J., 123.
Mount Marshall, near Georgetown, Colo., 214.
Mount McClellan, Clear Creek Co., Colo., 211.
Mount Prometheus, Nev., 232.
Mount Shasta, Cal., 241.
Mule Pass Mountains, Ariz., 148.
Mullica Hill, N. J., bog ore with vivianite, 74.
Munroe, H. S., citation from, 30.
    Cited on the origin of magnetite, 130.
    Scheme of classification of ore deposits, 52, 53.
Musconetcong iron belt, New Jersey, 124.
Nacemiento copper mines, New Mexico, 227.
Nason, F. L., cited on the geology of the Highlands, N. J., 123.
    On geology near Franklin Furnace, N. J., 175.
    On Missouri geology, 282.
    On Missouri iron ores, 118.
Neck of igneous rock defined, 11.
Neihart, Mont., 222.
Nelson Co., Va., tin, 274.
Neocomian beds, California, 267.
Nevada, antimony, 259.
    Geology, 230.
    Lead-silver mines, 196.
    Lovelock nickel mines, 271.
New Almaden, Cal., mercury, 267.
Newberry, J. S., cited on the Cave Mine, Utah, 195.
    On Clinton ore, 97.
    On Silver Reef, Utah, 226.
    Scheme of classification of ore deposits, 48, 50.
Newberry, W. E., cited on Aspen, Colo., 188, 190.
New Brunswick, antimony, 259.
    Manganese, 266.
New Brunswick, N. J., copper mines, 153.
Newburyport, Mass., lead mine, 157.
New Caledonia nickel, 271, 272.

INDEX. 297

New England, outline of geology, 7, 8.
New Hampshire tin, 274.
New Idria, Cal., mercury, 267, 268.
New Jersey copper ores, 153.
    Iron mines, 123, 124.
    Limonite, 81.
    Outline **of geology,** 7, 8.
    Titaniferous **ores,** 281.
    Zinc mines, **175.**
New Jersey Zinc and Iron Co.'s mine, Franklin Furnace, N. J., 176.
Newman Hill, Colo., banded veins, 35, 36.
    Figure of faulted vein at, 20.
    Lateral enrichments, 37.
    Mines **of,** 231.
New Mexico, geology, 202.
    Lead-silver mines, 181, **282.**
    Silver and gold mines, **202–204.**
    Triassic **copper ores,** 154.
Newton, Amador Co., Cal., copper, 136.
Newton Co., Mo., zinc mines, 167.
New York **copper** mine, Arizona, 150.
New York, **iron** (bog) ore, Staten Island, **74;** Rye, 74.
    Iron mines of the Highlands, 123.
        Literature, 124, 125.
    Iron ore of Adirondacks, **120, 121.**
        Literature, 122, 123.
    Lead mines, 157, 158.
    Limonite, 80, **85.**
    Outline of geology, 7, 8.
    Spathic ore **of** Burden **mines,** 89.
        at Warwarsing, 91.
    Clinton ore, **93.**
    Jefferson Co., 99.
Ney Co., Nev., 231.
Niccolite, 269.
Nickel **and cobalt, 269–272.**
Nickel, **statistics, 271, 272.**
Nicholas, W., cited **on precipitation** of gold, 251.
Nicholson, F., on the St. Genevieve, Mo., copper mines, 144.
Northampton, Mass., lead mine, 157.
North Carolina, gold mines, 252.
    Iron (bog) ore, 74.
    Limonite, 1, 84.
    Magnetite, 125.
        Literature, 125.
        Titaniferous magnetite, 281.
    'Nickel, **271.**
    Spathic ores, 89.
    Specular ores, 114.
    Tin, 274.
Norway, nickel, 272.
    Titaniferous ores, 281.
Nova Scotia, Clinton iron ore, 97.
    Gold, 255.

Nova Scotia, manganese, 266.
Oat Hill mercury mine, 267
Ogdensburg, **N. J.,** 174, 175–179.
Ohio, Clinton **ore, 92,** 93.
    Limonite, **78.**
    Spathic ores, **88.**
Okanogan Co., Wash., **238.**
Old Dominion **copper mine, Arizona, 150.**
Olmstead, I., cited **on the Burden** mines, New York, **90.**
Olympic Mountains, 238.
Oneida Co., Idaho, 223.
Ontario, arsenic mine at Deloro, **260.**
Ontario mine, Utah, 225.
Ontonagon, Mich., **141.**
Ophir Cañon, Utah, **194,** 225.
Oquirrh Mountains, **Utah,** 192, 225.
Orange Co., N. Y., iron **mines,** 123.
Oregon, geology, **239.**
    Mercury, 267.
Ore Knob, N. C., copper mines, 134, 135.
Ores, minerals forming, 23.
    Original source of, 23.
    Of iron, table of, 70.
Organic matter as a precipitating agent, 57.
Orton, E., cited on dolomitization, 22.
Osmium, with platinum, 272.
Ouachita uplift of Arkansas, 170, 201, 276.
Ouray Co., Colo., 205.
Owybee Co., Idaho, 223.
Oxford, **N. J.,** 123.
Ozark uplift, 7, 114, **170, 276.**
Pacific slope, geology, 10.
Pahranagat district, Nev., **231.**
Palmer Hill iron mines, N. Y., **122.**
Park Co., Colo., 212.
Passaic iron belt, New Jersey, 124.
Pearce, R., bismuth with gold **in** Colorado, 250.
    On gold ores of Gilpin Co., Colo., 214.
Penfield, S. F., on sperrylite, 272.
Peninsula **copper** mines, Michigan, 141.
Penrose, R. **A. F., on manganese, 264–** 266.
    On the iron ores **of Arkansas,** 279.
Pennsylvania, **brown hematites. See** under Iron, **brown hematite.**
    Chromite, 261.
    Clinton ore, 93.
    Lead mines, 157.
    Limonite, 76, 77, 81, 84, 85.
        For index of counties see under Iron **ores,** limonite.

Pennsylvania, Mansfield ore, 98.
   Nickel, 270.
   Spathic ores, 87.
Penokee-Gogebic district, Lake Superior, 108.
Pequest iron belt, New Jersey, 124.
Perkiomen copper mine, Pennsylvania, 153.
Phillips, J. A., scheme of classification of ore deposits, 49.
Phœnix copper mine, Michigan, 141.
Phosphorus in iron ores, 71, 72.
Pilot Knob, Mo., 114.
   Bibliography, 118, 119.
Pima Co., Ariz., **229**.
Pinal Co., Ariz., **228**.
Pinches and swells in a vein, 37.
Pioche, Nev., 230.
Pipe clays in California gravels, 246.
Pitkin, Colo., 211.
Piute Co., Utah, 226.
Placer Co., Cal., iron ores, 126.
   Chromite, 261.
Placers, 243–248.
Plateau region of Wyoming, 216.
Platinum, 272.
Platoro, Conejos Co., Colo., 212.
Point Orford, Ore., 240.
Poorman Lode, Idaho.
Portage Lake, Mich., 141.
Porter, J. B., cited on Clinton ore, 97.
Port Henry, N. Y., iron mines near, 121, 122.
Posepny, F., on origin of ore deposits, 282, 283.
   Cited on replacement theory for Raibl, 33.
Prairie region, 9.
   Of Wyoming, 216.
Prescott, Ariz., 151.
Prickly Pear Gulch, near Helena, Mont., 221, 245.
Pride of the West mine, Eagle Co., Colo., 211.
Prime, F., and Von Cotta, scheme of classification of ore deposits, 46, 47.
Printer Boy mine, Leadville, Colo., 183.
Psilomelane, 262.
Puget Sound, 238.
Pumpelly, R., cited on replacement, 33.
   On subaërial decay, 59.
   On Keweenaw Point copper mines, 142.
   Scheme of classification of ore deposits, 51, 52.
Putnam Co., N. Y., iron mines, 123.
Pyrite, 131, 132.

Pyrite, literature, 132.
   Origin of, 132.
Pyrite ore bodies, general relations, 276.
Pyrolusite, 262.
Pyrrhotite with nickel, 269.
Quaquaversal defined, 11.
Quaco Head, N. B., manganese, 266.
Quebec, bog ores, 279.
   Iron (bog) ore. See Addenda.
Quincy copper mine, Mich., 141.
Quicksilver (mercury), 267–269.
Quogue, Long Island, magnetite sands, 128.
Radnor ores, i.e., Three Rivers, Can., 279.
Raibl, Austria, silver-lead deposits, originating by replacement, 33.
Rainbow Lode, Butte, Mont., 220.
Rainier Mountain, 238.
Ramapo iron belt, N. J., 124.
Ramshorn mine, Custer Co., Idaho, 223.
Ranges of ore deposits, 275, 276.
Raymond & Ely mine, Pioche, Nev., 230.
Raymond, R. W., 277, 282.
Red Cliff Eagle Co., Colo., 211.
Red Mountain, Mont., 220.
Red Mountain, Ouray Co., Colo., 190.
Red River region, Kentucky, 77.
Red Rock, San Francisco, manganese, 266.
Reese River district, Nev., banded veins, 35.
   Geology of, 232.
Replacement of limestone by iron ores, 278, 279.
Replacement, method of vein filling by, 33.
Reyer, E., on the Marquette ores, 106.
Rico, Dolores Co., Colo., lead-silver mines, 190.
Riddle, Ore., nickel mines, 271.
Rifting in granite at Cape Ann, Mass., 12.
Rim of deep gravels, California, 245.
Rio Grande Co., Colo., 212.
River gravels, with gold, 243.
Roaring Fork Creek, Colo., 188.
Robert E. Lee mine, Leadville, Colo., 183.
Robinson mine, Summit Co., Colo., 186.
Rockbridge Co., Va., tin, 274.
Rock Point, Ore., nickel, 271.
Rocks, classification of, 6.
   General percentage of iron oxide, 72.

INDEX.   299

Rocky Mountains, geology, 9.
  Zinc ores, 179.
Rogers, H. D., on geology at Franklin Furnace, N. J., 175.
Rolker, C. M., cited on Colorado iron ores, 126.
  On Silver Reef, Utah, 226.
Rominger, C., on the Marquette district, 101, 102.
Ropes mine, gold, Michigan, 253.
Rosenbusch, H., cited on succession of rock-forming minerals, 24.
Rosita, Colo., 212, 213.
Rothpletz, A., cited on oölites, 57.
Rothwell, R. P., on Silver Reef, Utah, 226.
Roxbury, Conn., spathic ore, 91.
Ruby, Colo., 211.
Russell, J. C., cited on Clinton ore, 97.
Russia, platinum, 272.
Sacramento Valley, 243.
Saguache Co., Colo., 207-209.
Saline Co., Ark., nickel, 271.
Salt Lake Co., Utah, 225.
San Benito Co., Cal., antimony, 259.
San Bernardino Co., Cal., iron ore, 127.
Sandberger, F., cited on source of ores, 25.
  Barite in limestone, 26.
  On lateral secretion, 30.
San Diego Co., Cal., iron ores, 127.
San Emigdio, Kern Co., Cal., antimony, 259.
Sangre de Cristo range, Colorado, 212.
Sandia Mountains, N. M., 204.
Sanford Lake, titaniferous ores, New York, 280.
San Joaquin Co., Cal., 266.
San Juan Co., Colo., 205
  Bismuth, 260.
San Juan Mountains, 205.
San Juan region, Colorado, 205.
San Luis Obispo Co. (Cal.) chromite, 261.
San Miguel Co., Colo., 206, 245.
Santa Fé Co., N. M., 204.
  Placers of, 245.
Santa Rita, N. M., copper mines, 150, 151.
Santa Rita Mountains, 203.
Saucon Valley zinc mines, Pennsylvania, 174, 175.
Schmitt, A., on Missouri iron ores, 118.
  On replacement theory as applied to Missouri iron ores, 33.
  On zinc mines of southwestern Missouri, 167.

Schapbach in the Black Forest, veins at, 31.
Schooley's Mountain, N. J., iron ores, 281.
Schuyler copper mine, New Jersey, 153.
Sebenius, U., on Adirondack titaniferous ores, 280.
Sedimentary rocks defined and roughly classified, 6.
  Forms assumed by, 10, 11.
  Segregated veins, as applied to magnetite, 281.
  Described, 60.
Senarmontite, 259.
Sevier Co., Ark., antimony, 259.
Shasta Co., Cal., chromite, 261.
Shear zones, defined, 13.
Sheet of igneous rock, defined, 11.
Shepherd Mountain, Mo., 116.
Shumagin, Alaska, 254.
Siderite, iron ore, 86-91.
Siegenite, Mine la Motte, Mo., 271.
Sierra Co., Cal., iron ore, 126.
Sierra Nevada, chromite, 261.
Sierra Nevada in California, 241.
  Recent geological history, 282.
Sigmoid fold, defined, 11.
Silliman, B., on gold quartz, 248.
Siluro-Cambrian limonites, 276.
Silver and gold, mode of occurrence, 199, 200.
Silver belt of Utah, 275.
Silver Bow Co., Mont., 220.
Silver Bow Creek, Butte, Mont., 136.
Silver, California, 242.
Silver City, Idaho, 223.
Silver Cliff, Colo., 212, 213.
Silver Islet, Lake Superior, vein illustrating lateral secretion, 31.
  Literature, 31.
  Mines, 201.
Silver King mine, Arizona, 228.
Silver minerals, 200.
Silver Plume, Colo., 214.
Silver Reef, Utah, 154, 226.
Silverton, Colo., 207
Silver, Washington, 239.
Slayback Lode, New Mexico, 204.
Slickensides, or slips, defined, 18, 19.
Smyth, C. H., Jr., on Clinton ore, 97.
Smithfield iron mine, Colorado, 126.
Smuggler Mountain, near Aspen, Colo., 189.
Snake River, Idaho, basalt, 222.
  Placers, Idaho, 245.
Socorro Co., N. M., 204.
Sonora, Mex., antimony, 260.
South Dakota, lead-silver ores, 191.
  Geology, 216.

300 INDEX.

South Dakota, source of ore, 277.
Southern States, gold, 252.
South Mountain, Penn., iron mines, 123, 125.
   Literature, 124, 125.
South Park, Colo., 212.
South Wallingford, Vt., manganese, 263.
Spanish Peaks, Colo., 212.
Spathic iron ore, 86–91.
Spenceville, Cal., copper mines, 136.
Sperry, F. L., discovered sperrylite, 272.
Sperrylite, 272.
St. Clair limestone, Arkansas, 265.
St. François Co., Mo., gash veins, 165.
St. Genevieve, Mo., copper mines, 143.
   Geology, 282.
St. Lawrence Co., N. Y., lead mines, 156.
Stanley-Browne, J., cited on gold in sea beaches, 240.
Stannite, 273.
Star District, Utah, iron mines, 128.
Steamboat Springs, Nev., illustrate infiltration by ascension, 32.
   Mercury mines, 267.
Stein Mountains, Ore., 239.
Stelzner, A. W., cited on source of ores, 25.
Sterling Hill, zinc mines, Ogdensburg, N. J., 174–178.
Stevens Co., Wash., gold, 238.
Stibnite, 259.
Stillwater Co., Wyo., gold mines, 216.
Stobie nickel mine, Sudbury, Ont., 270.
Stokes Co., N. C., magnetite, 125.
Storey Co., Nev., 232.
Strahan, A., cited on explosive slickensides, 19.
Stream tin, 273.
Structure of veins, 35.
Sudbury, Ontario, 131.
   Nickel, 270.
Sullivan Co., N. Y., lead mines, 158.
Sullivan, Me., silver mines, 201.
Sulphur Bank, Cal., mercury mine, 267.
Summit Co., Colo, 211.
   Ten Mile district, 185, 186.
Summit Co, Utah, lead-silver, 195.
   Silver mines, 225.
Summit district, Rio Grande Co., Colo., 212.
Sunrise, Wyo., copper, 152.
Sweden, lake ore, 75.
   Nickel, 272.
   Titaniferous ores, 281.

Sweetwater district, Wyo., 245.
Sylvanite mine, Gothic, Colo., 211.
Syncline defined, 11.
   Cause of cavities, 14.
Tacoma Mountain, 238.
Tamarack copper mines, Michigan, 141.
Tarr, R. S., on rifting, 12.
Telegraph mine, Utah, 194.
Telluride ores, Colorado, 214.
Tellurides in gold quartz, 248.
Temescal tin mines, California, 274.
Tem Pahute district, Nev., 231.
Ten Mile district, Colo., 185, 211.
Tennessee, Clinton ore, 94.
Tennessee limonite, 78, 82.
Tenny Cape, N. S., manganese, 266.
Terrane defined, 5.
Texas, copper, Llano Co., 139.
   Limonite, 78.
   Triassic or Peruvian copper ores, 154.
Thiess-Hutchins antimony mines, Nev., 259.
Three Rivers, Can., bog ores, 279.
Thunder Bay, Lake Superior, 202.
Tilly Foster iron mine, N. Y., 123, 124.
Tin, 273, 274.
Tin Cup, Colo., 211.
Tintic district copper mines, Utah, 152.
   Lead-silver, 194.
   Other silver mines, 225.
Titaniferous magnetite, 122, 280, 281.
Tombstone district, Arizona, 229.
Tooele Co., Utah, 194, 225.
Torrington, Conn., nickel mine, 270.
Tourtelotte Park, near Aspen, Colo., 188.
Toyabe range, Nev., 232.
Treadwell mine, Alaska, 255.
Triassic copper ores, 152–154.
   Copper mines, general relations, 276.
Trotter zinc mine, Franklin Furnace, N. J., 176.
Tucson, Arizona, 229.
Tuolumne Co., Cal., 249.
Tuscarora district, Nev., 232.
Tybo, Nev., 231.
Ueberroth zinc mine, Pennsylvania, 174.
Uintah Mountains, 10, 224.
Ulster Co., N. Y., lead mines, 158.
United States Antimony Co., Philadelphia, 259.
United States Geological Survey, terms used by in geological classification, 4.

United States, general geology, 7-10.
   General topography, 6, 7.
Uralitization in the rocks of the Comstock Lode, 236.
Utah, antimony, 259.
   Copper, 152.
   Geology, 224.
   Lead-silver ores, 192-196.
   Limonite, 79.
   Magnetite, 128.
   Silver mines, 275.
   Triassic copper ores, 154.
Vadose circulations, 283.
Van Diest, P. H., on Boulder Co., Colo., 214.
Van Dyck, F. C., analysis by, 177.
Van Hise, C. R., cited on replacement, 33.
   On the Marquette district, 101, 102.
   On the Penokee Gogebic district, Lake Superior, 109.
Veins, methods of filling, 28.
Verde River, Ariz., 151.
Vermilion district, Lake Superior, 110.
Vermilion Lake iron ores, origin of, 60, 61.
Vermont, copper ores, 134, 135.
   Limonite, 80.
Vershire, Vt., mines of pyrite, 131.
   Chalcopyrite, 134.
Virginia, Clinton ore, 94.
   Limonite or brown hematite, 77, 82.
   Louisa Co., pyrite, 131.
   Magnetite, 125.
   Manganese, 263.
   Titaniferous ores, 281.
   Wythe Co., zinc and lead, 172.
Virginia City, Mont., 219.
Von Herder, on filling of veins, 28.
Von Richthofen, on the Comstock Lode, 233-236.
Vuggs of a vein, 35.
Wadsworth, M E., on the Marquette district, 101, 102.
   On Silver Islet, 202.
   On the classification of ore deposits, 278.
   Origin of the ores, 104.
Wagon Wheel Gap, Colo., 209.
Wall rock of veins, precipitating influence of, 31.
Wardner, Idaho, lead-silver ores, 192.
Warren or Bisbee copper district, Arizona, 148.
Wasatch Mountains, 9, 224.
Washington Co., Mo., lead and zinc mines, 165.

Washington, geology, 238.
Washoe Co., Nev., 232.
Webb City, Mo., zinc and lead mines near, 166-171.
Webster, Jackson Co., N. C., nickel, 271.
Weed, W. H., cited on siliceous sinters, 57.
Weissenbach, Von, scheme of classification of ore deposits, 44.
Wells, H. L., on sperrylite, 272.
Wendt, A. F., on Arizona copper mines, 147.
Westport, N. Y., iron mines, 120.
West Virginia, Clinton ore, 94.
   Oriskany, red hematite, 98.
   Spathic ores, 88.
Wet Mountain Valley, Custer Co., Colo., 212.
Wheatley lead mine, Pennsylvania, 157.
White Pine Co., Nev., 231.
Whitney, J. D., cited on Missouri ores, 118.
   On California gravels, 247.
   On gold veins, 250.
   On lead and zinc veins of the Mississippi Valley, 163.
   On Washington Co., Missouri, lead and zinc mines, 165.
   Scheme of classification of ore deposits, 48.
Wickes, Mont., lead-silver mines, 192, 220.
Williams, J. F., on Arkansas bauxite, 258.
Willow Creek, near Creede, Colo., 209.
Wiltsee, E., on the Half Moon mine, Pioche, Nev., 230.
Winchell, N. H. and H. V., cited on origin of Vermilion Lake iron ores, 61.
Winchell, N. H., cited on the Vermilion Lake district, 110.
Winchell, H. V., on the origin of the ore, 111.
   On the Mesabi ores, 113.
Winslow, A., on Missouri geology, 282.
Winslow, tin, 274.
Wisconsin, Clinton ore, Dodge Co., 92.
   Lead and zinc mines, 161.
   Menominee district, 107, 108.
   Penokee-Gogebic district, 108, 109.
Wood River mines, Idaho, lead-silver, 191, 223.
Woods mine, chromite, 261.

Wright, C. E., origin of Lake Superior ores, 106.
Wyoming copper mines, 152.
  Geology of, 216.
  Iron mines, 126.
  Iron ore near Fort Laramie, 114.
  Spathic iron ore, 89.
  Specular ore near Fort Laramie, **114**.
  Tin ores, 274.
  Titaniferous ores, 281.
Wythe Company zinc mines, **282**.
Whyte Co., Va., zinc and lead mines, 172.
Xenogenites of Posepny, 283.

Yakima Co., Wash., gold, 238.
Yakutat Bay, Alaska, 240.
Yarmouth, N. S., 255.
Yavapai Co., Ariz., 228.
York Co., N. B., antimony, 259.
York Co., Penn., iron mines, 127.
Yuma Co., Ariz., 228, 229.
Yuba River, California, 282.
Yukon River, Alaska, 253.
Zinc minerals, 174.
  Statistics, 180.
Zirkel, F., on the rocks of the Comstock Lode, 236.
Zone of oxidized ores in a vein, 38.
  Of sulphides, **38**.

BOOKS

ON

Metallurgy,

Electricity,   Mining,

Assaying,   Chemistry,

Geology,   Mineralogy,

Railroading,   Mechanics,

Surveying,   Hydraulics,

Architecture,   Roads and Bridges,

Mathematics,   Sanitary Engineering,

27 PARK PLACE.
(Copyrighted.)

and Other Subjects.

Magazines and Papers Furnished.

☞ SPECIAL DISCOUNTS GIVEN

to Libraries, Educational Institutions, and on Important Cash Orders.

Those desiring to investigate any scientific or technical subject can learn which are the best books and sources of information on the same by writing to

The Scientific Publishing Company.

Publishers and Booksellers.
New York.

GEMS
AND
PRECIOUS STONES
OF
NORTH AMERICA

A POPULAR DESCRIPTION

OF THEIR OCCURRENCE, VALUE, HISTORY,
ARCHÆOLOGY, AND OF THE COLLECTIONS IN
WHICH THEY EXIST, ALSO A CHAPTER ON
PEARLS AND ON REMARKABLE FOREIGN GEMS
OWNED IN THE UNITED STATES . . .

ILLUSTRATED
WITH EIGHT COLORED PLATES AND NUMEROUS
MINOR ENGRAVINGS

BY
GEORGE FREDERIC KUNZ,

Price, - - $10.00

SCIENTIFIC PUBLISHING COMPANY,
PUBLISHERS,
27 PARK PLACE, NEW YORK.

MANUAL

OF

QUALITATIVE BLOWPIPE ANALYSIS

AND

DETERMINATIVE MINERALOGY.

BY

F. M. ENDLICH, S.N.D.,

MINING ENGINEER AND METALLURGIST,
LATE MINERALOGIST SMITHSONIAN INSTITUTION, AND UNITED STATES GEOLOGICAL
AND GEOGRAPHICAL SURVEY OF THE TERRITORIES.

Bound in Cloth. Illustrated. Price $4.00.

This work has been specially prepared for the use of all students in this great department of chemical science. The difficulties which beset beginners are borne in mind, and detailed information has been given concerning the various manipulations. All enumerations of species as far as possible have been carried out in alphabetical order, and in the determinative tables more attention has been paid to the physical characteristics of substances under examination than has ever yet been done in a work of this kind. To a compilation of all the blowpipe reactions heretofore recognized as correct the author has added a number of new ones not previously published. The entire arrangement of the volume is an original one, and to the knowledge born of an extensive practical experience the author has added everything of value that could be gleaned from other sources. The book cannot fail to find a place in the library or workshop of almost every student and scientist in America.

TABLE OF CONTENTS.

Chapter I.—Appliances and Reagents required for Qualitative Blowpipe Analysis.
Chapter II.—Methods of Qualitative Blowpipe Analysis.
Chapter III.—Tables giving Reactions for the Oxides of Earth and Minerals.
Chapter IV.—Prominent Blowpipe Reactions for the Elements and their Principal Mineral Compounds.
Chapter V.—Systematic Qualitative Determination of Compounds.
Chapter VI.—Determinative Tables and their Application.

THE SCIENTIFIC PUBLISHING COMPANY.

PUBLISHERS,

27 PARK PLACE, NEW YORK.

THE METALLURGY OF STEEL,

BY

HENRY M. HOWE, A.M., S.B.

Royal Quarto, Handsomely Bound, Printed on Superfine Paper, and Profusely Illustrated.

Price, - - - - - $10.00.

    This work is the most notable contribution to the literature of iron and steel metallurgy ever published. The series of papers on the subject which have appeared as supplements to the "Engineering and Mining Journal" attracted world-wide attention, and have received the heartiest commendation from all quarters. The volume now published presents this material in much more convenient shape, with considerable additional matter, giving the results of the most recent research, experiment and practice. Mr. Howe also presents a complete review of all important conclusions reached by earlier investigators, and his masterly discussion of them renders the work classic. Every statement and citation has been carefully weighed and verified and the references to the literature of the subject are given minutely, the book thus furnishing in itself a key to the whole range of steel metallurgy. It also furnishes the results of much new and original investigation, specially undertaken for the present work.

    Every metallurgist, every manufacturer of steel in any form, and all who are interested in the iron or steel industries, and all engineers who use iron or steel should have this standard work and cannot afford to be without it.

SCIENTIFIC PUBLISHING COMPANY,

PUBLISHERS,

27 PARK PLACE, NEW YORK.

BASIC BESSEMER PROCESS.

By Dr. H. WEDDING.

The Scientific Publishing Company has secured the rights of publication in the United States of a translation of this, the acknowledged authoritative work on the Basic Bessemer or Thomas Process, which is now for the first time placed before American metallurgists.

Translated from the German by

WILLIAM B. PHILLIPS, Ph. D.,
Professor of Chemistry and Metallurgy in the University of Alabama,

and

ERNST PROCHASKA, Met. E.,
Late Engineer at the Basic Steel Works, Teplitz, Bohemia, and at the Works of the Pottstown Iron Co., Pottstown, Pa.

With supplementary chapter on Dephosphorization in the Basic Open Hearth Furnace, by ERNST PROCHASKA.

The Standard Work, and the Only Book in English on this Subject.

THE SCIENTIFIC PUBLISHING COMPANY,
PUBLISHERS,
27 PARK PLACE, NEW YORK.
(OUT OF PRINT.)

MODERN AMERICAN METHODS OF COPPER SMELTING,

BY

EDWARD D. PETERS, Jr., M. E., M. D.

No one who has a copy of the First Edition of this great work should fail to secure the Fourth Edition, Revised and Enlarged.

Profusely Illustrated.   Price $4.00.

This is the Best Book on Copper Smelting in the language.

It contains a record of practical experience, with directions how to build furnaces and how to overcome the various metallurgical difficulties met with in copper smelting.

TABLE OF CONTENTS.

Chapter I.—Description of the Ores of Copper.
Chapter II.—Distribution of the Ores of Copper.
Chapter III.—Methods of Copper Assaying.
Chapter IV.—The Roasting of Copper Ores in Lump Form.
Chapter V.—Stall Roasting.
Chapter VI.—The Roasting of Ores in Lump Form in Kilns.
Chapter VII.—Calcination of Ore and Matte in Finely Divided Condition.
Chapter VIII.—The Chemistry of the Calcining Process.
Chapter IX.—The Smelting of Copper.
Chapter X.—Blast Furnaces Constructed of Brick.
Chapter XI.—General Remarks on Blast Furnace Smelting.
Chapter XII.—Late Improvements in Blast Furnaces.
Chapter XIII—The Smelting of Pyritous Ores Containing Copper and Nickel.
Chapter XIV.—Reverberatory Furnaces.
Chapter XV.—Refining Copper Gas in Sweden.
Chapter XVI.—Treatment of Gold and Silver Bearing Copper Ores.
Chapter XVII.—The Bessemerizing of Copper Mattes.
General Index, Etc.

THE SCIENTIFIC PUBLISHING COMPANY,
PUBLISHERS,
27 PARK PLACE, NEW YORK.

THE METALLURGY OF LEAD

AND THE

DESILVERIZATION OF BASE BULLION

BY

H. O. HOFMAN, E.M., Ph.D.,

Associate Professor of Mining and Metallurgy, Massachusetts Institute of Technology.

Bound in Cloth.   Illustrated with Working Drawings.

PRICE, $6.00.

TABLE OF CONTENTS.

PART I. INTRODUCTION.

Chapter I   Historical and Statistical Notices.
"   II   Properties of Lead and some of its Compounds.
"   III   Lead Ores.
"   IV   Distribution of Lead Ores.
"   V   Receiving, Sampling and Purchasing of Ores, Flues and Fuel.

PART II. METALLURGICAL TREATMENT OF LEAD ORES.

Chapter VI   Smelting in the Reverberatory Furnace.
"   VII   "   "   Ore   "
"   VIII   "   "   Blast   "

PART III. DESILVERIZATION OF BASE BULLION.

Chapter IX   Pattinson Process.
"   X   Parkes   "
"   XI   Cupellation   "

THE SCIENTIFIC PUBLISHING COMPANY,

PUBLISHERS,

27 PARK PLACE, NEW YORK.

Florida, South Carolina AND Canadian

PHOSPHATES.

Giving a Complete Account of Their Occurrence, Methods and Cost of Production, Quantities Raised, and Commercial Importance.

—BY—

C. C. HOYER MILLAR.

The author gives his experience and investigations during the last few years in the Phosphate Fields of Florida, South Carolina and Canada.

Bound in Cloth, 223 pages ; Price, $2.50.

TABLE OF CONTENTS.

Chapter I. Introduction.
    Chapter II. Florida Phosphates.
        Chapter III. South Carolina Phosphates.
            Chapter IV. Canadian Phosphates.

Appendix—Analysis of Various Phosphates.

THE SCIENTIFIC PUBLISHING COMPANY,
27 Park Place, New York.

THE PHOSPHATES OF AMERICA

Where and How They Occur; How They Are Mined; and What They Cost.

With Practical Treatises on the Manufacture of Sulphuric Acid, Acid Phosphate, Phosphoric Acid and Concentrated Superphosphates, and Select Methods of Chemical Analysis.

BY

FRANCIS WYATT, Ph. D.

BOUND IN CLOTH. PROFUSELY ILLUSTRATED. PRICE ONLY $4.00.

It is the first work of its kind ever published, and cannot fail to be welcomed by every one at all interested in the subject of phosphates or phosphate mining.

TO THE MINER

This remarkable book is indispensable, describing and fully illustrating the geological occurrence; the various modern methods of mining; the actual mining cost and the chemical composition of all the workable phosphate deposits of the American continent.

TO THE FERTILIZER MANUFACTURER

It is, as the title implies, a standard authority in every phase of his industry. It gives methods, resulting from long practical experience, by which the most satisfactory results are obtained from all grades of phosphates, including those chemically defective and heretofore unused. It also contains valuable working tables whereby perfect products can be produced by those unfamiliar with chemistry.

TO THE CHEMIST

It must prove a veritable guide and reference book, since he will find in it every detail required for reliable, rapid, and strictly scientific analyses of all material used in the industries indicated by the title. The methods are those used in the author's own laboratory, and are supplemented by detailed descriptions of manipulation and apparatus; factors for facilitating calculations; formulæ for preparation of all reagents, and instructions for estimating and stating results of analyses.

TO THE GENERAL READER

It affords an opportunity, hitherto lacking, of studying a concise treatise on phosphates and fertilizers, by a practical and conscientious expert in everyday language. The capitalist, banker, merchant, and intelligent farmer may all derive from it with facility a liberal education on the important subjects treated, as the use of scientific terms is restricted to the technical pages alone.

THE SCIENTIFIC PUBLISHING COMPANY,

PUBLISHERS,

27 PARK PLACE, NEW YORK.

CHEMICAL AND GEOLOGICAL ESSAYS

— BY —

THOMAS STERRY HUNT, M. A., LL. D.,

Author of "Mineral Physiology and Physiography," "A New Basis for Chemistry," "Systematic Mineralogy," etc.

FOURTH EDITION. REVISED AND ENLARGED.

PRICE, $2.50.

TABLE OF CONTENTS.

Preface;
- I. Theory of Igneous Rocks and Volcanoes;
- II. Some Points in Chemical Geology;
- III. The Chemistry of Metamorphic Rocks;
- IV. The Chemistry of the Primeval Earth;
- V. The Origin of Mountains;
- VI. The Probable Seat of Volcanic Action;
- VII. On Some Points in Dynamical Geology;
- VIII. On Limestone, Dolomites and Gypsums;
- IX. The Chemistry of Natural Waters;
- X. Petroleum, Asphalt, Pyroschists and Coal;
- XI. Granites and Granitic Vein stones;
- XII. The Origin of Metalliferous Deposits;
- XIII. The Geognosy of the Appalachians and the Origin of Crystalline Rocks;
- XIV. The Geology of the Alps;
- XV. History of the Names Cambrian and Silurian in Geology;
- XVI. Theory of Chemical Changes and Equivalent Volumes;
- XVII. The Constitution and Equivalent Volume of Mineral Species;
- XVIII. Thoughts on Solution and the Chemical Process;
- XIX. On the Objects and Method of Mineralogy;
- XX. The Theory of Types in Chemistry.

Appendix and Index.

THE SCIENTIFIC PUBLISHING COMPANY,
PUBLISHERS,
27 PARK PLACE, NEW YORK.

MINERAL PHYSIOLOGY AND PHYSIOGRAPHY.

A SECOND SERIES OF

CHEMICAL AND GEOLOGICAL ESSAYS,

WITH

A GENERAL INTRODUCTION.

BY

THOMAS STERRY HUNT, M. A., LL. D.,

Author of "Chemical and Geological Essays," "A New Basis for Chemistry," "Systematic Mineralogy," etc.

PRICE, $5.00.

TABLE OF CONTENTS.

PREFACE.
Chapter I.—Nature in Thought and Language.
Chapter II.—The Order of the Natural Sciences.
Chapter III.—Chemical and Geological Relations of the Atmosphere.
Chapter IV.—Celestial Chemistry from the Time of Newton.
Chapter V.—The Origin of Crystalline Rocks.
Chapter VI.—The Genetic History of Crystalline Rocks.
Chapter VII.—The Decay of Crystalline Rocks.
Chapter VIII.—A Natural System in Mineralogy, with a Classification of Silicates.
Chapter IX.—History of Pre-Cambrian Rocks.
Chapter X.—The Geological History of Serpentine, with Studies of Pre Cambrian Rocks.
Chapter XI.—The Taconic Question in Geology.
Appendix and Index.

THE SCIENTIFIC PUBLISHING COMPANY,

PUBLISHERS,

27 PARK PLACE, NEW YORK.

A NEW BASIS FOR CHEMISTRY.

A CHEMICAL PHILOSOPHY

BY

THOMAS STERRY HUNT, M. A., LL. D.,

Author of "Chemical and Geological Essays," "Mineral Physiology and Physiography," "Systematic Mineralogy," etc.

PRICE, $2.00.

TABLE OF CONTENTS.

I. Introduction.
II. Nature of the Chemical Process.
III. Genesis of the Chemical Elements.
IV. Gases, Liquids and Solids.
V. The Law of Numbers.
VI. Equivalent Weights.
VII. Hardness and Chemical Indifference.
VIII. The Atomic Hypothesis.
IX. The Law of Volumes.
X. Metamorphosis in Chemistry.
XI. The Law of Densities.
XII. Historical Retrospect.
XIII. Conclusions.
XIV. Supplement.
Appendix and Index.

THE SCIENTIFIC PUBLISHING COMPANY,
PUBLISHERS,
27 PARK PLACE, NEW YORK.

SYSTEMATIC MINERALOGY

BASED ON A

NATURAL CLASSIFICATION.

WITH A GENERAL INTRODUCTION.

—BY—

THOMAS STERRY HUNT, M.A., LL.D.,

Author of "Chemical and Geological Essays," "Mineral Physiology and Physiography," "A New Basis for Chemistry," etc.

BOUND IN CLOTH.   PRICE $5.00.

The aim of the author in the present treatise has been to reconcile the rival and hitherto opposed Chemical and Natural History methods in Mineralogy, and to constitute a new system of classification, which is "at the same time Chemical and Natural Historical," or, in the words of the preface, "to observe a strict conformity to chemical principles, and at the same time to retain all that is valuable in the Natural History method; the two opposing schools being reconciled by showing that when rightly understood, chemical and physical characters are really dependent on each other, and present two aspects of the same problem which can never be solved but by the consideration of both." He has, moreover, devised and adopted a Latin nomenclature and arranged the mineral kingdom in classes, orders, genera and species, the designations of the latter being binomial.

TABLE OF CONTENTS.

Chapter I. The Relations of Mineralogy;
    II. Mineralogical Systems;
    III. First Principles in Chemistry;
    IV. Chemical Elements and Notation;
    V. Specific Gravity;
    VI. The Coefficient of Mineral Condensation;
    VII. The Theory of Solution;
    VIII. Relations of Condensation to Hardness and Insolubility;
    IX. Crystallization and Its Relations;

Chapter X. The Constitution of Mineral Species;
    XI. A New Mineralogical Classification;
    XII. Mineralogical Nomenclature;
    XIII. Synopsis of Mineral Species;
    XIV. The Metallaceous Class;
    XV. The Halidaceous Class;
    XVI. The Oxydaceous Class;
    XVII. The Pyricaustaceous Class;
    XVIII. Mineral History of Waters;
    General Index;
    Index of Names of Minerals.

THE SCIENTIFIC PUBLISHING COMPANY,

PUBLISHERS,

27 PARK PLACE, NEW YORK.

MINING ACCIDENTS
AND
THEIR PREVENTION.

BY SIR FREDERICK AUGUSTUS ABEL.

With Discussion by Leading Experts. Also, the United States, British and Prussian Laws relating to the Working of Coal Mines.

Price, - - - $4.00 in Cloth.

CONTENTS:

Mining Accidents. By Sir Frederick A. Abel. With discussion by President Bruce, of the British Institute of Civil Engineers; and Prof. Arnold Lupton, C. Tylden Wright, Emerson Bainbridge, William Morgans, Sydney F. Walker, Col. Paget Mosley, Henry Hall, Col. J. D. Shakespear, Stephen Humble, Sir George Elliot, Sir Warington Smyth, A. R. Sawyer, A. Giles, R. Bedlington, Edward Combes, George Seymour, Henry Harries, William Cochrane, James Ashworth, J. B. Atkinson, W. N. Atkinson, Bennett H. Brough, T. Foster Brown, S. B. Coxon, C. Le Neve Foster, W. Galloway, Max Georgi, W. S. Gresley, J. A. Longdon, A. R. Sennett, M. H. N. Story Maskelyne, Arthur Sopwith, A. L. Steavenson, A. H. Stokes and others.

List of safety appliances, with description of detachment of mineral from its bed, carriage of mineral to the surface, difficulties attendant on the presence of gases, etc. Safety lamps (oil and spirit), safety lamps (electric) and other appliances.

The Mining Laws of Colorado, Illinois, Indiana, Iowa, Kansas, Kentucky, Maryland, Missouri, Ohio, Pennsylvania, Washington, West Virginia and Wyoming; also those of Great Britain and Prussia add a feature of great value—for these laws have never before been collected or published in accessible form.

Of the unanimously favorable criticisms of this book, we have only space to quote one:

"It is a work that should be in the hands of every intelligent man connected with a colliery, no matter what his position. It is as valuable to the intelligent miner as it is to the mining engineer or the colliery official."—*Colliery Engineer.*

SCIENTIFIC PUBLISHING COMPANY,
27 PARK PLACE, NEW YORK.

THE
MINING LAWS
OF THE
REPUBLIC OF COLOMBIA

WITH

A SHORT EXPLANATION OF THEIR APPLICATION

AND

OFFICIAL FORMS FOR NOTICES, DENOUNCEMENTS

AND APPLICATIONS FOR TITLE.

TRANSLATED AND EDITED

BY

CHARLES BULLMAN, M.E.

BOUND IN CLOTH, 113 PAGES,   -   PRICE, $1.50.

THE SCIENTIFIC PUBLISHING COMPANY,
PUBLISHERS,
27 PARK PLACE, NEW YORK.

Universal Bimetallism

AND

An International Monetary Clearing House,

TOGETHER WITH

A RECORD OF THE WORLD'S MONEY,
STATISTICS OF GOLD AND SILVER, ETC.

BY

RICHARD P. ROTHWELL, M.E., C.E.,

Editor of the "Engineering and Mining Journal,"
Ex-President American Institute of Mining Engineers,
Special Agent of the 11th United States Census on Gold and Silver, Etc., Etc.

PRICE, 75 CENTS.

This important contribution is a practical business man's solution of the great question which has brought upon the entire world the most acute financial crisis of modern times.

It proposes a simple, absolute, and permanent solution of this problem by the well-tried and universally successful agencies, International Arbitration and Clearing House Methods.

THE SCIENTIFIC PUBLISHING CO., Publishers,
NEW YORK.

Parliamentary Tactics

FOR THE USE OF THE

PRESIDING OFFICER

AND

PUBLIC SPEAKERS.

BY

HARRY W. HOOT.

    This little manual on Parliamentary Tactics is arranged in a way especially adapted for use by "the Chair" from such standard authorities as Roberts, Cushing, Matthias, Jefferson, and Crocker.

    Where the authorities differ, the view most conformable to the latest and most thoroughly established usages of Parliamentary Law has been accepted.

    The table of motions has been arranged by placing on the margin the names of such motions as are used in Common Parliamentary Practice, with their classification and order of procedure in a deliberative body, being placed at the top of the book, and descending in regular consecutive order will be found those having the next highest order of precedence, viz.: Privileged questions, Incidental questions, Subsidiary motions, and Main questions. Thus can be seen without turning a page and in a moment's time whether a motion is in order.

    Between each marginal reference will be found in a condensed form all the rules relating to that particular question.

    The book is so arranged that it can be carried in the pocket. Bound in cloth. It will be found to be of invaluable aid to those who are required to decide questions of Parliamentary Law without reference to larger text-books.

PRICE, 50 CENTS.

THE SCIENTIFIC PUBLISHING CO., Publishers,

NEW YORK.

THE MINERAL INDUSTRY,

Its Statistics, Technology, and Trade

IN THE

UNITED STATES AND OTHER COUNTRIES FROM THE EARLIEST TIMES TO THE END OF 1892.

BY

R. P. ROTHWELL, C.E., M.E.,

Editor of the "Engineering and Mining Journal."

This is the most thorough and exhaustive work on the mineral productions and industries of the world that has ever been issued, and no person at all interested in mining or metallurgy can afford to be without it.

PRICE, BOUND IN CLOTH, $2.50.

THE SCIENTIFIC PUBLISHING COMPANY,

PUBLISHERS,

27 PARK PLACE, NEW YORK.

The Engineering and Mining Journal

RICHARD P. ROTHWELL C.E., M.E., Editor.
ROSSITER W. RAYMOND, Ph.D., M.E., Special Contributor.
SOPHIA BRAEUNLICH, Business Manager.

ESTABLISHED 1866.

THE ENGINEERING AND MINING JOURNAL is conceded to be "the best mining paper in the world." Its reports, criticisms on the technical information it gives, its market reports, and its peerless and impartial criticisms of things calculated to injure legitimate mining investments have gained the approval and confidence of the entire mining industry.

The Largest Circulation of any Technical Paper in America.

BEST ADVERTISING MEDIUM.

Circulating Among the Largest Buyers of Tools, Hardware, Machinery of all kinds, Railway and Mining Supplies in every part of the World.

PUBLISHED WEEKLY.—ILLUSTRATED.

SUBSCRIPTION PRICE: For United States, Mexico, and Canada, $5.00 per annum; $2.50 for six months; all other countries in the Postal Union, $7.00.

THE SCIENTIFIC PUBLISHING CO., Publishers,
NEW YORK.

www.ingramcontent.com/pod-product-compliance
Lightning Source LLC
Chambersburg PA
CBHW031853220426
43663CB00006B/610